REA's Books Are The Best...
They have rescued lots of grades and more!

D1128172

(continued from previous page)

"Your books have saved my GPA, and quite possibly my sanity. My course grade is now an 'A', and I couldn't be happier."

Student, Winchester, IN

"These books are the best review books on the market. They are fantastic!"

Student, New Orleans, LA

"Your book was responsible for my success on the exam. . . I will look for REA the next time I need help."

Student, Chesterfield, MO

"I think it is the greatest study guide I have ever used!"

Student, Anchorage, AK

"I encourage others to buy REA because of their superiority. Please continue to produce the best quality books on the market."

Student, San Jose, CA

"Just a short note to say thanks for the great support your book gave me in helping me pass the test . . . I'm on my way to a B.S. degree because of you!"

Student, Orlando, FL

GEOLOGY

**By the Staff of
Research & Education Association
Carl Fuchs, Chief Editor**

Research & Education Association
61 Ethel Road West
Piscataway, New Jersey 08854

Dr. M. Fogiel, Director

SUPER REVIEW®
OF GEOLOGY

Printed in the United States of America

Library of Congress Control Number 2002111537

International Standard Book Number 0-87891-388-2

SUPER REVIEW is a registered trademark of Research & Education Association, Piscataway, New Jersey 08854

WHAT THIS Super Review WILL DO FOR YOU

This **Super Review** provides all that you need to know to do your homework effectively and succeed on exams and quizzes.

The book focuses on the core aspects of the subject, and helps you to grasp the important elements quickly and easily.

Outstanding **Super Review** features:

- Geological periods are covered in chronological sequence

- Topics are reviewed in a concise and comprehensive manner

- The material is presented in student-friendly language that makes it easy to follow and understand

- Individual topics are fully indexed, making them easy to find

- Provides excellent preparation for midterms, finals, and in-between quizzes

- Written by professionals and experts who function as your very own tutors

Dr. Max Fogiel
Program Director

Carl Fuchs
Chief Editor

v

CONTENTS

Chapter   **Page**

1   GEOLOGIC TIME 1

   The Relative Time Scale 2
   The Radiometric Time Scale 12
   The Age of the Earth 19

2   FOSSILS, ROCKS, AND TIME 23

   Putting Events in Order 24
   The Relative Time Scale 26
   Rocks and Layers 28
   Fossils and Rocks 32
   Fossil Succession 34
   The Numeric Time Scale 40

3   THE EARTH'S INTERIOR 45

   The Crust 46
   The Mantle 48
   The Core 48
   The Structure of the Moon 51

4   ROCKS AND MINERALS 53

   Rocks 53
   Igneous Rocks 54
   Naming Igneous Rocks 54
   Sedimentary Rocks 56
   Metamorphic Rocks 57
   Useful Terms 61

5   VOLCANOES 63

   The Nature of Volcanoes 65
   Principal Types of Volcanoes 68
   Other Volcanic Structures 75
   Types of Volcanic Eruptions 78
   Submarine Volcanoes 82
   Geysers, Fumaroles, and Hot Springs 83
   Volcano Environments 85
   Plate-Tectonics Theory 90
   Extraterrestrial Volcanism 92
   Volcano Monitoring and Research 94
   Volcanoes and People 98

6 GLACIERS 101

Glacial Hazards 102
Glacial Advances 103
Glaciers and Climatic Change 105

7 DESERTS 113

Overview 113
How the Atmosphere Influences Aridity 116
Where Deserts Form 119
Types of Deserts 120
Desert Features 125
Eolian Processes 129
Types of Dunes 133
Remote Sensing of Arid Lands 135
Mineral Resources in Deserts 138
Desertification 140

8 EARTHQUAKES 147

Earthquakes in History 148
Where Earthquakes Occur 150
How Earthquakes Happen 151
Measuring Earthquakes 156
Volcanoes and Earthquakes 160
Predicting Earthquakes 161

9 HISTORICAL GEOLOGY 165

Fossils 166
Geological Chronology 168
Divisions of Geological Time 171

10 EARTH BEFORE THE CAMBRIAN 173

Pre-Cambrian Eras 173
The Archaeozoic Era 174
The Proterozoic Era 180

11 THE CAMBRIAN PERIOD 189

Overview 189
Life of the Cambrian 198
Plants 199
Animals 200
Summary 208

12 THE ORDOVICIAN PERIOD 211

Overview 211
Petroleum and Natural Gas 216
Life of the Ordovician 220
Plants 231
Summary 232

13 THE SILURIAN PERIOD 235

Overview 235
Life of the Silurian 240
Summary 248

14 THE DEVONIAN PERIOD 251

Overview 251
Life of the Devonian 256
Plants 269
Summary 270

15 THE CARBONIFEROUS PERIODS 273

Mississippian or Lower Carboniferous 273
Pennsylvanian or Upper Carboniferous 275
Coal Fields of North America 277
Summary of the Pennsylvanian 278
Permian 280
Invertebrates of the Carboniferous 285
Vertebrates of the Carboniferous 291
Carboniferous Plants 298
Coal 305
Problems of the Permian 311
Summary of the Paleozoic Era 313

16 THE MESOZOIC ERA: AGE OF REPTILES 317

Triassic 317
Jurassic 321
Lower Cretaceous (Comanchean) 324
Upper Cretaceous (Cretaceous) 326
Life of the Mesozoic 332
Invertebrates 334
Fish and Amphibians 346
Reptiles 350
Dinosaurs 353
Toothed Birds 372
Mammals 375

Plants 376
Climate 381
Coal 383

17 THE CENOZOIC ERA: AGE OF MAMMALS 385

Tertiary Period 385
Physical Geography of the Tertiary—Eocene 387
Oligocene 392
Miocene 393
Pliocene 401
Life of the Tertiary 406
Factors in the Evolution of Mammals 417
Odd-Toed Mammals (Perissidactyls) 420
Even-Toed Hoofed Mammals (Artiodactyls) 433
Other Animals 438
Invertebrates 445
Vegetation 449
Climate 454
Effects of Isolation and Migration 456

18 THE QUATERNARY PERIOD 465

Changes at the Close of the Tertiary 465
Distribution of the Ice Sheets 467
Development of the Ice Sheets 470
Glacial and Interglacial Stages 471
History of the Great Lakes 474
Other Pleistocene Lakes 479
Loess 481
Duration 482
Causes of Glaciation 484
Effects of Glaciation 486
Life of the Pleistocene 487
Prehistoric Man 498
Future Habitability of the Earth 509

APPENDIX: COMMON MINERALS 511

Iron Minerals 512
Zinc Minerals 513
Calcium Minerals 513
Copper Minerals 515
Lead Minerals 515
Silica Minerals 515
Silicate Minerals 516

ix

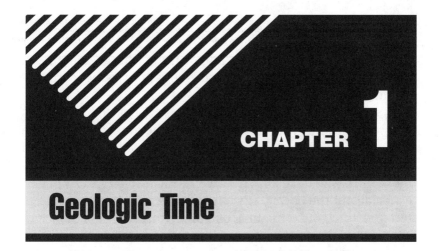

CHAPTER 1

Geologic Time

The Earth is very old—$4^1/_2$ billion years or more, according to recent estimates. This vast span of time, called geologic time by earth scientists, is difficult to comprehend in the familiar time units of months and years, or even centuries. How then do scientists reckon geologic time, and why do they believe the Earth is so old? A great part of the secret of the Earth's age is locked up in its rocks, and our centuries-old search for the key led to the beginning and nourished the growth of geologic science.

Mankind's speculations about the nature of the Earth inspired much of the lore and legend of early civilizations, but at times there were flashes of insight. The ancient historian Herodotus, in the 5th century B.C., made one of the earliest recorded geological observations. After finding fossil shells far inland in what are now parts of Egypt and Libya, he correctly inferred that the Mediterranean Sea had once extended much farther to the south. Few believed him, however, nor did the idea catch on. In the 3rd century B.C., Eratosthenes depicted a spherical Earth and even calculated its diameter and circumference, but the concept of a spherical Earth was beyond the imagination of most men. Only 500 years ago, sailors aboard the *Santa Maria* begged Columbus to turn back lest they sail off the Earth's "edge." Similar opinions and prejudices about the nature and age of the Earth have waxed and waned through the centuries. Most people, however, appear to have traditionally believed the Earth to be quite young—that its age might be measured in terms of thousands of years, but certainly not in millions.

The evidence for an ancient Earth is concealed in the rocks that form the Earth's crust and surface. The rocks are not all the same age—or even nearly so—but, like the pages in a long and complicated history, they record the Earth-shaping events and life of the past. The record, however, is incomplete. Many pages, especially in the early parts, are missing and many others are tattered, torn, and difficult to decipher. But enough of the pages are preserved to reward the reader with accounts of astounding episodes which certify that the Earth is billions of years old.

Two scales are used to date these episodes and to measure the age of the Earth: a *relative* time scale, based on the sequence of layering of the rocks and the evolution of life, and the *radiometric* time scale, based on the natural radioactivity of chemical elements in some of the rocks. An explanation of the relative scale highlights events in the growth of geologic science itself; the radiometric scale is a more recent development borrowed from the physical sciences and applied to geologic problems.

The Relative Time Scale

At the close of the 18th century, the haze of fantasy and mysticism that tended to obscure the true nature of the Earth was being swept away. Careful studies by scientists showed that rocks had diverse origins. Some rock layers, containing clearly identifiable fossil remains of fish and other forms of aquatic animal and plant life, originally formed in the ocean. Other layers, consisting of sand grains winnowed clean by the pounding surf, obviously formed as beach deposits that marked the shorelines of ancient seas. Certain layers are in the form of sand bars and gravel banks—rock debris spread over the land by streams. Some rocks were once lava flows or beds of cinders and ash thrown out of ancient volcanoes; others are portions of large masses of once-molten rock that cooled very slowly far

In places where layers of rocks are contorted, the relative ages of the layers may be difficult to determine. View near Copiapo, Chile.

beneath the Earth's surface. Other rocks were so transformed by heat and pressure during the heaving and buckling of the Earth's crust in periods of mountain building that their original features were obliterated.

Between the years of 1785 and 1800, James Hutton and William Smith advanced the concept of geologic time and strengthened the belief in an ancient world. Hutton, a Scottish geologist, first proposed formally the fundamental principle used to classify rocks according to their relative ages. He concluded, after studying rocks at many outcrops, that each layer represented a specific interval of geologic time. Further, he proposed that wherever uncontorted layers were exposed, the bottom layer was deposited first and was, therefore, the oldest layer exposed; each succeeding layer, up to the topmost one, was progressively younger.

Today, such a proposal appears to be quite elementary but, nearly 200 years ago, it amounted to a major breakthrough in scientific reasoning by establishing a rational basis for relative time measurements. However, unlike tree-ring dating—in which each ring is a measure of one year's growth—no precise rate of deposition can be determined for most of the rock layers. Therefore, the actual length of geologic time represented by any given layer is usually unknown or, at best, a matter of opinion.

William "Strata" Smith, a civil engineer and surveyor, was well acquainted with areas in southern England where "limestone and shales are layered like slices of bread and butter." His hobby of collecting and cataloging fossil shells from these rocks led to the discovery that certain layers contained fossils unlike those in other layers. Using these key or *index* fossils as markers, Smith could identify a particular layer of rock wherever it was exposed. Because fossils actually record the slow but progressive development of life, scientists use them to identify rocks of the same age throughout the world.

From the results of studies on the origins of the various kinds of rocks (petrology), coupled with studies of rock layering (stratigraphy) and the evolution of life (paleontology), geologists reconstruct the sequence of events that has shaped the Earth's surface. Their studies show, for example, that during a particular episode the land surface was raised in one part of the world to form high plateaus and mountain ranges. After the uplift of the land, the forces of erosion attacked the highlands and the eroded rock debris was transported and redeposited in the lowlands. During the same interval of time in another part of the world, the land surface subsided and was covered by the seas. With the sinking of the land surface, sediments were deposited on the ocean floor. The evidence for the preexistence of ancient mountain ranges lies in the nature of the eroded rock debris, and the evidence of the seas' former presence is, in part, the fossil forms of marine life that accumulated with the bottom sediments.

Such recurring events as mountain building and sea encroachment, of which the rocks themselves are records, comprise units of geologic time even though the actual dates of the events are unknown. By comparison, the history of mankind is similarly organized into relative units of time. We speak of human events as occurring either B.C. or A.D.—broad divisions of time. Shorter spans are measured by the dynasties of ancient Egypt or by the reigns of kings and queens in Europe. Geologists have done the same thing to geologic time by dividing the Earth's history into *Eras*—broad spans based on the general character of life that existed during these times, and *Periods*—shorter spans based partly on evidence of major disturbances of the Earth's crust.

The names used to designate the divisions of geologic time are a fascinating mixture of words that mark highlights in the historical development of geologic science over the past 200 years. Nearly every name signifies the acceptance of a new scientific concept—a new rung in the ladder of geologic knowledge.

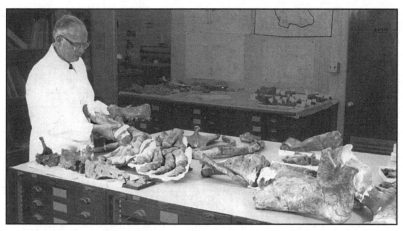

A paleontologist of the U.S. Geological Survey examining the fossil bones of *Paleoparadoxia*, an aquatic mammal that lived about 14 million years ago.

The major divisions, with brief explanations of each, are shown in the following scale of relative geologic time, which is arranged in chronological order with the oldest divisions at the bottom, the youngest at the top.

Major Divisions of Geologic Time

CENOZOIC ERA (Age of Recent Life)	Quaternary Period	The several geologic eras were originally named Primary, Secondary, Tertiary, and Quaternary. The first two names are no longer used; Tertiary and Quaternary have been retained but used as period designations.
	Tertiary Period	
MESOZOIC ERA (Age of Medieval Life)	Cretaceous Period	From the Latin word for chalk (creta) and first applied to extensive deposits that form white cliffs along the English Channel.
	Jurassic Period	Named for the Jura Mountains, located between France and Switzerland, where rocks of this age were first studied.
	Triassic Period	Taken from word "trias" in recognition of the threefold character of these rocks in Europe.
PALEOZOIC ERA (Age of Ancient Life)	Permian Period	Named after the province of Perm, Russia, where these rocks were first studied.
	Pennsylvanian Period	Named for the State of Pennsylvania where these rocks have produced much coal.
	Mississippian Period	Named for the Mississippi River valley where these rocks are well exposed.
	Devonian Period	Named after Devonshire, England, where these rocks were first studied.
	Silurian Period	Named after Celtic tribes, the Silures and the Ordovices, that lived in Wales during the Roman Conquest.
	Ordovician Period	
	Cambrian Period	Roman for Wales (Cambria) where rocks containing the earliest evidence of complex forms of life were first studied.
PRECAMBRIAN	— — —	Time between the birth of the planet and the appearance of complex life forms. More than 80% of Earth's est. 4.5 billion years falls here.

Keyed to the relative time scale are examples of index fossils, the forms of life which existed during limited periods of geologic time and thus are used as guides to the age of the rocks in which they are preserved.

Index Fossils

Pecten gibbus	Neptunea tabulata
Calyptraphorus velatus	Venericardia planicosta
Scaphites hippocrepis	Inoceramus labiatus
Perisphinctes tiziani	Nerinea trinodosa
Tropites subbullatus	Monotis subcircularis
Leptodus americanus	Parafusulina bosei
Dictyoclostus americanus	Lophophyllidium proliferum
Cactocrinus multibrachiatus	Prolecanites gurleyi
Mucrospirifer mucronatus	Palmatolepus unicornis
Cystiphyllum niagarense	Hexamoceras hertzeri
Bathyurus extans	Tetragraptus fructicosus
Paradoxides pinus	Billingsella corrugata

The following examples show how the rock layers themselves are used as a relative time scale:

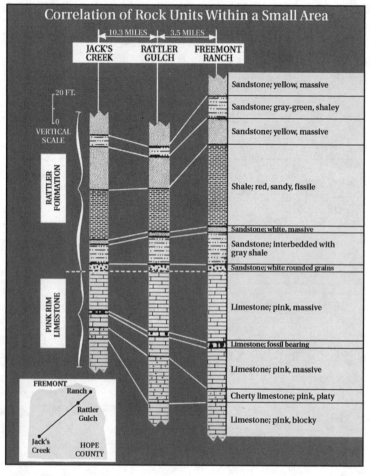

The diagram above *correlates* or matches rock units from
three localities within a small area by means of geologic sections compiled from results of field studies.

These sections are typical of the ones geologists prepare
when studying the relationships of layers of rocks (beds)
throughout a region. Each column represents the sequence of
beds at a specific locality. The same beds, which in places may
thicken or thin (some may pinch out entirely) according to the

The Wingate Sandstone is a reddish-brown formation consisting largely of wind-blown sand believed to have accumulated as desert dunes in the Four Corners region of the Southwestern United States about 200 million years ago. Erosion of this formation commonly produces vertical cliffs. View near Gateway, Colorado.

local environment of deposition, are bracketed within the lines connecting the three columns.

For convenience, geologists commonly group adjoining beds that possess similar or related features (including fossils) into a single, more conspicuous unit called a *formation*. The component beds of each formation are described, the formation is named, and the information is published for the use of all geologists. Formation names comprise two or more words, the first part usually taken from a geographic feature near which the rocks are prominently displayed. The last word indicates the principal rock type, or if of mixed rock types, the word formation is used: The *Morrison Formation*—the *Wingate Sandstone*—the *Todilto Limestone*—the *Mancos Shale*.

The diagram on pages 10-11 is a composite *geologic section*, greatly simplified. This example correlates formations throughout a large region in the southwestern part of the United States noted for its spectacular scenery. The composite section can be thought of as representing a large, although incomplete, segment of relative geologic time. A global time scale has been derived by extending the correlations from one place to another and reaching from one continent to another. These correlations are based in large part on fossil evidence.

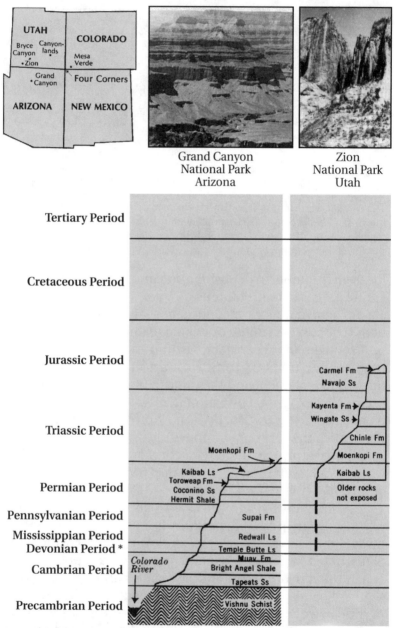

Grand Canyon
National Park
Arizona

Zion
National Park
Utah

Period	Grand Canyon	Zion
Tertiary Period		
Cretaceous Period		
Jurassic Period		Carmel Fm / Navajo Ss
Triassic Period	Moenkopi Fm	Kayenta Fm / Wingate Ss / Chinle Fm / Moenkopi Fm
Permian Period	Kaibab Ls / Toroweap Fm / Coconino Ss / Hermit Shale	Kaibab Ls / Older rocks not exposed
Pennsylvanian Period	Supai Fm	
Mississippian Period	Redwall Ls	
Devonian Period *	Temple Butte Ls	
Cambrian Period	Muav Fm / Bright Angel Shale / Tapeats Ss	
Precambrian Period	Vishnu Schist	

Colorado River

* Rocks of Ordovician and Silurian age are not present in the Grand Canyon.
Ss=Sandstone; Fm=Formation; Ls=Limestone

Canyonlands
National Park
Utah

Mesa Verde
National Park
Colorado

Bryce Canyon
National Park
Utah

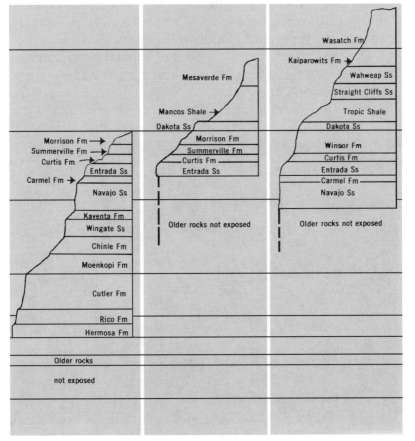

Wasatch Fm

Kaiparowits Fm →

Mesaverde Fm

Wahweap Ss

Straight Cliffs Ss

Mancos Shale →

Tropic Shale

Dakota Ss

Dakota Ss

Morrison Fm →
Summerville Fm →
Curtis Fm →

Morrison Fm
Summerville Fm
Curtis Fm

Winsor Fm
Curtis Fm

Entrada Ss

Entrada Ss

Entrada Ss

Carmel Fm →

Carmel Fm

Navajo Ss

Navajo Ss

Kayenta Fm
Wingate Ss

Older rocks not exposed

Older rocks not exposed

Chinle Fm

Moenkopi Fm

Cutler Fm

Rico Fm
Hermosa Fm

Older rocks

not exposed

The Radiometric Time Scale

The discovery of the natural radioactive decay of uranium in 1896 by Henry Becquerel, the French physicist, opened new vistas in science. In 1905, the British physicist Lord Rutherford—after defining the structure of the atom—made the first clear suggestion for using radioactivity as a tool for measuring geologic time directly; shortly thereafter, in 1907, Professor B. B. Boltwood, radiochemist of Yale University, published a list of geologic ages based on radioactivity. Although Boltwood's ages have since been revised, they did show correctly that the duration of geologic time would be measured in terms of hundreds-to-thousands of millions of years.

The next 40 years was a period of expanding research on the nature and behavior of atoms, leading to the development of nuclear fission and fusion as energy sources. A by-product of this atomic research has been the development and

A technician of the U.S. Geological Survey uses a mass spectrometer to determine the proportions of neodymium isotopes in a sample of igneous rock.

continuing refinement of the various methods and techniques used to measure the age of Earth materials. Precise dating has been accomplished since 1950.

A chemical element consists of atoms with a specific number of protons in their nuclei but different atomic weights owing to variations in the number of neutrons. Atoms of the same element with differing atomic weights are called *isotopes.* Radioactive decay is a spontaneous process in which an isotope (the parent) loses particles from its nucleus to form an isotope of a new element (the daughter). The rate of decay is conveniently expressed in terms of an isotope's *half-life,* or the time it takes for one-half of a particular radioactive isotope in a sample to decay. Most radioactive isotopes have rapid rates of decay (that is, short half-lives) and lose their radioactivity within a few days or years. Some isotopes, however, decay slowly, and several of these are used as geologic clocks. The parent isotopes and corresponding daughter products most commonly used to determine the ages of ancient rocks are listed below:

Parent Isotope	Stable Daughter Product	Currently Accepted Half-Life Values
Uranium-238	Lead-206	4.5 billion years
Uranium-235	Lead-207	704 million years
Thorium-232	Lead-208	14.0 billion years
Rubidium-87	Strontium-87	48.8 billion years
Potassium-40	Argon-40	1.25 billion years
Samarium-147	Neodymium-143	106 billion years

The mathematical expression that relates radioactive decay to geologic time is called the age equation and is:

$$t = \frac{1}{\lambda} \ln\left(1 + \frac{D}{P}\right)$$

where: t is the age of a rock or mineral specimen,
D is the number of atoms of a daughter product today,
P is the number of atoms of the parent isotope today,
ln is the natural logarithm (logarithm to base e), and
λ is the appropriate decay constant.

(The decay constant for each parent isotope is related to its half-life, $t\frac{1}{2}$, by the following expression: $t\frac{1}{2} = \dfrac{ln2}{\lambda}$)

Dating rocks by these radioactive timekeepers is simple in theory, but the laboratory procedures are complex. The numbers of parent and daughter isotopes in each specimen are determined by various kinds of analytical methods. The principal difficulty lies in measuring precisely very small amounts of isotopes.

The potassium-argon method can be used on rocks as young as a few thousand years as well as on the oldest rocks known. Potassium is found in most rock-forming minerals, the half-life of its radioactive isotope potassium-40 is such that measurable quantities of argon (daughter) have accumulated in potassium-bearing minerals of nearly all ages, and the amounts of potassium and argon isotopes can be measured accurately, even in very small quantities. Where feasible, two or more methods of analysis are used on the same specimen of rock to confirm the results.

Another important atomic clock used for dating purposes is based on the radioactive decay of the isotope carbon-14, which has a half-life of 5,730 years. Carbon-14 is produced continuously in the Earth's upper atmosphere as a result of the bombardment of nitrogen by neutrons from cosmic rays. This newly formed radiocarbon becomes uniformly mixed with the nonradioactive carbon in the carbon dioxide of the air, and it eventually finds its way into all living plants and animals. In effect, all carbon in living organisms contains a constant proportion of radiocarbon to nonradioactive carbon. After the death of the organism, the amount of radiocarbon gradually decreases as it reverts to nitrogen-14 by radioactive decay. By measuring the amount of radioactivity remaining in organic materials, the amount of carbon-14 in the materials can be calculated and the time of death can be determined. For example, if carbon from a sample of wood is found to contain only half as much carbon-14 as that from a living plant, the estimated age of the old wood would be 5,730 years.

The radiocarbon clock has become an extremely useful and efficient tool in dating the important episodes in the recent prehistory and history of man, but because of the relatively short half-life of carbon-14, the clock can be used for dating events that have taken place only within the past 50,000 years.

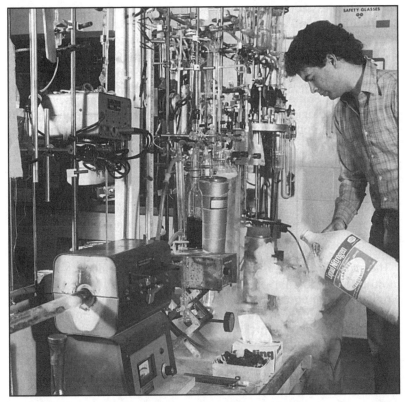

Carbon samples are converted to acetylene gas by combustion in a vacuum line. The acetylene gas is then analyzed in a mass spectrometer to determine its carbon isotopic composition.

The following is a group of rocks and materials that have dated by various atomic clock methods:

	Approximate Age in Years
Cloth wrappings from a mummified bull	2,050
Samples taken from a pyramid in Dashur, Egypt. This date agrees with the age of the pyramid as estimated from historical records.	
Charcoal	6,640
Sample, recovered from bed of ash near Crater Lake, Oregon, is from a tree burned in the violent eruption of Mount Mazama which created Crater Lake. This eruption blanketed several states with ash, providing geologists with an excellent time zone.	
Charcoal	10,130
Sample collected from the "Marmes Man" site in southeastern Washington. This rock shelter is believed to be among the oldest known inhabited sites in North America.	
Spruce wood	11,640
Sample from the Two Creeks forest bed near Milwaukee, Wisconsin, dates one of the last advances of the continental ice sheet into the United States.	
Bishop Tuff	700,000
Samples collected from volcanic ash and pumice that overlie glacial debris in Owens Valley, California. This volcanic episode provides an important reference datum in the glacial history of North America.	
Volcanic ash	1,750,000
Samples collected from strata in Olduvai Gorge, East Africa, which sandwich the fossil remains of Zinjanthropus and Homo habilis— possible precursors of modern man.	
Monzonite	37,500,000
Samples of copper-bearing rock from vast open-pit mine at Bingham Canyon, Utah.	
Quartz monzonite	80,000,000
Samples collected from Half Dome, Yosemite National Park, California.	
Conway Granite	180,000,000
Samples collected from Redstone Quarry in the White Mountains of New Hampshire.	
Rhyolite	820,000,000
Samples collected from Mount Rogers, the highest point in Virginia.	
Pikes Peak Granite	1,030,000,000
Samples collected on top of Pikes Peak, Colorado.	
Gneiss	2,700,000,000
Samples from outcrops in the Karelian area of eastern Finland are believed to represent the oldest rocks in the Baltic region.	
The Old Granite	3,200,000,000
Samples from outcrops in the Transvaal, South Africa. These rocks intrude even older rocks that have not been dated.	
Morton Gneiss	3,600,000,000
Samples from outcrops in southwestern Minnesota are believed to represent some of the oldest rocks in North America.	

Interweaving the relative time scale with the atomic time scale poses certain problems because only certain types of rocks, chiefly the igneous variety, can be dated directly by radiometric methods; but these rocks do not ordinarily contain fossils. Igneous rocks are those such as granite and basalt which crystallize from molten material called *magma.*

When igneous rocks crystallize, the newly formed minerals contain various amounts of chemical elements, some of which have radioactive isotopes. These isotopes decay within the rocks according to their half-life rates, and by selecting the appropriate minerals (those that contain potassium, for instance) and measuring the relative amounts of parent and daughter isotopes in them, the date at which the rock crystallized can be determined. Most of the large igneous rock masses of the world have been dated in this manner.

Most sedimentary rocks such as sandstone, limestone, and shale are related to the radiometric time scale by bracketing them within time zones that are determined by dating appropriately selected igneous rocks, as shown in the hypothetical example on page 18.

The age of the volcanic ash bed and the igneous dike are determined directly by radiometric methods. The layers of sedimentary rocks below the ash bed are obviously older than the ash, and all the layers above the ash are younger. The igneous dike is younger than the Mancos Shale and Mesaverde Formation but older than the Wasatch Formation because the dike does not intrude the Tertiary rocks.

From this kind of evidence, geologists estimate that the end of the Cretaceous Period and the beginning of the Tertiary Period took place between 63 and 66 million years ago. Similarly, a part of the Morrison Formation of Jurassic age was deposited about 160 million years ago as indicated by the age of the ash bed.

Literally thousands of dated materials are now available for use to bracket the various episodes in the history of the Earth within specific time zones. Many points on the time scale are being revised, however, as the behavior of isotopes in the Earth's crust is more clearly understood. Thus the graphic illustration of the geologic time scale, showing both relative time and radiometric time, represents only the present state of knowledge. Certainly, revisions and modifications will be forthcoming as research continues to improve our knowledge of Earth history.

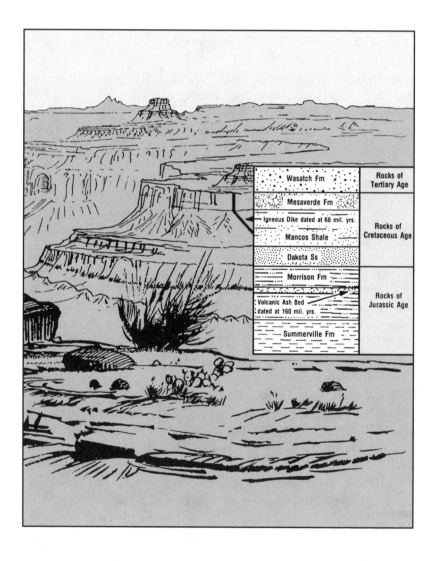

Wasatch Fm	Rocks of Tertiary Age
Mesaverde Fm	Rocks of Cretaceous Age
Igneous Dike dated at 66 mil. yrs.	
Mancos Shale	
Dakota Ss	
Morrison Fm	Rocks of Jurassic Age
Volcanic Ash Bed dated at 160 mil. yrs.	
Summerville Fm	

The Age of the Earth

The oldest known formation in North America, the Morton Gneiss of Minnesota, appears to have formed about $3\frac{1}{2}$ billion years ago. Other rocks of similar or somewhat younger age occur in South Africa, Finland, Australia, and elsewhere. Upon what evidence is a minimum age for the Earth of $4\frac{1}{2}$ billion years based?

Some estimates of the age of the Earth are based on the decay rates of radioactive isotopes. One such estimate is based on two long-lived isotopes, uranium-238 and uranium-235, and places the maximum age of the chemical elements that make up the Earth at about 6 billion years. The reasoning behind this estimate involves theoretical concepts about the origin of uranium which stipulate that not more than half the first-formed uranium consisted of the uranium-235 isotope. But, because of the changes in isotopic composition that have taken place throughout geologic time as a result of radioactive decay, uranium today contains only about 0.7 percent of the uranium-235 isotope. Calculations based on the known decay rates for these two isotopes (see table, page 13) indicate that a span of about 6 billion years is required for the isotopic composition of uranium to have changed from 50 percent to 0.7 percent uranium-235. Had uranium originally consisted of *more than half* of the uranium-235 isotope, the calculated age of the Earth would exceed 6 billion years.

Another estimate of the Earth's age is based on the progressive buildup in the Earth's crust of lead isotopes that have been derived from radioactive decay of uranium-238, uranium-235, and thorium-232. Studies of the relative abundance of these "radiogenic" leads suggest that the Earth is not older than 5.5 billion years.

An approximate age for the Earth has been determined from studies of meteorites. Meteorites are specimens of matter from space; they are classified according to their chemical composition into three main groups: the irons, the stony irons, and the stones. Because of chemical similarities between meteorites and the rocks of the Earth (these three groups may correspond to the three major parts of the Earth—the core, the mantle, and the crust), it is generally assumed that meteorites and the Earth have a common origin and were formed as solid objects at about the same time during the evolution of the Solar System. Thus, what can be learned from meteorites can be used to interpret the Earth. Chemical analyses of iron meteorites show that the uranium content is so low that the isotopic composition of lead has not been changed significantly by the addition of radiogenic lead since the formation of meteorites. Thus, the lead in iron meteorites is considered to be primordial lead. Stony meteorites, however, contain sufficient uranium to have produced appreciable quantities of radiogenic lead since the meteorites were formed. When the measured amounts of radiogenic lead in stony meteorites are corrected by subtracting the amount of primordial lead in iron meteorites, calculations yield an age of about $4^1/_2$ billion years for stony meteorites. If the Earth and meteorites have a common or a similar origin, then it seems reasonable to assume that the age of the Earth is about the same as the age of the meteorites.

Thus, as a current working hypothesis based on several kinds of evidence, geologists consider the Earth to be at least $4^1/_2$ billion years old.

Figure at right: Most of the evidence for an ancient Earth is contained in the rocks that form the Earth's crust. The rock layers themselves record the surface-shaping events of the past, and buried within them are traces of life—the plants and animals that evolved from organic structures that existed perhaps 3 billion years ago.

Also contained in rocks once molten are radioactive elements whose isotopes provide Earth scientists with an atomic clock. Within these rocks, *parent* isotopes decay at a predictable rate to form *daughter* isotopes. By determining the relative amounts of parent and daughter isotopes, the age of these rocks can be calculated.

Thus, the results of studies of rock layers (stratigraphy), and of fossils (paleontology), coupled with the ages of certain rocks as measured by atomic clocks (geochronology), attest to a very old Earth!

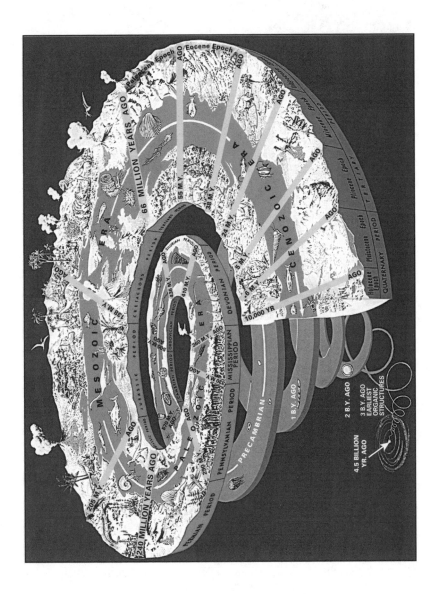

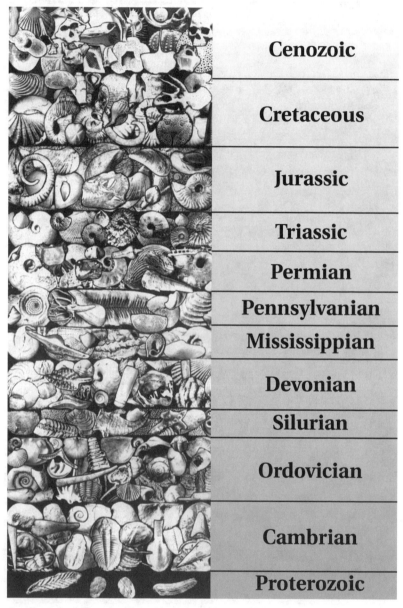

Cenozoic

Cretaceous

Jurassic

Triassic

Permian

Pennsylvanian

Mississippian

Devonian

Silurian

Ordovician

Cambrian

Proterozoic

A collage depicting the diversity and evolution of life on Earth through the last 600 million years. The oldest fossils are at the bottom and the youngest at the top. The size of each time interval is proportional to its duration.

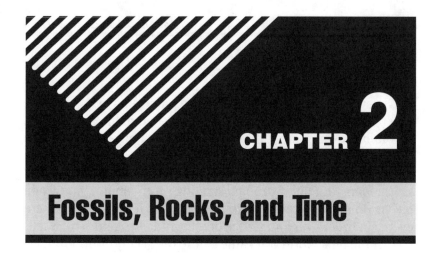

CHAPTER 2

Fossils, Rocks, and Time

We study our Earth for many reasons: to find water to drink or oil to run our cars or coal to heat our homes, to know where to expect earthquakes or landslides or floods, and to try to understand our natural surroundings. Earth is constantly changing—nothing on its surface is truly permanent. Rocks that are now on top of a mountain may once have been at the bottom of the sea. Thus, to understand the world we live on, we must add the dimension of time. We must study Earth's history.

When we talk about recorded history, time is measured in years, centuries, and tens of centuries. When we talk about Earth history, time is measured in millions and billions of years.

Time is an everyday part of our lives. We keep track of time with a marvelous invention, the calendar, which is based on the movements of Earth in space. One spin of Earth on its axis is a day, and one trip around the Sun is a year. The modern calendar is a great achievement, developed over many thousands of years as theory and technology improved.

People who study Earth's history also use a type of calendar, called the geologic time scale. It looks very different from the familiar calendar. In some ways, it is more like a book, and the rocks are its pages. Some of the pages are torn or missing, and the pages are not numbered, but geology gives us the tools to help us read this book.

Putting Events in Order

Scientists who study the past try to put events in their proper order. When we discuss events that happened in historical times, we often use dates or numbers, but we do not have to do so.

Consider six historical events: the Wright brothers' flight, the bicentennial of American independence, the First and Second World Wars, the first astronaut landing on the moon, and when television became common in homes. First, let's try to put these events in order. Our knowledge of the words *first* and *second* tells us that the First World War came before the Second World War. We may know or may have been told that the landing of Neil Armstrong on the moon was seen by many people on television, but there was no television around when the Wright brothers flew at Kitty Hawk. Thus, we can order these three events: first Wright brothers'

In layered rocks like these at Saint Stephens, Alabama, geologists can easily determine the order in which the rocks were formed.

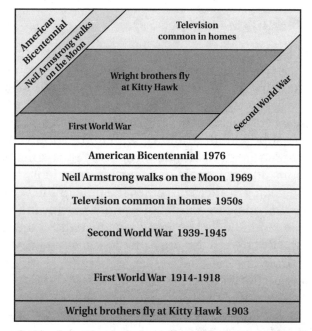

The box at the top shows six events that occurred during the twentieth century. The bottom shows these events in relative order and numeric order.

flight, then television common in homes, then the landing on the moon. By a process of gathering evidence and making comparisons, we can eventually put all six events in the complete proper order: Wright brothers' flight, First World War, Second World War, television common in homes, landing on the moon, and American bicentennial.

Because we have written records of the time each of these events happened, we can also put them in order by using numbers. The Wright brothers' flight occurred in 1903, the First World War lasted from 1914 to 1918, and the Second World War lasted from 1939 to 1945. Televisions became part of our homes in the 1950s, Neil Armstrong walked on the moon in 1969, and America celebrated 200 years of independence in 1976.

Written records are available for only a tiny fraction of the history of Earth. Understanding the rest of the history requires detective work: gathering the evidence and making comparisons.

The Relative Time Scale

Long before geologists had the means to recognize and express time in numbers of years before the present, they developed the geologic time scale. This time scale was developed gradually, mostly in Europe, over the eighteenth and nineteenth centuries. Earth's history is subdivided into eons, which are subdivided into eras, which are subdivided into periods, which are subdivided into epochs. The names of these subdivisions, like Paleozoic or Cenozoic, may look daunting, but to the geologist there are clues in some of the words. For example, *zoic* refers to animal life, and *paleo* means ancient, *meso* means middle, and *ceno* means recent. So the relative order of the three youngest eras, first *Paleozoic*, then *Mesozoic*, then *Cenozoic*, is straightforward.

Fossils are the recognizable remains, such as bones, shells, or leaves, or other evidence, such as tracks, burrows, or impressions, of past life on Earth. Scientists who study fossils are called *paleontologists*. Remember that *paleo* means ancient; so a paleontologist studies ancient forms of life. Fossils are fundamental to the geologic time scale. The names of most of

the eons and eras end in *zoic*, because these time intervals are often recognized on the basis of animal life. Rocks formed during the Proterozoic Eon may have fossils of relative simple organisms, such as bacteria, algae, and wormlike animals. Rocks formed during the Phanerozoic Eon may have fossils of complex animals and plants such as dinosaurs, mammals, and trees.

This rock sample will be taken to the laboratory where tiny fossils will be extracted for further study.

Eon	Era	Period	Epoch
Phanerozoic	Cenozoic	Quaternary	Holocene Pleistocene
		Tertiary	Pliocene Miocene Oligocene Eocene Paleocene
	Mesozoic	Cretaceous	Late Early
		Jurassic	Late Middle Early
		Triassic	Late Early
	Paleozoic	Permian	Late Early
		Pennsylvanian	Late Middle Early
		Mississippian	Late Early
		Devonian	Late Middle Early
		Silurian	Late Middle Early
		Ordovician	Late Middle Early
		Cambrian	Late Middle Early
Proterozoic	Late Proterozoic Middle Proterozoic Early Proterozoic		
Archean	Late Archean Middle Archean Early Archean		
pre-Archean			

The relative geologic time scale. The oldest time interval is at the bottom and the youngest is at the top.

Rocks and Layers

We study Earth's history by studying the record of past events that is preserved in the rocks. The layers of the rocks are the pages in our history book.

Most of the rocks exposed at the surface of Earth are *sedimentary*—formed from particles of older rocks that have been broken apart by water or wind. The gravel, sand, and mud settle to the bottom in rivers, lakes, and oceans. These sedimentary particles may bury living and dead animals and plants on the lake or sea bottom. With the passage of time and the accumulation of more particles, and often with chemical changes, the sediments at the bottom of the pile become rock. Gravel becomes a rock called conglomerate, sand becomes sandstone, mud becomes mudstone or shale, and the animal skeletons and plant pieces can become fossils.

An idealized view of a modern landscape and some of the plants and animals that could be preserved as fossils.

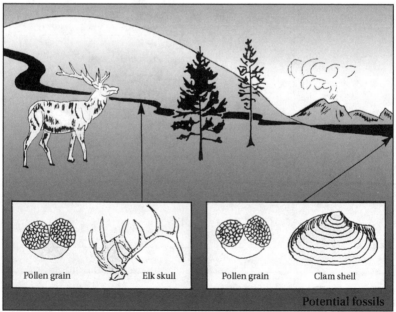

Pollen grain   Elk skull   Pollen grain   Clam shell

Potential fossils

EON	FORMS OF LIFE	
Phanerozoic	Animals with shells or bones; land animals and plants	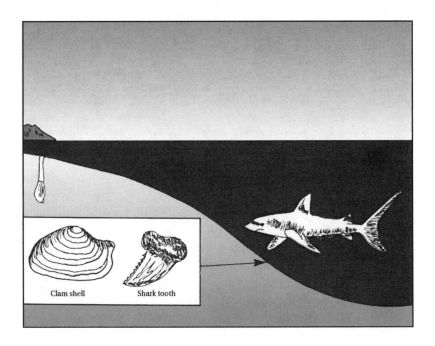
Proterozoic	Simple and complex single-celled organisms, algae, wormlike organisms	
Archean	Microscopic single-celled or filament-shaped organisms	
pre-Archean	No record of life	

Originations of major life forms.

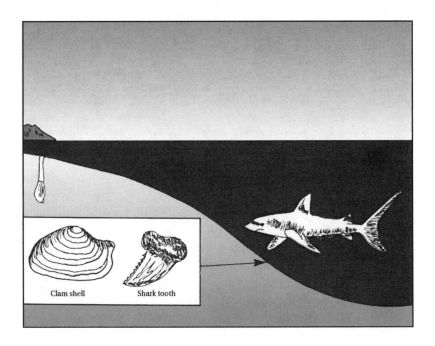

Clam shell   Shark tooth

Outcrop of the Ordovician Lexington Limestone, which is rich in fossil shells, near Lexington, Kentucky. These horizontally layered beds were deposited about 450 million years ago. The dark stains on the rocks are formed by water seeping from springs. The vertical marks on the rocks are drill holes in which dynamite charges were exploded to remove the rock so that an interstate highway could be built.

As early as the mid-1600s, the Danish scientist Nicholas Steno studied the relative positions of sedimentary rocks. He found that solid particles settle from a fluid according to their relative weight or size. The largest, or heaviest, settle first, and the smallest, or lightest, settle last. Slight changes in particle size or composition result in the formation of layers, also called beds, in the rock. Layering, or bedding, is the most obvious feature of sedimentary rocks.

Sedimentary rocks are formed particle by particle and bed

by bed, and the layers are piled one on top of the other. Thus, in any sequence of layered rocks, a given bed must be older than any bed on top of it. This *Law of Superposition* is fundamental to the interpretation of Earth history, because at any one location it indicates the relative ages of rock layers and the fossils in them.

Layered rocks form when particles settle from water or air. Steno's *Law of Original Horizontality* states that most sediments, when originally formed, were laid down horizontally. However, many layered rocks are no longer horizontal. Because of the *Law of Original Horizontality*, we know that sedimentary rocks that are not horizontal either were formed in special ways or, more often, were moved from their horizontal position by later events, such as tilting during episodes of mountain building.

Rock layers are also called *strata* (the plural form of the Latin word *stratum*), and stratigraphy is the science of strata. Stratigraphy deals with all the characteristics of layered rocks; it includes the study of how these rocks relate to time.

Nearly vertical limestone beds that were disturbed from their original horizontal position by mountain building. The men are collecting Silurian fossil shells from rocks in the Arbuckle Mountains, near Ardmore, Oklahoma.

Fossils and Rocks

To tell the age of most layered rocks, scientists study the fossils these rocks contain. Fossils provide important evidence to help determine what happened in Earth history and when it happened.

The word *fossil* makes many people think of dinosaurs. Dinosaurs are now featured in books, movies, and television programs, and the bones of some large dinosaurs are on display in many museums. These reptiles were dominant animals on Earth for well over 100 million years from the Late Triassic through the Late Cretaceous. Many dinosaurs were quite small, but by the middle of the Mesozoic Period, some species weighed as much as 80 tons. By around 65 million years ago, all dinosaurs were extinct. The reasons for and the rapidity of their extinction are a matter of intense debate among scientists.

In spite of all the interest in dinosaurs, they form only a small fraction of the millions of species that live and have lived on Earth. The great bulk of the fossil record is dominated by fossils of animals with shells and microscopic remains of plants and animals, and these remains are widespread in sedimentary rocks. It is these fossils that are studied by most paleontologists.

In the late eighteenth and early nineteenth centuries, the English geologist and engineer William Smith and the French paleontologists Georges Cuvier and Alexandre Brongniart discovered that rocks of the same age may contain the same fossils even when the rocks are separated by long distances. They published the first geologic maps of large areas on which rocks containing similar fossils were shown. By careful observation of the rocks and their fossils, these men and other geologists were able to recognize rocks of the same age on opposite sides of the English Channel.

William Smith was able to apply his knowledge of fossils in a very practical way. He was an engineer building canals in England, which has lots of vegetation and few surface exposures of rock. He needed to know what rocks he could expect to find on the hills through which he had to build a canal. Often he could tell what kind of rock was likely to be below the surface by examining the fossils that had eroded from the rocks of the hillside or by digging a small hole to find fossils. Knowing what rocks to expect allowed Smith to estimate costs and determine what tools were needed for the job.

Smith and others knew that the succession of life forms preserved as fossils is useful for understanding how and when the rocks formed. Only later did scientists develop a theory to explain that succession.

Collecting samples from the bottom of the Mississippi Sound to look for the kinds of microorganisms that are preserved as fossils. The metal box is lowered overboard, scoops up bottom sediments, and then is raised onto the ship by a winch and pulleys.

Fossil Succession

Three concepts are important in the study and use of fossils: (1) Fossils represent the remains of once-living organisms. (2) Most fossils are the remains of extinct organisms; that is, they belong to species that are no longer living anywhere on Earth. (3) The kinds of fossils found in rocks of different ages differ because life on Earth has changed through time.

If we begin at the present and examine older and older layers of rock, we will come to a level where no fossils of humans are present. If we continue backwards in time, we will successively come to levels where no fossils of flowering plants are present, no birds, no mammals, no reptiles, no four-footed vertebrates, no land plants, no fishes, no shells, and no animals. The three concepts are summarized in the general principle called the *Law of Fossil Succession:* The kinds of animals and plants found

Stratigraphic ranges and origins of some major groups of animals and plants.

Period	Animals							
Quaternary								
Tertiary								
Cretaceous								Birds
Jurassic							Mammals	
Triassic								
Permian					Amphibians	Reptiles		
Pennsylvanian		Animals with shells						
Mississippian								
Devonian				Fishes				
Silurian								
Ordovician								
Cambrian								

as fossils change through time. When we find the same kinds of fossils in rocks from different places, we know that the rocks are the same age.

How do scientists explain the changes in life forms, which are obvious in the record of fossils in rocks? Early explanations were built around the idea of successive natural disasters or catastrophes that periodically destroyed life. After each catastrophe, life began anew. In the mid-nineteenth century, both Charles Darwin and Alfred Wallace proposed that older species of life give rise to younger ones. According to Darwin, this change or *evolution* is caused by four processes: variation, over-reproduction, competition, and survival of those best adapted to the environment in which they live. Darwin's theory accounts for all of the diversity of life, both living and fossil. His explanation gave scientific meaning to the observed succession of once-living species seen as fossils in the record of Earth's history preserved in the rocks.

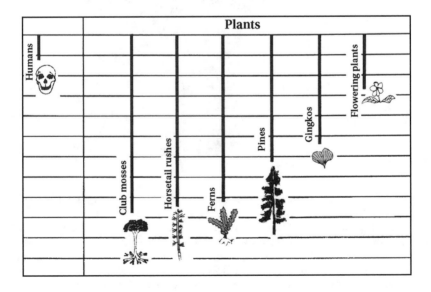

Scientific theories are continually being corrected and improved, because theory must always account for known facts and observations. Therefore, as new knowledge is gained, a theory may change. Application of theory allows us to develop new plants that resist disease, to transplant kidneys, to find oil, and to establish the age of our Earth. Darwin's theory of evolution has been refined and modified continuously as new information has accumulated. All of the new information has supported Darwin's basic concept—that living beings have changed through time and older species are ancestors of younger ones.

Scientists look for ancestors and descendants through geologic time. The fossil *Archaeopteryx lithographica* was a Jurassic animal with the skeleton of a reptile, including fingers with claws on the wings (solid arrows), backbone extending into the tail (open arrow), and teeth, but it was covered with feathers. We can see fossils of many other reptiles in rock of the same age and even older, but *Archaeopteryx lithographica* is the oldest known fossil to have feathers. We conclude that this animal is a link between reptiles and birds and that birds are descended from reptiles. The specimen is about 45 centimeters long.

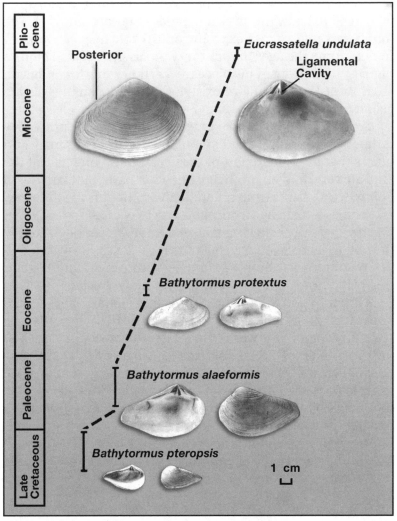

A species is the most basic unit of classification for living things. This group of fossil clams shows likely ancestor-descendant relationships at the species level. These fossils from the Mid-Atlantic States show the way species can change through time. Notice how the shape of the posterior (rear) end of these clams becomes more rounded in the younger species, and the area where the two shells are held together (ligamental cavity) gets larger. Paleontologists pay particular attention to the shape of the shells and the details of the anatomy preserved as markings on the shells.

The *Law of Fossil Succession* is very important to geologists who need to know the ages of the rocks they are studying. The fossils present in a rock exposure or in a core hole can be used to determine the ages of rocks very precisely. Detailed studies of many rocks from many places reveal that some fossils have a short, well-known time of existence. These useful fossils are called *index fossils*.

Today the animals and plants that live in the ocean are very different from those that live on land, and the animals and plants that live in one part of the ocean or on one part of the land are very different from those in other parts. Similarly, fossil animals and plants from different environments are different. It becomes a challenge to recognize rocks of the same age when one rock was deposited on land and another was deposited in the deep ocean. Scientists must study the fossils from a variety of environments to build a complete picture of the animals and plants that were living at a particular time in the past.

The study of fossils and the rocks that contain them occurs both out of doors and in the laboratory. The field work can take place anywhere in the world. In the laboratory, rock saws, dental drills, pneumatic chisels, inorganic and organic acids, and other mechanical and chemical procedures may be used to prepare samples for study. Preparation may take days, weeks, or months—large dinosaurs may take years to prepare. Once the fossils are freed from the rock, they can be studied and interpreted. In addition, the rock itself provides much useful information about the environment in which it and the fossils were formed.

Fossils can be used to recognize rocks of the same or different ages. The fossils in the figure at right are the remains of microscopic algae. The pictures shown were made with a scanning electron microscope and have been magnified about 250 times. In South Carolina, three species are found in a core of rock. In Virginia, only two of the species are found. We know from the species that do occur that the rock record from the early part of the middle Eocene is missing in Virginia. We also use these species to recognize rocks of the same ages (early Eocene and latter part of the middle Eocene) in both South Carolina and Virginia. The study of layered rocks and the fossils they contain is called biostratigraphy; the prefix *bio* is Greek and means *life.*

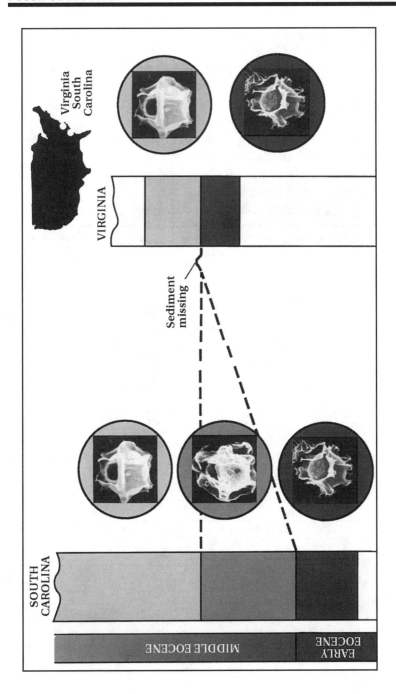

The Numeric Time Scale

Thus far we have been discussing the relative time scale. How can we add numbers to our time scale? How have geologists determined that

—Earth is about 4.6 billion years old?

—the oldest known fossils are from rocks that were deposited about 3.5 billion years ago?

—the first abundant shelly fossils occur in rocks that are about 570 million years old?

—the last ice age ended about 10,000 years ago?

Eon	Era	Period	Millions of years ago
Phanerozoic	Cenozoic	Quaternary	— 1.6 —
		Tertiary	
			— 66 —
	Mesozoic	Cretaceous	— 138 —
		Jurassic	— 205 —
		Triassic	— 240 —
	Paleozoic	Permian	— 290 —
		Pennsylvanian	— 330 —
		Mississippian	
		Devonian	— 360 —
		Silurian	— 410 —
		Ordovician	— 435 —
		Cambrian	— 500 —
Proterozoic	Late Proterozoic Middle Proterozoic Early Proterozoic		— 570 —
			—2500—
Archean	Late Archean Middle Archean Early Archean		
pre-Archean			—3800?—

Geologic time scale showing both relative and numeric ages. Ages in millions of years are approximate.

Nineteenth-century geologists and paleontologists believed that Earth was quite old, but they had only crude ways of estimating just how old. The assignment of ages of rocks in thousands, millions, and billions of years was made possible by the discovery of radioactivity. Now we can use minerals that contain naturally occurring radioactive elements to calculate the numeric age of a rock in years.

The basic unit of each chemical element is the atom. An atom consists of a central nucleus, which contains protons and neutrons, surrounded by a cloud of electrons. *Isotopes* of an element are atoms that differ from one another only in the number of neutrons in the nucleus. For example, radioactive atoms of the element potassium have 19 protons and 21 neutrons in the nucleus (potassium 40); other atoms of potassium have 19 protons and 20 or 22 neutrons (potassium 39 and potassium 41). A radioactive isotope (the parent) of one chemical element naturally converts to a stable isotope (the daughter) of another chemical element by undergoing changes in the nucleus.

Isotopes		Half-life of Parent (years)	Useful range (years)
Parent	Daughter		
Carbon 14	Nitrogen 14	5,730	100 to 30,000
Potassium 40	Argon 40	1.3 billion	100,000 to 4.5 billion
Rubidium 87	Strontium 87	47 billion	10 million to 4.5 billion
Uranium 238	Lead 206	4.5 billion	10 million to 4.6 billion
Uranium 235	Lead 207	710 million	

Parents and daughters for some isotopes commonly used to establish numeric ages of rocks.

The change from parent to daughter happens at a constant rate, called the *half-life*. The half-life of a radioactive isotope is the length of time required for exactly one-half of the parent atoms to decay to daughter atoms. Each radioactive isotope has its own unique half-life. Precise laboratory measurements of the number of remaining atoms of the parent and the number of atoms of the new daughter produced are used to compute the age of the rock. For dating geologic materials, four parent/daughter decay series are especially useful: carbon to nitrogen, potassium to argon, rubidium to strontium, and uranium to lead.

Age determinations using radioactive isotopes are subject to relatively small errors in measurement—but errors that look small can mean many years or millions of years. If the measurements have an error of 1 percent, for example, an age determination of 100 million years could actually be wrong by a million years too low or too high.

Isotopic techniques are used to measure the time at which a particular mineral within a rock was formed. To allow us to assign numeric ages to the geologic time scale, a rock that can be dated isotopically is found together with rocks that can be assigned relative ages because of their fossils. Many samples, usually from several different places, must be studied before assigning a numeric age to a boundary on the geologic time scale.

The geologic time scale is the product of many years of detective work, as well as a variety of dating techniques not discussed here. The details will change as more and better information and tools become available. Many scientists have contributed and continue to contribute to the refinement of the geologic time scale as they study the fossils and the rocks, and the chemical physical properties of the materials of which Earth is made. Just as in the time of William Smith, knowing what kinds of rocks are found below the soil can help people to make informed judgments about the uses of the resources of the planet.

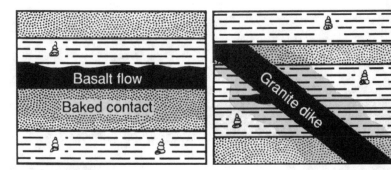

The basalt *flow* shown here obeys the *Law of Superposition*. It is younger than the beds below it and older than the beds above it. Note that the molten rock of the volcanic flow has baked the rock underneath it. The bed above was deposited long after the flow had cooled and hardened and has not been baked. Numeric ages from the *flow* and *dike* and relative ages from the fossils in the surrounding rocks contribute to the geologic time scale.

The granite *dike* (a mass of rock that cuts across the structure of the rocks around it) shown here illustrates the *Law of Cross-Cutting Relations*. The dike is younger than all the rocks that it cuts across and older than the rocks above it that it does not cut. Note that the contact between the dike and the rocks around it has been baked by the heat of the molten granite.

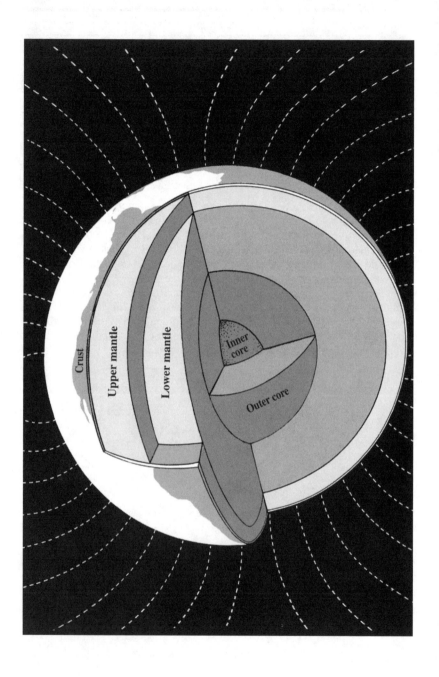

CHAPTER 3

The Earth's Interior

Three centuries ago, the English scientist Isaac Newton calculated, from his studies of planets and the force of gravity, that the average density of the Earth is twice that of surface rocks and therefore that the Earth's interior must be composed of much denser material. Our knowledge of what's inside the Earth has improved immensely since Newton's time, but his estimate of the density remains essentially unchanged. Our current information comes from studies of the paths and characteristics of earthquake waves travelling through the Earth, as well as from laboratory experiments on surface minerals and rocks at high pressure and temperature. Other important data on the Earth's interior come from geological observation of surface rocks and studies of the Earth's motions in the Solar System, its gravity and magnetic fields, and the flow of heat from inside the Earth.

The planet Earth is made up of three main shells: the very thin, brittle *crust*, the *mantle*, and the *core*; the mantle and core are each divided into two parts. A table on page 50 lists the thicknesses of all the parts. Although the core and mantle are about equal in thickness, the core actually forms only 15 percent of the Earth's volume, whereas the mantle occupies 84 percent. The crust makes up the remaining 1 percent. Our knowledge of the layering and chemical composition of the Earth is steadily being improved by earth scientists doing laboratory experiments on rocks at high pressure and analyzing earthquake records on computers.

The Crust

Because the crust is accessible to us, its geology has been extensively studied, and therefore much more information is known about its structure and composition than about the structure and composition of the mantle and core. Within the crust, intricate patterns are created when rocks are redistributed and deposited in layers through the geologic processes of eruption and intrusion of lava, erosion and consolidation of rock particles, and solidification and recrystallization of porous rock.

By the large-scale process of plate tectonics, about twelve plates, which contain combinations of continents and ocean basins, have moved around on the Earth's surface through much of geologic time. The edges of the plates are marked by concentrations of earthquakes and volcanoes. Collisions of plates can produce mountains like the Himalayas, the tallest range in the world. The plates include the crust and part of the upper mantle, and they move over a hot, yielding upper mantle zone at very slow rates of a few centimeters per year, slower than the rate at which fingernails grow. The crust is much thinner under the oceans than under continents (see figure at right).

The boundary between the crust and mantle is called the Mohorovicic discontinuity (or Moho); it is named in honor of the man who discovered it, the Croatian scientist Andrija Mohorovicic. No one has ever seen this boundary, but it can be detected by a sharp increase downward in the speed of earthquake waves there. The explanation for the increase at the Moho is presumed to be a change in rock types. Drill holes to penetrate the Moho have been proposed, and a Soviet hole on the Kola Peninsula was drilled to a depth of 12 kilometers before the project was abandoned in 1989. However, drilling expense increases enormously with depth, and Moho penetration is not likely very soon.

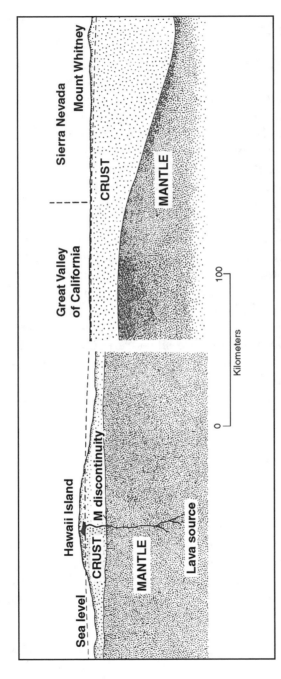

The oceanic crust at Hawaii Island is about 5 kilometers thick. The thickness of the continental crust under eastern California ranges from a normal 25 kilometers under the Great Valley to 60 kilometers under the Sierra Nevada.

The Mantle

Our knowledge of the upper mantle, including the tectonic plates, is derived from analyses of earthquake waves (see figure at right for paths); heat flow, magnetic, and gravity studies; and laboratory experiments on rocks and minerals. Between 100 and 200 kilometers below the Earth's surface, the temperature of the rock is near the melting point; molten rock erupted by some volcanoes originates in this region of the mantle. This zone of extremely yielding rock has a slightly lower velocity of earthquake waves and is presumed to be the layer on which the tectonic plates ride.

Below this low-velocity zone is a transition zone in the upper mantle; it contains two discontinuities caused by changes from less dense to more dense minerals. The chemical composition and crystal forms of these minerals have been identified by laboratory experiments at high pressure and temperature. The lower mantle, below the transition zone, is made up of relatively simple iron and magnesium silicate minerals, which change gradually with depth to very dense forms. Going from mantle to core, there is a marked decrease (about 30 percent) in earthquake wave velocity and a marked increase (about 30 percent) in density.

The Core

The core was the first internal structural element to be identified. It was discovered in 1906 by R.D. Oldham from his study of earthquake records, and it helped to explain Newton's calculation of the Earth's density. The outer core is presumed to be liquid because it does not transmit shear (S) waves and because the velocity of compressional (P) waves that pass through it is sharply reduced. The inner core is considered to be solid because of the behavior of P and S waves passing through it.

Data from earthquake waves, rotations and inertia of the

whole Earth, magnetic-field dynamo theory, and laboratory experiments on melting and alloying of iron all contribute to the identification of the composition of the inner and outer core. The core is presumed to be composed principally of iron, with about 10 percent alloy of oxygen or sulfur or nickel, or perhaps some combination of these three elements.

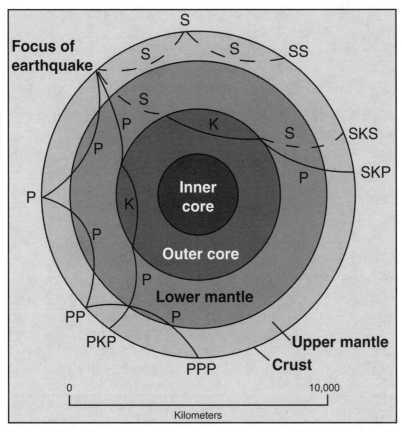

Cross section of the whole Earth, showing the complexity of paths of earthquake waves. The paths curve because the different rock types found at different depths change the speed at which the waves travel. Solid lines marked P are compressional waves; dashed lines marked S are shear waves. S waves do not travel through the core but may be converted to compressional waves (marked K) on entering the core (PKP, SKS). Waves may be reflected at the surface (PP, PPP, SS).

Data on the Earth's Interior

	Thickness (km)	Top	Bottom	Types of rock found
Crust	30	2.2	2.9	Silicic rocks, Andesite, basalt at base
Upper mantle	720	3.4	4.4	Peridotite, eclogite, olivine, spinel, garnet, pyroxene Perovskite, oxides
Lower mantle	2171	4.4	5.6	Magnesium and silicon oxides
Outer core	2259	9.9	12.2	Iron, oxygen, sulfer, nickel alloy
Inner core	1221	12.8	13.1	Iron, oxygen, sulfer, nickel alloy
Total thickness	6371			

Scientists are continuing to refine the chemical and mineral composition of the Earth's interior by laboratory experiments, using pressures two million times the pressure of the atmosphere at the surface and temperatures as high as 2,000°C.

The Structure of the Moon

The Moon, our fellow-traveler in space, has a diameter half that of the Earth's core, and it rotates around the Earth, as all the planets rotate around the Sun, under the force of gravity. Moonquakes of very low energy are caused by land tides produced by the pull of Earth's gravity, and, from analysis of moonquake data, scientists believe the Moon has two layers: a crust, from the surface to 65 kilometers depth, and an inner, more dense mantle from the crust to the center at 3,700 kilometers. The crust is presumed to be composed primarily of rocks containing feldspar, calcium aluminum silicate, and lesser pyroxene, iron and magnesium silicate; the crust also contains basalt in the mares, which contains less iron and more titanium than earth basalt. The mantle is thought to be made up of calcic peridotite, containing both pyroxene and feldspar.

Quartz

Quartz is one of the most common minerals in the Earth's crust. It is made up of silicon dioxide (SiO_2), otherwise known as silica. Quartz in its natural state is shown below. The photo at left is a quartz cluster.

Iron Pyrite

Iron sulfide mineral (FeS) forms silvery to brassy metallic cubes or masses. Known as fool's gold, it is common in many rocks. Weathered pyrite produces limonite (iron oxide) that stains rock brown or yellow.

Calcite

Calcite is a mineral made of calcium carbonate ($CaCO3$). It is generally white and easily scratched with a knife. Most seashells are made of calcite or related minerals. This is the lime of limestone. Here, it appears on fluorite.

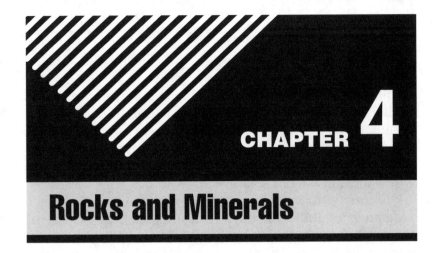

CHAPTER 4

Rocks and Minerals

Rocks

Rocks are all around us. They make up the backbones of hills and mountains and the foundations of plains and valleys. Beneath the soil you walk on and the deep layers of soft mud that cover the ocean basins is a basement of hard rock.

Rocks are made up mostly of crystals of different kinds of minerals, or broken pieces of crystals, or broken pieces of rocks. Some rocks are made of the shells of once-living animals, or of compressed pieces of plants.

We can learn something about the way a rock formed from looking carefully at the evidence preserved inside. What a rock is made of, the shapes of the grains or crystals within the rock, and how the grains or crystals fit together all provide valuable clues to help us unlock the rock's history hidden within.

Rocks are divided into three basic types—*igneous, sedimentary,* and *metamorphic*—depending upon how they were formed. Plate tectonics provides an explanation for how rocks are recycled from igneous to sedimentary to metamorphic and back to igneous again.

Igneous Rocks

Igneous rocks (from the Greek word for fire) form when hot, molten rock (magma) crystallizes and solidifies. The melt originates deep within the Earth near active plate boundaries or hot spots, then rises toward the surface. Igneous rocks are divided into two groups, intrusive or extrusive, depending upon where the molten rock solidifies.

Extrusive, or volcanic, igneous rock is produced when magma exits and cools outside of, or very near the Earth's surface. These are the rocks that form at erupting volcanoes and oozing fissures. The magma, called lava when molten rock erupts on the surface, cools and solidifies almost instantly when it is exposed to the relatively cool temperature of the atmosphere.

Quick cooling means that mineral crystals don't have much time to grow, so these rocks have a very fine-grained or even glassy texture. Hot gas bubbles are often trapped in the quenched lava, forming a bubbly, vesicular texture. Pumice, obsidian, and basalt are all extrusive igneous rocks.

Intrusive, or plutonic igneous rock forms when magma is trapped deep inside the Earth. Great globs of molten rock rise toward the surface. Some of the magma may feed volcanoes on the Earth's surface, but most remains trapped below, where it cools very slowly over many thousands or millions of years until it solidifies. Slow cooling means the individual mineral grains have a very long time to grow, so they grow to a relatively large size. Intrusive rocks have a coarse grained texture. Granite is an intrusive igneous rock.

Naming Igneous Rocks

The names we give igneous rocks are based on their chemical compositions. The relative amounts of just three main minerals–quartz, plagioclase feldspar, and potassium feldspar– are all you need to know to start naming igneous rocks. In addition to these three light-colored, felsic minerals, the

Intrusive Plutonic	Gabro	Diorite	Granodiorite	Granite
Extrusive Volcanic	Basalt	Andesite	Dacite	Rhyolite

Mafic ⟵⟶ Felsic

Increasing differentiation
Increasing SIO_2, NA, K
Decreasing FE, MG, CA

abundance of dark, mafic minerals can also help you distinguish one type of igneous rock from another.

The chart above will help you get started. Once you know you have an igneous rock, look at the texture to decide if it is intrusive or extrusive. Then use this chart to make your first guess based on how dark (*mafic*) or light (*felsic*) your rock appears.

To be sure you've named your rock correctly you need to compare the amounts of plagioclase feldspar, potassium feldspar, and quartz and plot it on the chart below. It takes a little practice to get used to a triangular graph.

Try this: Suppose your rock is coarse-grained, so you know it's intrusive. It has 40% quartz, 30% potassium feldspar, and 30% plagioclase feldspar—it's called granite.

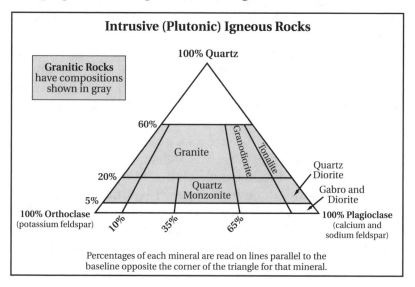

Intrusive (Plutonic) Igneous Rocks

Granitic Rocks have compositions shown in gray

100% Quartz

60%

Granite

Granodiorite

Tonalite

20%

Quartz Monzonite

5%

Quartz Diorite

Gabro and Diorite

100% Orthoclase (potassium feldspar)

10%   35%   65%

100% Plagioclase (calcium and sodium feldspar)

Percentages of each mineral are read on lines parallel to the baseline opposite the corner of the triangle for that mineral.

Sedimentary Rocks

Sedimentary rocks are formed from preexisting rocks or pieces of once-living organisms. They form from deposits that accumulate on the Earth's surface. Sedimentary rocks often have distinctive layering or bedding. Many of the picturesque views of the desert Southwest show mesas and arches made of layered sedimentary rock.

Clastic sedimentary rocks are the group of rocks most people think of when they think of sedimentary rocks. Clastic sedimentary rocks are made up of pieces (clasts) of preexisting rocks. Pieces of rock are loosened by weathering, then transported to some basin or depression where sediment is trapped. If the sediment is buried deeply, it becomes compacted and cemented, forming sedimentary rock.

Clastic sedimentary rocks may have particles ranging in size from microscopic clay to huge boulders. Their names are based on their clast or grain size. The smallest grains are called clay, then silt, then sand. Grains larger that 2 millimeters are called pebbles. Shale is a rock made mostly of clay, siltstone is made up of silt-sized grains, sandstone is made of sand-sized clasts, and conglomerate is made of pebbles surrounded by a matrix of sand or mud.

Biologic sedimentary rocks form when large numbers of living things die, pile up, and are compressed and cemented to form rock. Accumulated carbon-rich plant material may form coal. Deposits made mostly of animal shells may form limestone, coquina, or chert.

Chemical sedimentary rocks are formed by chemical precipitation. The stalactites and stalagmites you see in caves form this way; so does the rock salt that table salt comes from. This process begins when water traveling through rock dissolves some of the minerals, carrying them away from their source. Eventually these minerals can be redeposited, or precipitated, when the water evaporates away or when the water becomes oversaturated with minerals.

Metamorphic Rocks

Metamorphic rocks started out as some other type of rock, but have been substantially changed from their original igneous, sedimentary, or earlier metamorphic form. Metamorphic rocks form when rocks are subjected to high heat, high pressure, hot, mineral-rich fluids or, more commonly, some combination of these factors. Conditions like these are found deep within the Earth or where tectonic plates meet.

In metamorphic rocks some or all of the minerals in the original rock are replaced, atom by atom, to form new minerals.

Foliation forms when pressure squeezes the flat or elongate minerals within a rock so they become aligned. These rocks develop a platy or sheetlike structure that reflects the direction that pressure was applied in. Slate, schist, and gneiss (pronounced "nice") are all foliated metamorphic rocks.

Non-foliated metamorphic rocks do not have a platy or sheetlike structure. There are several ways that non-foliated rocks can be produced. Some rocks, such as limestone, are made of minerals that are not flat or elongate. No matter how much pressure you apply, the grains will not align. Another type of metamorphism, contact metamorphism, occurs when hot igneous rock intrudes into some preexisting rock. The preexisting rock is essentially baked by the heat, changing the mineral structure of the rock without addition of pressure.

IGNEOUS

Igneous rocks form when molten rock (magma) originating from deep within the Earth solidifies. The chemical composition of the magma and its cooling rate determine the final igneous rock type.

Intrusive (plutonic)

Intrusive igneous rocks are formed from magma that cools and solidifies deep beneath the Earth's surface. The insulating effect of the surrounding rock allows the magma to solidify very slowly. Slow cooling means the individual mineral grains have a long time to grow, so they grow to a relatively large size. Intrusive rocks have a characteristically coarse grain size.

Extrusive (volcanic)

Extrusive igneous rocks are formed from magma that cools and solidifies at or near the Earth's surface. Exposure to the relatively cool temperature of the atmosphere or water makes the erupted magma solidify very quickly. Rapid cooling means the individual mineral grains have only a short time to grow, so their final size is very tiny, or fine-grained. Sometimes the magma is quenched so rapidly that individual minerals have no time to grow. This is how volcanic glass forms.

SEDIMENTARY

Sedimentary rocks are formed from preexisting rocks or pieces of once-living organisms. They form from deposits that accumulate on the Earth's surface.

Clastic

Clastic sedimentary rocks are made up of pieces (clasts) of preexisting rocks. Pieces of rock are loosened by weathering, then transported to some basin or depression where sediment is trapped. If the sediment is buried deeply, it becomes compacted and cemented, forming sedimentary rock.

Clastic sedimentary rocks may have particles ranging in size from microscopic clay to huge boulders. Their names are based on their grain size.

Chemical

Chemical sedimentary rocks are formed by chemical precipitation. This process begins when water traveling through rock dissolves some of the minerals, carrying them away from their source. Eventually these minerals are redeposited when the water evaporates away or when the water becomes oversaturated.

Biologic

Biologic sedimentary rocks form from once-living organisms. They may form from accumulated carbon-rich plant material or from deposits of animal shells.

METAMORPHIC

Metamorphic rocks are rocks that have been substantially changed from their original igneous, sedimentary, or earlier metamorphic form. Metamorphic rocks form when rocks are subjected to high heat, high pressure, hot, mineral-rich fluids or, more commonly, some combination of these factors.

Foliated

Foliation forms when pressure squeezes the flat or elongate minerals within a rock so they become aligned. These rocks develop a platy or sheetlike structure that reflects the direction that pressure was applied.

Non-foliated

Non-foliated metamorphic rocks do not have a platy or sheetlike structure. There are several ways that non-foliated rocks can be produced. Some rocks, such as limestone, are made of minerals that are not flat or elongate. No matter how much pressure you apply, the grains will not align. Another type of metamorphism, contact metamorphism, occurs when hot igneous rock intrudes into some preexisting rock. The preexisting rock is essentially baked by the heat, changing the mineral structure of the rock without addition of pressure.

Useful Terms

Basalt

A dark, fine-grained, extrusive (volcanic) igneous rock with a low silica content (40% to 50%). Rich in iron, magnesium, and calcium. Most of the ocean floor is made up of basalt and it is the most abundant volcanic rock in the Earth's crust.

Sediment

Sediment is the word geologists use for loose pieces of minerals and rock. Of course, pieces of rock come in all sizes so we have developed a way to classify them (see chart below). Most of the names are ones you already use; sand, boulders, and clay, for example.

Sediment Sizes		
256 mm and up	Boulders	Gravel
64-256 mm	Cobbles	Gravel
2-64 mm	Pebbles	Gravel
0.0625-2 mm	Sand	
0.002-0.0625 mm	Silt	
0.002 mm	Clay	

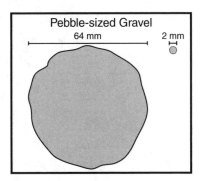

Pebble-sized Gravel

Tuff

Tuff is a volcanic rock made up of a mixture of volcanic rock and mineral fragments in a volcanic ash matrix. Wherever there are explosive volcanic eruptions you can expect to find tuff.

Tuff forms when some combination of ash, rock, and mineral fragments (pyroclastics or tephra) are blasted into the air, then fall to the ground as a mixed deposit. Most of the rock fragments tend to be volcanic rocks that were once solidified parts of the volcano that erupted to produce the tuff, but sometimes other types of rock are blasted out and incorporated into the tuff as well. Sometimes erupted material is so hot when it reaches the ground that it fuses together to produce a welded tuff.

Mount St. Helens, about noon, May 18, 1980.

CHAPTER 5

Volcanoes

Volcanoes destroy and volcanoes create. The catastrophic eruption of Mount St. Helens on May 18, 1980, made clear the awesome destructive power of a volcano. Yet, over a time span longer than human memory and record, volcanoes have played a key role in forming and modifying the planet upon which we live. More than 80 percent of the Earth's surface—above and below sea level—is of volcanic origin. Gaseous emissions from volcanic vents over hundreds of millions of years formed the Earth's earliest oceans and atmosphere, which supplied the ingredients vital to evolve and sustain life. Over geologic eons, countless volcanic eruptions have produced mountains, plateaus, and plains, which subsequent erosion and weathering have sculpted into majestic landscapes and formed fertile soils.

Ironically, these volcanic soils and inviting terranes have attracted, and continue to attract, people to live on the flanks of volcanoes. Thus, as population density increases in regions of active or potentially active volcanoes, mankind must become increasingly aware of the hazards and learn not to "crowd" the volcanoes. People living in the shadow of volcanoes must live in harmony with them and expect, and should plan for, periodic violent unleashings of their pent-up energy.

This chapter presents a generalized summary of the nature, workings, products, and hazards of the common types of volcanoes around the world, along with a brief introduction to the techniques of volcano monitoring and research.

On August 24, A.D. 79, Vesuvius Volcano suddenly exploded and destroyed the Roman cities of Pompeii and Herculaneum. Although Vesuvius had shown stirrings of life when a succession of earthquakes in A.D. 63 caused some damage, it had been literally quiet for hundreds of years and was considered "extinct." Its surface and crater were green and covered with vegetation, so the eruption was totally unexpected. Yet in a few hours, hot volcanic ash and dust buried the two cities so thoroughly that their ruins were not uncovered for nearly 1,700 years, when the discovery of an outer wall in 1748 started a period of modern archaeology. Vesuvius has continued its activity intermittently ever since A.D. 79 with numerous minor eruptions and several major eruptions occurring in 1631, 1794, 1872, 1906, and in 1944 in the midst of the Italian campaign of World War II.

In the United States on March 27, 1980, Mount St. Helens Volcano in the Cascade Range, southwestern Washington, reawakened after more than a century of dormancy and provided a dramatic and tragic reminder that there are active volcanoes in the lower 48 states as well as in Hawaii and Alaska. The catastrophic eruption of Mount St. Helens on May 18, 1980, and related mudflows and flooding caused significant loss of life (57 dead or missing) and property damage (over $1.2 billion). Mount St. Helens is expected to remain intermittently active for months or years, possibly even decades.

The word "volcano" comes from the little island of Vulcano in the Mediterranean Sea off Sicily. Centuries ago, the people living in this area believed that Vulcano was the chimney of the forge of Vulcan—the blacksmith of the Roman gods. They thought that the hot lava fragments and clouds of dust erupting from Vulcano came from Vulcan's forge as he beat out thunderbolts for Jupiter, king of the gods, and weapons for Mars, the god of war. In Polynesia, the people attributed eruptive activity to the beautiful but wrathful Pele, Goddess of Volcanoes, whenever she was angry or spiteful. Today we know that volcanic eruptions are not supernatural but can be studied and interpreted by scientists.

The Nature of Volcanoes

Volcanoes are mountains, but they are very different from other mountains; they are not formed by folding and crumpling or by uplift and erosion. Instead, volcanoes are built by the accumulation of their own eruptive products—lava, bombs (crusted over lava blobs), ashflows, and tephra (airborne ash and dust). A *volcano* is most commonly a conical hill or mountain built around a vent that connects with reservoirs of molten rock below the surface of the Earth. The term *volcano* also refers to the opening or vent through which the molten rock and associated gases are expelled.

Driven by buoyancy and gas pressure, the molten rock, which is lighter than the surrounding solid rock, forces its way upward and may ultimately break through zones of weaknesses in the Earth's crust. If so, an eruption begins, and the molten rock may pour from the vent as nonexplosive lava flows, or it may shoot violently into the air as dense clouds of lava fragments. Larger fragments fall back around the vent, and accumulations of fall-back fragments may move downslope as ash flows under the force of gravity. Some of the finer ejected materials may be carried by the wind only to fall to the ground many miles away. The finest ash particles may be injected miles into the atmosphere and carried many times around the world by stratospheric winds before settling out.

Molten rock below the surface of the Earth that rises in volcanic vents is known as *magma*, but after it erupts from a volcano it is called *lava*. Originating many tens of miles beneath the ground, the ascending magma commonly contains some crystals, fragments of surrounding (unmelted) rocks, and dissolved gases, but it is primarily a liquid composed principally of oxygen, silicon, aluminum, iron, magnesium, calcium, sodium, potassium, titanium, and manganese. Magmas also contain many other chemical elements in trace quantities. Upon cooling, the liquid magma may precipitate crystals of various minerals until solidification is complete to form an *igneous* or *magmatic rock*.

The diagram below shows that heat concentrated in the Earth's upper *mantle* raises temperatures sufficiently to melt the rock locally by fusing the materials with the lowest melting temperatures, resulting in small, isolated blobs of magma. These blobs then collect, rise through conduits and fractures, and some ultimately may re-collect in larger pockets or reservoirs ("holding tanks") a few miles beneath the Earth's surface. Mounting pressure within the reservoir may drive the magma further upward through structurally weak zones to erupt as lava at the surface. In a continental environment, magmas are generated in the Earth's crust as well as at varying depths in the upper mantle. The variety of molten rocks in the crust, plus the possibility of mixing with molten materials from the underlying mantle, leads to the production of magmas with widely different chemical compositions.

If magmas cool rapidly, as might be expected near or on the Earth's surface, they solidify to form igneous rocks that are finely crystalline or glassy with few crystals. Accordingly, lavas, which of course are very rapidly cooled, form volcanic rocks typically characterized by a small percentage of crystals or fragments set in a matrix of *glass* (quenched or supercooled magma) or finer

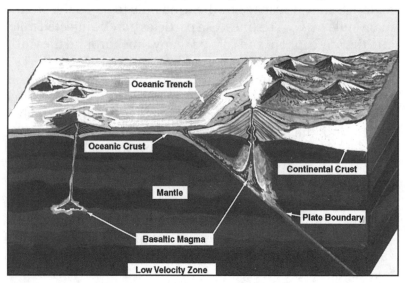

Two Polynesian terms are used to identify the surface character of Hawaiian lava flows. *Aa*, a basalt with a rough, blocky appearance, much like furnace slag, is shown at left. *Pahoehoe*, a more fluid variety with a smooth, satiny, and sometimes glassy appearance, is shown at right.

grained crystalline materials. If magmas never breach the surface to erupt and remain deep underground, they cool much more slowly and thus allow ample time to sustain crystal precipitation and growth, resulting in the formation of coarser grained, nearly completely crystalline, igneous rocks. Subsequent to final crystallization and solidification, such rocks can be exhumed by erosion many thousands or millions of years later and be exposed as large bodies of so-called *granitic* rocks, as, for example, those spectacularly displayed in Yosemite National Park and other parts of the majestic Sierra Nevada mountains of California.

Lava is red hot when it pours or blasts out of a vent but soon changes to dark red, gray, black, or some other color as it cools and solidifies. Very hot, gas-rich lava containing abundant iron and magnesium is fluid and flows like hot tar, whereas cooler, gas-poor lava high in silicon, sodium, and potassium flows sluggishly, like thick honey in some cases or in others like pasty, blocky masses.

All magmas contain dissolved gases, and as they rise to the surface to erupt, the confining pressures are reduced and the dissolved gases are liberated either quietly or explosively. If the lava is a thin fluid (not viscous), the gases may escape easily.

But if the lava is thick and pasty (highly viscous), the gases will not move freely but will build up tremendous pressure, and ultimately escape with explosive violence. Gases in lava may be compared with the gas in a bottle of a carbonated soft drink. If you put your thumb over the top of the bottle and shake it vigorously, the gas separates from the drink and forms bubbles. When you remove your thumb abruptly, there is a miniature explosion of gas and liquid. The gases in lava behave in somewhat the same way. Their sudden expansion causes the terrible explosions that throw out great masses of solid rock as well as lava, dust, and ashes.

The violent separation of gas from lava may produce rock froth called *pumice*. Some of this froth is so light—because of the many gas bubbles—that it floats on water. In many eruptions, the froth is shattered explosively into small fragments that are hurled high into the air in the form of volcanic cinders (red or black), volcanic ash (commonly tan or gray), and volcanic dust.

Principal Types of Volcanoes

Geologists generally group volcanoes into four main kinds— *cinder cones, composite volcanoes, shield volcanoes,* and *lava domes.*

Cinder cones

Cinder cones are the simplest type of volcano. They are built from particles and blobs of congealed lava ejected from a single vent. As the gas-charged lava is blown violently into the air, it breaks into small fragments that solidify and fall as *cinders* around the vent to form a circular or oval cone. Most cinder cones have a bowl-shaped *crater* at the summit and rarely rise more than a thousand feet or so above their surroundings. Cinder cones are numerous in western North America as well as throughout other volcanic terrains of the world.

In 1943, a cinder cone started growing on a farm near the

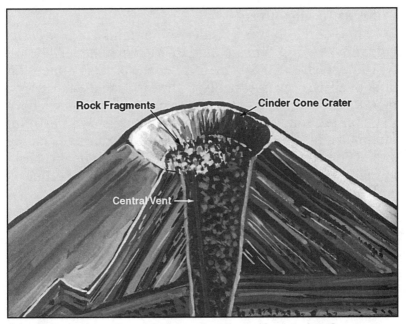

Schematic representation of the internal structure of a typical cinder cone.

village of Parícutin in Mexico. Explosive eruptions caused by gas rapidly expanding and escaping from molten lava formed cinders that fell back around the vent, building up the cone to a height of 1,200 feet. The last explosive eruption left a funnel-shaped crater at the top of the cone. After the excess gases had largely dissipated, the molten rock quietly poured out on the surrounding surface of the cone and moved downslope as lava flows. This order of events—eruption, formation of cone and crater, lava flow—is a common sequence in the formation of cinder cones.

During nine years of activity, Parícutin built a prominent cone, covered about 100 square miles with ashes, and destroyed the town of San Juan. Geologists from many parts of the world studied Parícutin during its lifetime and learned a great deal about volcanism, its products, and the modification of a volcanic landform by erosion.

Composite volcanoes

Some of the Earth's grandest mountains are *composite* volcanoes—sometimes called *stratovolcanoes*. They are typically steep-sided, symmetrical cones of large dimension built of alternating layers of lava flows, volcanic ash, cinders, blocks, and bombs and may rise as much as 8,000 feet above their bases. Some of the most conspicuous and beautiful mountains in the world are composite volcanoes, including Mount Fuji in Japan, Mount Cotopaxi in Ecuador, Mount Shasta in California, Mount Hood in Oregon, and Mount St. Helens and Mount Rainier in Washington.

Most composite volcanoes have a crater at the summit which contains a central vent or a clustered group of vents.

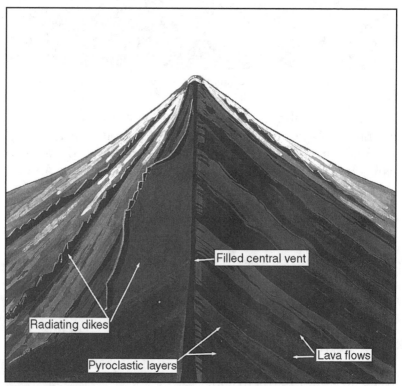

Schematic representation of the internal structure of a typical composite volcano.

Lavas either flow through breaks in the crater wall or issue from fissures on the flanks of the cone. Lava, solidified within the fissures, forms dikes that act as ribs which greatly strengthen the cone.

The essential feature of a composite volcano is a conduit system through which magma from a reservoir deep in the Earth's crust rises to the surface. The volcano is built up by the accumulation of material erupted through the conduit and increases in size as lava, cinders, ash, etc., are added to its slopes.

When a composite volcano becomes dormant, erosion begins to destroy the cone. As the cone is stripped away, the hardened magma filling the conduit (the volcanic plug) and fissures (the dikes) becomes exposed, and it too is slowly reduced by erosion. Finally, all that remains is the plug and dike complex projecting above the land surface—a telltale remnant of the vanished volcano.

An interesting variation of a composite volcano can be seen at Crater Lake in Oregon. From what geologists can interpret of its past, a high volcano—called Mount Mazama—probably similar in appearance to present-day Mount Rainier was once located at this spot. Following a series of tremendous explosions about 6,800 years ago, the volcano lost its top. Enormous volumes of volcanic ash and dust were expelled and swept down the slopes as ash flows and avalanches. These large-volume explosions rapidly drained the lava beneath the mountain and weakened the upper part. The top then collapsed to form a large depression, which later filled with water and is now completely occupied by beautiful Crater Lake. A last gasp of eruptions produced a small cinder cone, which rises above the water surface as Wizard Island near the rim of the lake. Depressions such as Crater Lake, formed by the collapse of volcanoes, are known as *calderas*. They are usually large, steep-walled, basin-shaped depressions formed by the collapse of a large area over, and around, a volcanic vent or vents. Calderas range in form and size from roughly circular depressions 1 to 15 miles in diameter to huge elongated depressions as much as 60 miles long.

**The Evolution
of a Composite
Volcano**

a. Magma, rising upward through a conduit, erupts at the Earth's surface to form a volcanic cone. Lava flows spread over the surrounding area.

b. As volcanic activity continues, perhaps over spans of hundreds of years, the cone is built to a great height and lava flows form an extensive plateau around its base. During this period, streams enlarge and deepen their valleys.

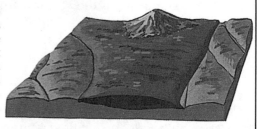

c. When volcanic activity ceases, erosion starts to destroy the cone. After thousands of years, the great cone is stripped away to expose the hardened "volcanic plug" in the conduit. During this period of inactivity, streams broaden their valleys and dissect the lava plateau to form isolated lava-capped mesas.

d. Continued erosion removes all traces of the cone and the land is worn down to a surface of low relief. All that remains is a projecting plug or "volcanic neck," a small lava-capped mesa, and vestiges of the once lofty volcano and its surrounding lava plateau.

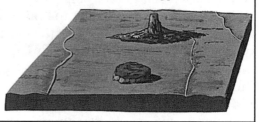

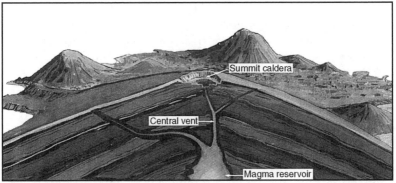

The internal structure of a typical shield volcano.

Shield volcanoes

Shield volcanoes, the third type of volcano, are built almost entirely of fluid lava flows. Flow after flow pours out in all directions from a central summit vent, or group of vents, building a broad, gently sloping cone of flat, domical shape, with a profile much like that of a warrior's shield. They are built up slowly by the accretion of thousands of highly fluid lava flows called basalt lava that spread widely over great distances, and then cool as thin, gently dipping sheets. Lavas also commonly erupt from vents along fractures (rift zones) that develop on the flanks of the cone. Some of the largest volcanoes in the world are shield volcanoes. In northern California and Oregon, many shield volcanoes have diameters of three or four miles and heights of 1,500 to 2,000 feet. The Hawaiian Islands are composed of linear chains of these volcanoes including Kilauea and Mauna Loa on the island of Hawaii—two of the world's most active volcanoes. The floor of the ocean is more than 15,000 feet deep at the bases of the islands. As Mauna Loa, the largest of the shield volcanoes (and also the world's largest active volcano), projects 13,677 feet above sea level, its top is over 28,000 feet above the deep ocean floor.

In some eruptions, basaltic lava pours out quietly from long fissures instead of central vents and floods the surrounding countryside with lava flow upon lava flow, forming broad plateaus. Lava plateaus of this type can be seen in Iceland, southeastern Washington, eastern Oregon, and southern Idaho. Along the Snake River in Idaho, and the Columbia River in Washington and Oregon, these lava flows are beautifully exposed and measure more than a mile in total thickness.

Lava domes

Volcanic or lava domes are formed by relatively small, bulbous masses of lava too viscous to flow any great distance; consequently, on extrusion, the lava piles over and around its vent. A dome grows largely by expansion from within. As it grows, its outer surface cools and hardens, then shatters, spilling loose fragments down its sides. Some domes form craggy knobs or spines over the volcanic vent, whereas others form short, steep-sided lava flows known as *coulees*. Volcanic domes commonly occur within the craters or on the flanks of large composite volcanoes. The nearly circular Novarupta Dome that formed during the 1912 eruption of Katmai Volcano, Alaska, measures 800 feet across and 200 feet high. The internal structure of this dome—defined by layering of lava fanning upward and outward from the center—indicates that it grew

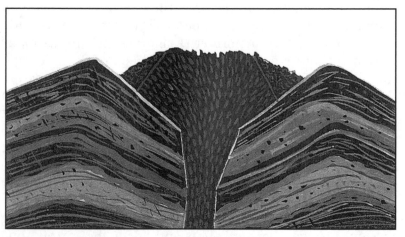

Schematic representation of the internal structure of a typical volcanic dome.

largely by expansion from within. Mont Pelée in Martinique, Lesser Antilles, and Lassen Peak and Mono domes in California are examples of lava domes. An extremely destructive eruption accompanied the growth of a dome at Mont Pelée in 1902. The coastal town of St. Pierre, about four miles downslope to the south, was demolished and nearly 30,000 inhabitants were killed by an incandescent, high-velocity ash flow and associated hot gases and volcanic dust. Only two men survived; one because he was in a poorly ventilated, dungeon-like jail cell and the other who somehow made his way safely through the burning city.

Other Volcanic Structures

Plugs (necks)

Congealed magma, along with fragmental volcanic and wallrock materials, can be preserved in the feeding conduits of a volcano upon cessation of activity. These preserved rocks form crudely cylindrical masses, from which project radiating dikes; they may be visualized as the fossil remains of the innards of a volcano (the so-called "volcanic plumbing system") and are referred to as volcanic *plugs* or *necks*. The igneous material in a plug may have a range of composition similar to that of associated lavas or ash, but may also include fragments and blocks of denser, coarser grained rocks—higher in iron and magnesium, lower in silicon—thought to be samples of the Earth's deep crust or upper mantle plucked and transported by the ascending magma. Many plugs and necks are largely or wholly composed of fragmental volcanic material and of fragments of wallrock, which can be of any type. Plugs that bear a particularly strong imprint of explosive eruption of highly gas-charged magma are called *diatremes* or *tuff-breccia*.

Volcanic plugs are believed to overlie a body of magma which could be either still largely liquid or completely solid depending on the state of activity of the volcano. Plugs are known, or postulated, to be commonly funnel shaped and to taper downward into bodies increasingly elliptical in plan or elongated to dike-like forms. Typically, volcanic plugs and necks tend to be more resistant to erosion than their enclosing rock formations. Thus, after the volcano becomes inactive and deeply eroded, the exhumed plug may stand up in bold relief as an irregular, columnar structure. One of the best known and most spectacular diatremes in the United States is Ship Rock in New Mexico, which towers some 1,700 feet above the more deeply eroded surrounding plain. Volcanic plugs, including diatremes, are found elsewhere in the western United States and also in Germany, South Africa, Tanzania, and Siberia.

Ship Rock, San Juan County, New Mexico

Maars

Also called "tuff cones," *maars* are shallow, flat-floored craters that scientists interpret have formed above diatremes as a result of a violent expansion of magmatic gas or steam; deep erosion of a maar presumably would expose a diatreme. Maars range in size from 200 to 6,500 feet across and from 30 to 650 feet deep, and most are commonly filled with water to form natural lakes. Most maars have low rims composed of a mixture of loose fragments of volcanic rock and rocks torn from the walls of the diatreme.

Maars occur in the western United States, in the Eifel region of Germany, and in other geologically young volcanic regions of the world. An excellent example of a maar is Zuni Salt Lake in New Mexico, a shallow saline lake that occupies a flat-floored crater about 6,500 feet across and 400 feet deep. Its low rim is composed of loose pieces of basaltic lava and wallrocks (sandstone, shale, limestone) of the underlying diatreme, as well as random chunks of ancient crystalline rocks blasted upward from great depths.

Nonvolcanic craters

Some well-exposed, nearly circular areas of intensely deformed sedimentary rocks, in which a central vent-like feature is surrounded by a ring-shaped depression, resemble volcanic structures in gross form. As no clear evidence of volcanic origin could be found in or near these structures, scientists initially described them as *cryptovolcanic*, a term now rarely used. Recent studies have shown that not all craters are of volcanic origin. Impact craters, formed by collisions with the Earth of large meteorites, asteroids, or comets, share with volcanoes the imprints of violent origin, as evidenced by severe disruption, and even local melting, of rock. Fragments of meteorites or chemically detectable traces of extraterrestrial materials and indications of strong forces acting from above, rather than from below, distinguish impact from volcanic features.

Other possible explanations for these nonvolcanic craters

include subsurface salt-dome intrusion (and subsequent dissolution and collapse); collapse caused by subsurface limestone dissolution and/or groundwater withdrawal; and collapse related to melting of glacial ice. An impressive example of an impact structure is Meteor Crater, Arizona, which is visited by thousands of tourists each year. This impact crater, 4,000 feet in diameter and 600 feet deep, was formed in the geologic past (probably 30,000-50,000 years before present) by a meteorite striking the Earth at a speed of many thousands of miles per hour.

In addition to Meteor Crater, very fresh, morphologically distinct, impact craters are found at three sites near Odessa, Texas, as well as 10 or 12 other locations in the world. Of the more deeply eroded, less obvious, postulated impact structures, there are about ten well-established sites in the United States and perhaps 80 or 90 elsewhere in the world.

Meteor Crater, Arizona

Types of Volcanic Eruptions

During an episode of activity, a volcano commonly displays a distinctive pattern of behavior. Some mild eruptions merely discharge steam and other gases, whereas other eruptions quietly extrude quantities of lava. The most spectacular eruptions consist of violent explosions that blast great clouds of gas-laden debris into the atmosphere.

Parícutin Volcano, Mexico, 1947.

The type of volcanic eruption is often labeled with the name of a well-known volcano where characteristic behavior is similar—hence the use of such terms as *Strombolian, Vulcanian, Vesuvian, Peléan, Hawaiian,* and others. Some volcanoes may exhibit only one characteristic type of eruption during an interval of activity—others may display an entire sequence of types.

In a *Strombolian*-type eruption observed during the 1965 activity of Irazu Volcano in Costa Rica, huge clots of molten lava burst from the summit crater to form luminous arcs through the sky. Collecting on the flanks of the cone, lava clots combined to stream down the slopes in fiery rivulets.

In contrast, the eruptive activity of Parícutin Volcano in 1947 demonstrated a *Vulcanian*-type eruption, in which a dense cloud of ash-laden gas explodes from the crater and rises high above the peak. Steaming ash forms a whitish cloud near the upper level of the cone.

In a *Vesuvian* eruption, as typified by the eruption of Mount Vesuvius in Italy in A.D. 79, great quantities of ash-laden gas are violently discharged to form a cauliflower-shaped cloud high above the volcano.

In a *Peléan* or *Nuée Ardente* (glowing cloud) eruption, such as occurred on the Mayon Volcano in the Philippines in 1968, a large quantity of gas, dust, ash, and incandescent lava fragments are blown out of a central crater, fall back, and form tonguelike, glowing avalanches that move downslope at velocities as great as 100 miles per hour. Such eruptive activity can cause great destruction and loss of life if it occurs in populated areas, as demonstrated by the devastation of St. Pierre during the 1902 eruption of Mont Pelée on Martinique, Lesser Antilles.

Hawaiian eruptions may occur along fissures or fractures that serve as linear vents, such as during the eruption of Mauna Loa Volcano in Hawaii in 1950; or they may occur at a central vent such as during the 1959 eruption in Kilauea Iki Crater of

Mount Vesuvius Volcano, Italy, 1944.

Mauna Loa Volcano, Hawaii, 1950.

Kilauea Volcano, Hawaii. In fissure-type eruptions, molten, incandescent lava spurts from a fissure on the volcano's rift zone and feeds lava streams that flow downslope. In central-vent eruptions, a fountain of fiery lava spurts to a height of several hundred feet or more. Such lava may collect in old pit craters to form lava lakes, or form cones, or feed radiating flows.

Phreatic (or steam-blast) eruptions are driven by explosive expanding steam resulting from cold ground or surface water coming into contact with hot rock or magma. The distinguishing feature of phreatic explosions is that they only blast out fragments of preexisting solid rock from the volcanic conduit; no new magma is erupted. Phreatic activity is generally weak, but can be quite violent in some cases, such as the 1965 eruption of Taal Volcano, Philippines, and the 1975-76 activity at La Soufrière, Guadeloupe (Lesser Antilles).

The most powerful eruptions are called *plinian* and involve the explosive ejection of relatively viscous lava. Large plinian eruptions—such as on May 18, 1980 at Mount St. Helens or, more recently, on June 15, 1991 at Pinatubo in the Philippines—can send ash and volcanic gas tens of miles into the air. The resulting ash fallout can affect large areas hundreds of miles downwind. Fast-moving deadly pyroclastic flows (*nuées ardentes*) are also commonly associated with plinian eruptions.

Submarine Volcanoes

Submarine volcanoes and volcanic vents are common features on certain zones of the ocean floor. Some are active at the present time and, in shallow water, disclose their presence by blasting steam and rock-debris high above the surface of the sea. Many others lie at such great depths that the tremendous weight of the water above them results in high, confining pressure and prevents the formation and explosive release of steam and gases. Even very large, deep-water eruptions may not disturb the ocean surface.

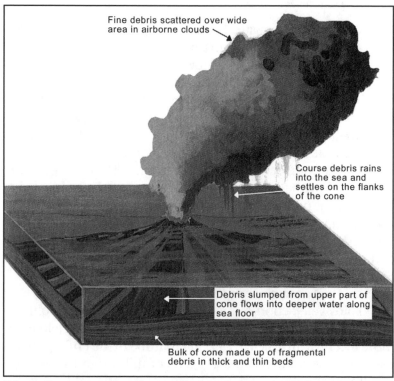

Fine debris scattered over wide area in airborne clouds

Course debris rains into the sea and settles on the flanks of the cone

Debris slumped from upper part of cone flows into deeper water along sea floor

Bulk of cone made up of fragmental debris in thick and thin beds

Schematic representation of a typical submarine eruption in the open ocean.

The unlimited supply of water surrounding submarine volcanoes can cause them to behave differently from volcanoes on land. Violent, steam-blast eruptions take place when sea water pours into active shallow submarine vents. Lava, erupting onto a shallow sea floor or flowing into the sea from land, may cool so rapidly that it shatters into sand and rubble. The result is the production of huge amounts of fragmental volcanic debris. The famous "black sand" beaches of Hawaii were created virtually instantaneously by the violent interaction between hot lava and sea water. On the other hand, recent observations made from deep-diving submersibles have shown that some submarine eruptions produce flows and other volcanic structures remarkably similar to those formed on land. Recent studies have revealed the presence of spectacular, high-temperature hydrothermal plumes and vents (called "smokers") along some parts of the mid-oceanic volcanic rift systems. In 1998, a device placed by the National Oceanic and Atmospheric Administration and Oregon State University directly recorded a deep submarine volcanic eruption for the first time.

During an explosive submarine eruption in the shallow open ocean, enormous piles of debris are built up around the active volcanic vent. Ocean currents rework the debris in shallow water, while other debris slumps from the upper part of the cone and flows into deep water along the sea floor. Fine debris and ash in the eruptive plume are scattered over a wide area in airborne clouds. Coarse debris in the same eruptive plume rains into the sea and settles on the flanks of the cone. Pumice from the eruption floats on the water and drifts with the ocean currents over a large area.

Geysers, Fumaroles, and Hot Springs

Geysers, fumaroles (also called *solfataras*), and hot springs are generally found in regions of young volcanic activity. Surface water percolates downward through the rocks below the Earth's surface to high-temperature regions surrounding a magma reservoir, either active or recently solidified but still hot. There the water is

Old Faithful Geyser, Yellowstone National Park, Wyoming.

heated, becomes less dense, and rises back to the surface along fissures and cracks. Sometimes these features are called "dying volcanoes" because they seem to represent the last stage of volcanic activity as the magma, at depth, cools and hardens.

Erupting geysers provide spectacular displays of underground energy suddenly unleashed, but their mechanisms are not completely understood. Large amounts of hot water are presumed to fill underground cavities. The water, upon further heating, is violently ejected when a portion of it suddenly flashes into steam. This cycle can be repeated with remarkable regularity, as for example, at Old Faithful Geyser in Yellowstone National Park, which erupts on an average of about once every 65 minutes.

Fumaroles, which emit mixtures of steam and other gases, are fed by conduits that pass through the water table before reaching the surface of the ground. Hydrogen sulfide (H_2S), one of the

Mammoth Hot Springs, Yellowstone National Park, Wyoming.

typical gases issuing from fumaroles, readily oxidizes to sulfuric acid and native sulfur. This accounts for the intense chemical activity and brightly colored rocks in many thermal areas.

Hot springs occur in many thermal areas where the surface of the Earth intersects the water table. The temperature and rate of discharge of hot springs depend on factors such as the rate at which water circulates through the system of underground channelways, the amount of heat supplied at depth, and the extent of dilution of the heated water by cool ground water near the surface.

Volcano Environments

There are more than 500 active volcanoes (those that have erupted at least once within recorded history) in the world—50 of which are in the United States (Hawaii, Alaska, Washington, Oregon, and California)—although many more are hidden under the seas. Most active volcanoes are strung like beads along, or near, the margins of the continents, and more than half encircle the Pacific Ocean as a "Ring of Fire."

Many volcanoes are in and around the Mediterranean Sea. Mount Etna in Sicily is the largest and highest of these mountains. Italy's Vesuvius is the only active volcano on the European mainland. Near the island of Vulcano, the volcano Stromboli has been in a state of nearly continuous, mild eruption since early Roman times. At night, sailors in the Mediterranean can see the glow from the fiery molten material that is hurled into the air. Very appropriately, Stromboli has

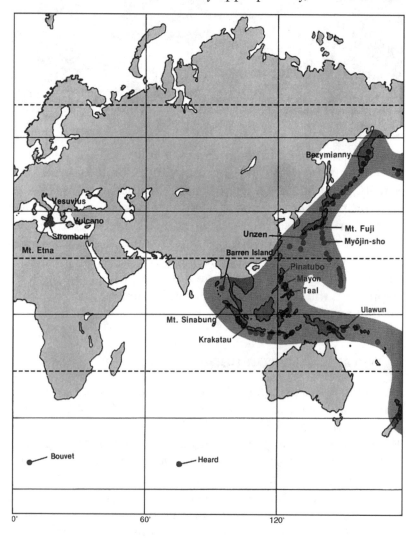

been called "the lighthouse of the Mediterranean."

Some volcanoes crown island areas lying near the continents, and others form chains of islands in the deep ocean basins. Volcanoes tend to cluster along narrow mountainous belts where folding and fracturing of the rocks provide channelways to the surface for the escape of magma.

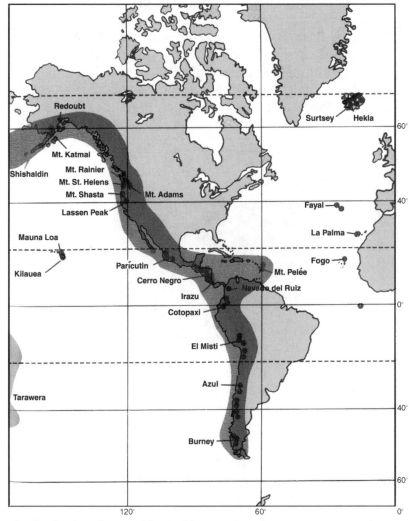

The distribution of some of the Earth's 500 active volcanoes.

Significantly, major earthquakes also occur along these belts, indicating that volcanism and seismic activity are often closely related, responding to the same dynamic Earth forces.

In a typical *island-arc* environment, volcanoes lie along the crest of an arcuate, crustal ridge bounded on its convex side by a deep oceanic trench. The granite or granitelike layer of the continental crust extends beneath the ridge to the vicinity of the trench. Basaltic magmas, generated in the mantle beneath the ridge, rise along fractures through the granitic layer. These magmas commonly will be modified or changed in composition during passage through the granitic layer and erupt on the surface to form volcanoes built largely of nonbasaltic rocks.

Island-arc
Mount Sinabung, Sumatra

In a typical *oceanic* environment, volcanoes are aligned along the crest of a broad ridge that marks an active fracture system in the oceanic crust. Basaltic magmas, generated in the upper mantle beneath the ridge, rise along fractures through the basaltic layer. Because the granitic crustal layer is absent, the magmas are not appreciably modified or changed in composition and they erupt on the surface to form basaltic volcanoes.

Oceanic
Mauna Kea Volcano, Hawaii

In the typical *continental* environment, volcanoes are located in unstable, mountainous belts that have thick roots of granite or granitelike rock. Magmas, generated near the base of the mountain root, rise slowly or intermittently along fractures in the crust. During passage through the granitic layer, magmas are commonly modified or changed in composition and erupt on the surface to form volcanoes constructed of nonbasaltic rocks.

Continental
Mount Adams, Washington

Plate-Tectonics Theory

According to the "plate-tectonics" theory, scientists believe that the Earth's surface is broken into a number of shifting slabs or plates, which average about 50 miles in thickness. These plates move relative to one another above a hotter, deeper, more mobile zone at average rates as great as a few inches per year. Most of the world's active volcanoes are located along or near the boundaries between shifting plates and are called "plate-boundary" volcanoes. However, some active volcanoes are not associated with plate boundaries, and many of these so-called "intra-plate" volcanoes form roughly linear chains in the interior of some oceanic plates. The Hawaiian Islands provide perhaps the best example of an "intra-plate" volcanic chain, developed by the northwest-moving Pacific plate passing over

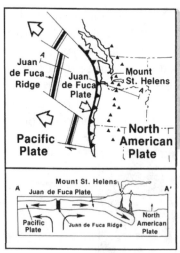

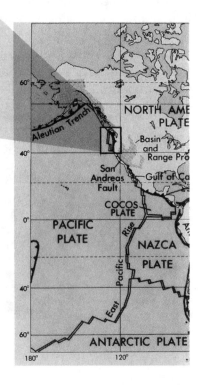

Major tectonic plates of the Earth.

an inferred "hot spot" that initiates the magma-generation and volcano-formation process. The peripheral areas of the Pacific Ocean Basin, containing the boundaries of several plates, are dotted by many active volcanoes that form the so-called "Ring of Fire." The "Ring" provides excellent examples of "plate-boundary" volcanoes, including Mount St. Helens.

The diagram below shows the boundaries of lithosphere plates that are presently active. The double lines indicate zones of spreading from which plates are moving apart. The lines with barbs show zones of underthrusting (subduction), where one plate is sliding beneath another. The barbs on the lines indicate the overriding plate. The single line defines a strike-slip fault along which plates are sliding horizontally past one another. The shaded areas indicate a part of a continent, exclusive of that along a plate boundary, which is undergoing active extensional, compressional, or strike-slip faulting.

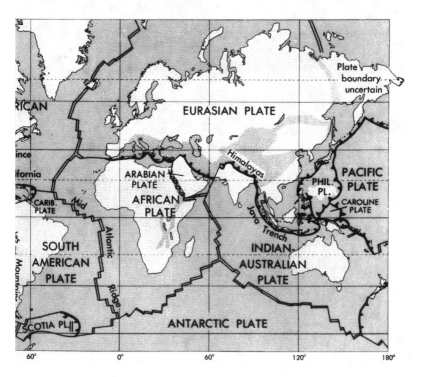

Extraterrestrial Volcanism

Volcanoes and volcanism are not restricted to the planet Earth. Manned and unmanned planetary explorations, beginning in the late 1960s, have furnished graphic evidence of past volcanism and its products on the Moon, Mars, Venus,

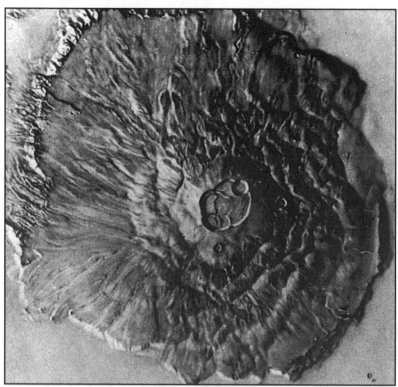

Mariner 9 imagery of Olympus Mons Volcano on Mars compared to the eight principal Hawaiian Islands at the same scale (*Mariner 9 Image Mosaic, NASA/JPL*).

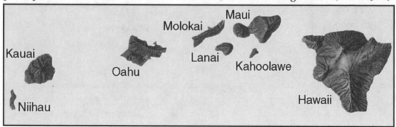

and other planetary bodies. Many pounds of volcanic rocks were collected by astronauts during the various Apollo lunar landing missions. Only a small fraction of these samples have been subjected to exhaustive study by scientists. The bulk of the material is stored under controlled-environment conditions at NASA's Lunar Receiving Laboratory in Houston, Texas for future study by scientists.

From the 1976-1979 Viking mission, scientists have been able to study the volcanoes on Mars, and their studies are very revealing when compared with those of volcanoes on Earth. For example, Martian and Hawaiian volcanoes closely resemble each other in form. Both are shield volcanoes, have gently sloping flanks, large multiple collapse pits at their centers, and appear to be built of fluid lavas that have left numerous flow

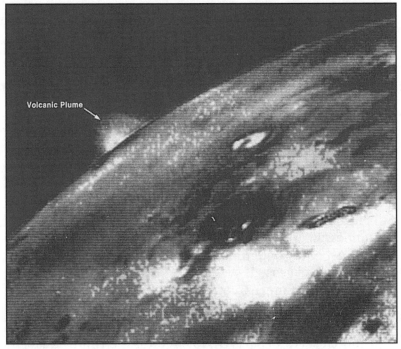

Volcanic Plume

Spacecraft image, made in July 1979, shows volcanic plume rising some 60 to 100 miles above the surface of Io, a moon of Jupiter (*Voyager 2 photo, NASA*).

features on their flanks. The most obvious difference between the two is size. The Martian shields are enormous. They can grow to over 17 miles in height and more than 350 miles across, in contrast to a maximum height of about six miles and width of 74 miles for the Hawaiian shields.

Voyager-2 spacecraft images taken of Io, a moon of Jupiter, captured volcanoes in the actual process of eruption. The volcanic plumes, shown in the image on page 93, rise some 60 to 100 miles above the surface of the moon. Thus, active volcanism is taking place, at present, on at least one planetary body in addition to our Earth.

Volcano Monitoring and Research

It has been said that the science of *volcanology* originated with the accurate descriptions of the eruption of Vesuvius in A.D. 79 contained in two letters from Pliny the Younger to the Roman historian Tacitus. Pliny's letters also described the death of his uncle, Pliny the Elder, who was killed in the eruption. Actually, however, it was not until the 19th century that serious scientific inquiry into volcanic phenomena flourished as part of the general revolution in the physical and life sciences, including the new science of "geology." In 1847, an observatory was established on the flanks of Vesuvius, upslope from the site of Herculaneum, for the more or less continuous recording of the activity of the volcano that destroyed the city in A.D. 79. Still, through the first decade of the 20th century, the study of volcanoes by and large continued to be of an expeditionary nature, generally undertaken after the eruption had begun or the activity had ceased.

Perhaps "modern" volcanology began in 1912, when Thomas A. Jaggar, Head of the Geology Department of the Massachusetts Institute of Technology, founded the Hawaiian Volcano Observatory (HVO), located on the rim of Kilauea's caldera. Initially supported by an association of Honolulu businessmen, HVO began to conduct systematic and continuous monitoring of seismic activity preceding, accompanying, and following eruptions, as well as a wide variety of other geological, geophysical, and geochemical observations and investigations. Between 1919 and 1948, HVO was administered by various Federal agencies (National Weather Service, U.S. Geological Survey, and National Park Service), and since 1948 it has been operated continuously by the Geological Survey as part of its Volcano Hazards Program. The more than 75 years of comprehensive investigations by HVO and other scientists in Hawaii have added substantially to our understanding of the eruptive mechanisms of Kilauea and Mauna Loa, two of the world's most active volcanoes. Moreover, the Hawaiian Volcano Observatory pioneered and refined most of the commonly used volcano-monitoring techniques presently employed by other observatories monitoring active volcanoes elsewhere, principally in Indonesia, Italy, Japan, Latin America, New Zealand, Lesser Antilles (Caribbean), Philippines, and Kamchatka (Russia).

What does "volcano monitoring" actually involve? Basically, it is the keeping of a detailed "diary" of the changes—visible and invisible—in a volcano and its surroundings. Between eruptions, visible changes of importance to the scientists would include marked increase or decrease of steaming from known vents; emergence of new steaming areas; development of new ground cracks or widening of old ones; unusual or inexplicable withering of plant life; changes in the color of mineral deposits encrusting fumaroles; and any other directly observable, and often measurable, feature that might reflect a change in the state of the volcano. Of course, the "diary" keeping during eruptive activity presents additional tasks. Wherever and whenever they can do so safely, scientists document, in words and on film, the course of the eruption in detail; make temperature measurements of lava and gas; collect the eruptive products and gases for subsequent laboratory analysis; measure the heights of lava fountains or ash plumes; gage the flow rate of ash ejection or lava flows; and carry out other necessary observations and measurements to fully document and characterize the eruption. For each eruption, such documentation and data collection and analysis provide another building block in constructing a model of the characteristic behavior of a given volcano or type of eruption.

Volcano monitoring also involves the recording and analysis of volcanic phenomena not visible to the human eye, but measurable by precise and sophisticated instruments. These phenomena include ground movements, earthquakes (particularly those too small to be felt by people), variations in gas compositions, and deviations in local electrical and magnetic fields that respond to pressure and stresses caused by the subterranean magma movements.

Some common methods used to study invisible, volcano-related phenomena are based on:

1. Measurement of changes in the shape of the volcano—volcanoes gradually swell or "inflate" in building up to an eruption because of the influx of magma into the volcano's reservoir or "plumbing system"; with the onset of eruption, pressure is immediately relieved and the volcano rapidly shrinks or "deflates." A wide variety of instruments, including precise spirit-levels, electronic "tiltmeters," and electronic-laser beam instruments, can measure changes in the slope or "tilt" of the volcano or in vertical and horizontal distances with a precision of only a few parts in a million.

2. Precise determination of the location and magnitude of earthquakes by a well-designed seismic network—as the volcano inflates by the rise of magma, the enclosing rocks are deformed to the breaking point to accommodate magma movement. When the rock ultimately fails to permit continued magma ascent, earthquakes result. By carefully mapping out the variations with time in the locations and depths of earthquake foci, scientists in effect can track the subsurface movement of magma, horizontally and vertically.

3. Measurement of changes in volcanic-gas composition and in magnetic field—the rise of magma high into the volcanic

edifice may allow some of the associated gases to escape along fractures, thereby causing the composition of the gases (measured at the surface) to differ from that usually measured when the volcano is quiescent and the magma is too deep to allow gas to escape. Changes in the Earth's magnetic field have been noted preceding and accompanying some eruptions, and such changes are believed to reflect temperature effects and/or the content of magnetic minerals in the magma.

Recording historic eruptions and modern volcano-monitoring in themselves are insufficient to fully determine the characteristic behavior of a volcano, because a time record of such information, though perhaps long in human terms, is much too short in geologic terms to permit reliable predictions of possible future behavior. A comprehensive investigation of any volcano must also include the careful, systematic mapping of the nature, volume, and distribution of the products of prehistoric eruptions, as well as the determination of their ages by modern isotopic and other dating methods. Research on the volcano's geologic past extends the data base for refined estimates of the recurrence intervals of active versus dormant periods in the history of the volcano. With such information in hand, scientists can construct so-called "volcanic hazards" maps that delineate the zones of greatest risk around the volcano and that designate which zones are particularly susceptible to certain types of volcanic hazards (lava flows, ash fall, toxic gases, mudflows and associated flooding, etc.).

A strikingly successful example of volcano research and volcanic-hazard assessment was the 1978 publication (Bulletin 1383-C) by two Geological Survey scientists, Dwight Crandell and Donal Mullineaux, who concluded that Mount St. Helens was the Cascade volcano most frequently active in the past 4,500 years and the one most likely to reawaken to erupt, "...perhaps before the end of this century." Their prediction came true when Mount St. Helens rumbled back to life in March of 1980. Intermittent explosions of ash and steam and periodic formation of short-lived lava domes continued throughout the decade. Analysis of the volcano's past behavior indicates that this kind of eruptive activity may continue for years or decades, but another catastrophic eruption like that of May 18, 1980, is unlikely to occur soon.

On May 18, 1982, the U.S. Geological Survey (USGS) formally dedicated the David A. Johnston Cascades Volcano Observatory (CVO) in Vancouver, Washington, in memory of the Survey volcanologist killed two years earlier. This facility—a sister observatory to the Hawaiian Volcano Observatory—facilitates the increased monitoring and research on not only Mount St. Helens but also the other volcanoes of the Cascade Range of the Pacific Northwest. More recently, in cooperation with the State of Alaska, the USGS established the Alaska Volcano Observatory in March 1988. The work being done at these volcano observatories provides important comparisons and contrasts between the behavior of the generally nonexplosive Hawaiian shield volcanoes and that of the generally explosive composite volcanoes of the Cascade and Alaskan Peninsula-Aleutian chains.

Volcanoes and People

Volcanoes both harass and help mankind. As dramatically demonstrated by the catastrophic eruption of Mount St. Helens in May 1980 and of Pinatubo in June 1991, volcanoes can wreak havoc and devastation in the short term. However, it should be emphasized that the short-term hazards posed by volcanoes are balanced by benefits of volcanism and related processes over geologic time. Volcanic materials ultimately break down to form some of the most fertile soils on Earth, cultivation of which fostered and sustained civilizations. People use volcanic products as construction materials, as abrasive and cleaning agents, and as raw materials for many chemical and industrial uses. The internal heat associated with some young volcanic systems has been harnessed to produce geothermal energy. For example, the electrical energy generated from The Geysers geothermal field in northern California can meet the present power consumption of the city of San Francisco.

The challenge to scientists involved with volcano research is to mitigate the short-term adverse impacts of eruptions, so that society may continue to enjoy the long-term benefits of volcanism. They must continue to improve the capability for predicting eruptions and to provide decision makers and the general public with the best possible information on high-risk volcanoes for sound decisions on land-use planning and public safety. Geoscientists still do not fully understand how volcanoes really work, but considerable advances have been made in recent decades. An improved understanding of volcanic phenomena provides important clues to the Earth's past, present, and possibly its future.

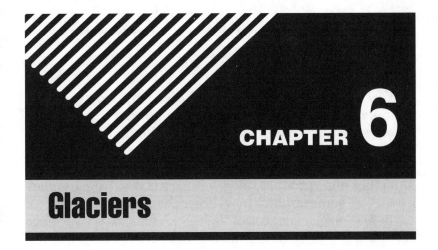

CHAPTER 6

Glaciers

A glacier is a large mass of ice having its genesis on land and represents a multiyear surplus of snowfall over snowmelt. At the present time, perennial ice covers about 10 percent of the land areas of the Earth. Although glaciers are generally thought of as polar entities, they also are found in mountainous areas throughout the world, on all continents except Australia, and even at or near the Equator on high mountains in Africa and South America.

Present-day glaciers and the deposits from more extensive glaciation in the geological past have considerable economic importance in many areas. In areas of limited precipitation during the growing season, for example, such as parts of the Western United States, glaciers are considered to be frozen freshwater reservoirs which release water during the drier summer months. In the Western United States, as in many other mountainous regions, they are of considerable economic importance in the irrigation of crops and to the generation of hydroelectric power. Lakes and ponds are numerous where continental ice sheets once covered New England and the upper Midwest, and the glacial deposits act as major groundwater reservoirs. The same deposits have substantial economic value as sand and gravel for building materials, and are often the basis, as in Massachusetts, for the largest mineral industry in a state.

Glacial Hazards

Glaciers can also pose dangers to ocean transportation. For example, the west side of Greenland produces a large number of icebergs which travel south into the shipping lanes of the North Atlantic. Some of Greenland's icebergs have been known to drift as far south as Delaware before melting completely. Antarctica often produces large tabular icebergs from its many ice shelves, some of which may exceed the area of Rhode Island.

In Alaska, glaciologists of the U.S. Geological Survey have carried out a long-term study of the rate of retreat of the Columbia Glacier. This large valley glacier's snout ends as a tidal glacier in Prince William Sound, a waterway traversed by large oil tankers en route to and from Valdez, the southern terminus

The terminus of the Columbia Glacier, Prince William Sound, Alaska.

of the trans-Alaskan pipeline. In 1980, U.S. Geological Survey glaciologists predicted that the 425-square-mile (1100-square-kilometer) Columbia Glacier would begin an accelerated retreat during the 1980s, thereby increasing the hazard to shipping because of an increased production of icebergs.

Other glacial hazards include "surging" glaciers. Such glaciers may suddenly advance several miles in a few months. Catastrophic outburst floods, resulting from the failure of ice-dammed lakes or from subglacial volcanic or geothermal activity, occur frequently in Iceland and in Alaska, less frequently in the State of Washington, and occasionally in other locales.

Glacial Advances

Within the past 3 million years, glaciers—in the form of icecaps and ice sheets—repeatedly advanced to cover as much as 30 percent of the total land area of the Earth. This land area includes most of Canada, all of New England, much of the upper Midwest, large areas in Alaska, all of Greenland, Svalbard, other Arctic islands, Scandinavia, most of Great Britain and Ireland, and much of the western part of northern Russia. Parts of southern South America, central Asia, the Alps, and many other mountainous areas in Asia also experienced an increase in glacier extent. Glaciers in Antarctica also expanded somewhat, but more in ice thickness than in ice area because of the limiting effect of the surrounding oceans.

Although four or five major glacial advances have been identified by geologists who have studied glacial deposits in Europe and North America, stratigraphic evidence from the Tjörnes area of Iceland suggests that as many as 10 major glacial advances may have occurred within the past 3 million years or during what is popularly referred to as the Ice Age.

Geologists recognized during the 19th century that the Earth has been repeatedly subjected to glaciation many times more

The South Cascade Glacier, Cascade Mountains, Washington.

extensive than the areal distribution of present-day glaciers. More recent evidence, from geochronological evidence (isotopic age dating), has shown that the last continental-size glaciers in Europe, Asia, and North America were still melting away only 10,000 years ago. These findings have caused scientists and nonscientists alike to ask themselves two questions with profound implications: What caused the continental glaciers to advance and retreat, and because there have been so many glacial advances within the past 3 million years, is the world still emerging from the last glacial retreat, or are we already heading back into the next glacial advance?

Glaciers and Climatic Change

Because variation in climate causes glaciers to advance or retreat, glaciers can serve as excellent indicators of climatic change. Geologists and other scientists study sedimentary deposits on land and in the sea to determine how long glacial intervals lasted, the frequency of repetition, the length of interglacial intervals, the variation in ocean surface temperature during the past 3 million years, and many other factors. These empirical or observational data are then compared with the various theories of glaciation to attempt to ascertain the actual causes of glaciation and to identify the clues to look for in determining whether we are still coming out of or going back into a glacial interval.

Among the more prominent theories of events that have triggered global climatic changes and lead to repeated glaciation are: (1) known astronomical variations in the orbital elements of the Earth (the so-called Milankovitch theory); (2) changes in energy output from the Sun; and (3) increases in volcanism that could have thrown more airborne volcanic material into the stratosphere, thereby creating a dust veil and lowered temperatures.

The years 1980, 1981, and 1982, for example, saw several major volcanic eruptions adding large quantities of particulate volcanic material and volatiles to the stratosphere, including the catastrophic eruption of Mount St. Helens, Washington, on May 18, 1980, and a large eruption of Mount Hekla, Iceland, on August 17, 1980. The 1982 series of eruptions from El Chichón volcano, Mexico, caused death and destruction in the populated area around the volcano, but a further reaching impact may result from the effect on Earth's climate because of the enormous ejection of volcanic material into the stratosphere.

The potential climatic effect of the Laki volcanic eruption in Iceland in 1783, the largest effusive (lava) volcanic eruption in historic time, was noted by the diplomat-scientist Benjamin Franklin in 1784, during one of his many sojourns in Paris.

Franklin concluded that the introduction of large quantities of volcanic particles into the Earth's upper atmosphere could cause a reduction in surface temperature, because the particles would lessen the amount of solar energy reaching the Earth's surface. The catastrophic eruption of the Tambora volcano, Indonesia, in 1815 was followed by a so-called "year-without-a-summer." In New England, for example, frost occurred during each of the summer months in 1816.

Some of the observational evidence matches certain aspects of these three theories and other theories as well. Whatever the actual cause or causes of climatic change, the irrefutable fact is that repeated periods of glaciation have occurred, and that the last such glacial interval ended in the geologically recent past— at the dawn of human civilization.

Long-term variations in climate are difficult to measure because of the rather short period of time in which scientifically valid meteorological observations have been made. For example, most weather stations in North America have observational records of less than 100 years. Long-term variations in climate, however, are measured in decades, centuries, and even millennia. This is one of the reasons why glaciers can be such valuable indicators of climate. Glaciers tend to "average" out the short term meteorological variations and reflect longer term variations which take place over several decades or centuries. Large ice sheets such as in Greenland and Antarctica have even greater response times, probably over several millennia (thousands of years) or tens of millennia.

Most glacial ice is encompassed in the two largest ice sheets, Antarctica and Greenland, which together contain an estimated 97 percent of all the glacial ice and 77 percent of the freshwater supply of the planet. During a maximum global advance of glaciers, however, it is estimated that North America contained volumetrically more ice than the combined present-day total of Antarctica and Greenland, and that sea level dropped by as much as 300 feet (90 meters). If all the present glacial ice were to melt from Antarctica and Greenland, the oceans would rise another 300 feet (90 meters) and inundate most of the coastal cities of the world. Some glaciologists have suggested that the West Antarctic Ice Sheet is inherently unstable and could suddenly surge forward under climatic conditions similar to the present time, resulting in a 7 to 20 feet (2-6 meters) global rise in sea level, depending on the size of the surge.

Until the launches in 1972, 1975, 1978, and 1982 of the Landsat series of spacecraft, glaciologists had no accurate means of measuring the areal extent of glacial ice on Earth. Landsat 1, 2, and 3 multispectral scanner (MSS) images, obtained from an orbital altitude of 570 miles (920 kilometers), covered about the same area, 10,000 square miles (34,000 square kilometers) every 18 days, and had a pixel (picture element) resolution of about 260 feet (80 meters). Such satellite images provide a means for delineating the areal extent of ice sheets and icecaps and for determining the position of the termini of valley, outlet, and tidal glaciers on a common base of data for the entire globe. To take advantage of this data, a "Satellite Image Atlas of Glaciers" is being prepared by the U.S. Geological Survey in association with a number of other U.S. and foreign scientific organizations. If Landsat-type surveys of the planet are continued for several decades, a means of monitoring long-term changes in glacial area will also become possible—thereby giving us a way of monitoring global climate change.

Iceland, an island in the North Atlantic at about 65° North latitude, is about the size of the Commonwealth of Virginia or about 40,000 square miles (103,000 square kilometers). About 10 percent of the area of Iceland is covered by glaciers, mostly occurring as icecaps of various sizes, the largest of which is Vatnajökull, with an area of about 3,200 square miles (8,300 square kilometers). Six other glaciers have areas in excess of 19 square miles (50 square kilometers). Many of the icecaps of Iceland, such as Vatnajökull, Langjökull, Hofsjökull, Myrdalsjökull, and Eyjafjallajökull, are extremely dynamic, with heavy annual accumulations of snow—over 20 feet (6 meters) per year in the interior of Vatnajökull. Rapid movement is nearly

6 feet (2 meters) per day in one of the outlet glaciers of Vatnajökull, and extensive melting occurs at lower elevations.

For a variety of reasons the glaciers of Iceland are well suited to serve as an indicator of changes in climate in at least the Northern Hemisphere. The historical record of pre-1900 observations has been well documented over the years by several Icelandic scientists such as Thoroddsen and Thorarinsson. The Danish Geodetic Survey began to survey and produce maps of Iceland in the early part of the 20th century just after the end of what is now called "The Little Ice Age," the cool climatic period between about the mid-1500s and the late 1800s. Many of the terminal moraines deposited by Iceland's glaciers most distant from the present-day ice margins (termini) were formed during the latter part of the 19th century. The warmer first half of the 20th century has seen all of Iceland's glaciers diminish in size in response to this period of climatic warming. For example, Glámujökull, an ice cap in northwest Iceland, which appeared on late 19th century and early 20th century maps with an area of 89 square miles (230 square kilometers) and 1.7 square miles (4.5 square kilometers), respectively, has subsequently completely disappeared and no longer appears on modern maps.

U.S. military maps were made during the 1940s, and a revised map series is presently in production. The Iceland Geodetic Survey has also published revised maps and has begun a new series of orthophotomaps in association with the United States. Landsat images of Iceland permit an excellent means of monitoring changes in the area of Icelandic glaciers at more frequent intervals. Already, Landsat images of Iceland have been used to good advantage by U.S. Geological Survey and Icelandic scientists to document changes resulting from two glacial surges; the velocity of an outlet glacier has been measured, and existing maps have been revised.

According to Milankovitch and other 20th century theoretical climatologists, the glaciers in the area around 65° North latitude are especially sensitive to astronomical variations

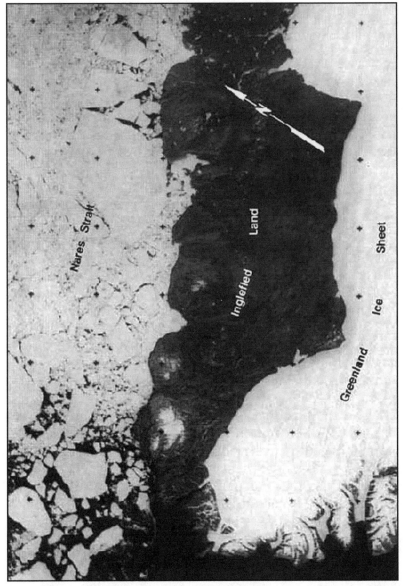

Landsat image of northwestern Greenland on July 29, 1976 shows the well-defined edge of the Greenland Ice Sheet in the Inglefield Land area and numerous outlet glaciers emptying into fiords. Sea ice fills the Nares Strait between Greenland and Ellesmere Island, Northwest Territories, Canada.

in the orbital elements of the Earth. In the Northern Hemisphere, glaciers on Baffin Island, Canada; in the Alaska Range, Alaska; in the southern tip of Greenland; in Iceland; and in Norway are at the right location. Except for Greenland, with its massive ice sheet, all the other geographic areas mentioned contain only remnants of a once far more extensive glacial ice cover during the Pleistocene era. The glaciers of Iceland, as with some of those in these other areas, are important as long-term indicators of climatic change because of their latitudinal location and because they are apparently just large enough not to be affected by short-term climatic variation yet are small and dynamic enough to respond to changes caused by climatic variation over several decades. Landsat images of Iceland and other areas at 65° North latitude are providing a practical and cost-effective means of making a rapid inventory of changes in glacial area.

The Earth can be considered a natural "spaceship" as it and other planets and stars of the galaxy travel through the universe. Climate is one aspect of the planetary environment which has an impact on every living thing; even a small change in climate can have a devastating effect on plant and animal life and on humans and their economies. Because climatic changes have global significance, scientists judiciously search for the correct clues to these changes. This search often requires making observations in remote areas—both on land and in the sea.

Only recently have scientific tools been developed that can make the necessary observations. With improved technology for extracting cores from the sea bottom and with existing and planned Earth resources survey satellites, geologists, glaciologists, and other scientists now have the tools to study and monitor a variety of environmental factors and phenomena. Many decades of observations by many scientists will be necessary, however, to begin to unravel the climate-glacier puzzle. But a start has been made to try to understand the enigma of climatic change and to understand the response of glaciers to such change. In turn, by studying the response and changes in glaciers, we can better understand—and anticipate—the range of past and possible future climatic changes. Scientists of the U.S. Geological Survey are among those who are deeply involved in the study of the many facets of an incredibly complex environmental problem, one which has extraordinary significance to the economic well-being of human societies.

Changes in climate over time produce variations in seasonal and annual temperature and precipitation at different latitudes, a fact confirmed by the geologic record and by meteorological observations during the past 200 years or so. Floods, droughts, higher space heating and cooling costs, variations in agricultural yield, higher snow removal costs, changes in sea level, and variation in length of time of ice-free conditions in navigable waterways have an impact on the economics of various human endeavors. The monitoring of glaciers gives scientists one of the most important indicators in determining whether observed climatic change is regional or global. This knowledge can then be used by governments to make long-range plans to better cope with the economic impact of such climatic changes.

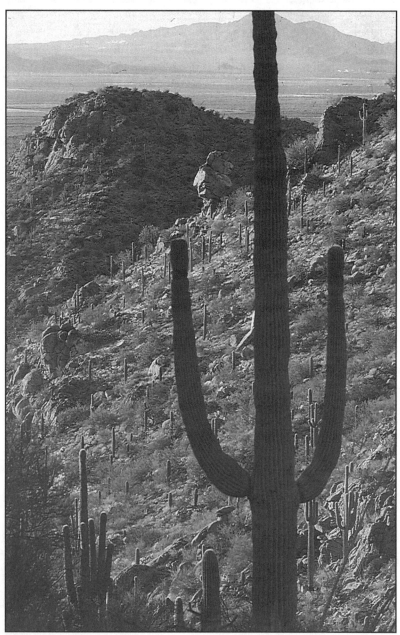
Cacti dominate the Sonoran Desert vegetation near Tucson, Arizona.

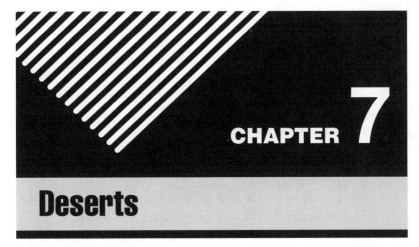

CHAPTER 7

Deserts

Approximately one-third of the Earth's land surface is desert, arid land with meager rainfall that supports only sparse vegetation and a limited population of people and animals. Deserts—stark, sometimes mysterious worlds—have been portrayed as fascinating environments of adventure and exploration from narratives such as that of *Lawrence of Arabia* to movies such as *Dune*. These arid regions are called *deserts* because they are dry. They may be hot, or they may be cold. They may be regions of sand or vast areas of rocks and gravel peppered with occasional plants. But deserts are always dry.

Ripples on a dune in the Gran Desierto, Mexico.

Deserts are natural laboratories in which to study the interactions of wind and sometimes water on the arid surfaces of planets. They contain valuable mineral deposits that were formed in the arid environment or that were exposed by erosion. Because deserts are dry, they are ideal places for human artifacts and fossils to be preserved. Deserts are also fragile environments. The misuse of these lands is a serious and growing problem in parts of our world.

There are almost as many definitions of deserts and classification systems as there are deserts in the world. Most classifications rely on some combination of the number of days

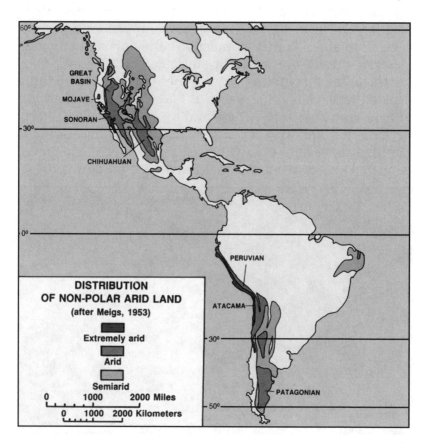

DISTRIBUTION
OF NON-POLAR ARID LAND
(after Meigs, 1953)

Extremely arid

Arid

Semiarid

GREAT BASIN
MOJAVE
SONORAN
CHIHUAHUAN
PERUVIAN
ATACAMA
PATAGONIAN

0   1000   2000 Miles
0   1000   2000 Kilometers

of rainfall, the total amount of annual rainfall, temperature, humidity, or other factors. In 1953, Peveril Meigs divided desert regions on Earth into three categories according to the amount of precipitation they received. In this now widely accepted system, extremely arid lands have at least 12 consecutive months without rainfall, arid lands have less than 250 millimeters of annual rainfall, and semiarid lands have a mean annual precipitation of between 250 and 500 millimeters. Arid and extremely arid land are deserts, and semiarid grasslands generally are referred to as steppes.

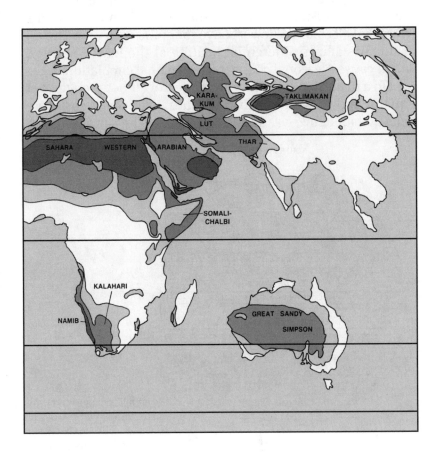

How the Atmosphere Influences Aridity

We live at the bottom of a gaseous envelope—the atmosphere—that is bound gravitationally to the planet Earth. The circulation of our atmosphere is a complex process because of the Earth's rotation and the tilt of its axis. The Earth's axis is inclined $23^{1}/_{2}°$ from the ecliptic, the plane of the Earth's orbit around the Sun. Due to this inclination, vertical rays of the Sun strike $23^{1}/_{2}°$ N. latitude, the Tropic of Cancer, at summer solstice in late June. At winter solstice, the vertical rays strike $23^{1}/_{2}°$ S. latitude, the Tropic of Capricorn. In the Northern Hemisphere, the summer solstice day has the most daylight hours, and the winter solstice has the fewest daylight hours each year. The tilt of the axis allows differential heating of the Earth's surface, which causes seasonal changes in the global circulation.

On a planetary scale, the circulation of air between the hot Equator and the cold North and South Poles creates pressure belts that influence weather. Air warmed by the Sun rises at the

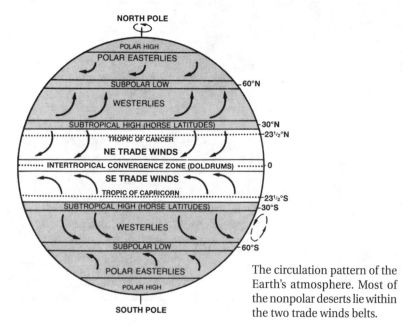

The circulation pattern of the Earth's atmosphere. Most of the nonpolar deserts lie within the two trade winds belts.

Equator, cools as it moves toward the poles, descends as cold air over the poles, and warms again as it moves over the surface of the Earth toward the Equator. This simple pattern of atmospheric convection, however, is complicated by the rotation of the Earth, which introduces the Coriolis Effect.

To appreciate the origin of this effect, consider the following. A stick placed vertically in the ground at the North Pole would simply turn around as the Earth rotates. A stick at the Equator would move in a large circle of almost 40,000 kilometers with the Earth as it rotates.

The Coriolis Effect illustrates Newton's first law of motion— a body in motion will maintain its speed and direction of motion unless acted on by some outside force. Thus, a wind travelling north from the equator will maintain the velocity acquired at the equator while the Earth under it is moving slower. This effect accounts for the generally east-west direction of winds, or streams of air, on the Earth's surface. Winds blow between areas of different atmospheric pressures.

The Coriolis Effect influences the circulation pattern of the Earth's atmosphere. In the zone between about 30° N. and 30° S., the surface air flows toward the Equator and the flow aloft is poleward. A low-pressure area of calm, light variable winds near the equator is known to mariners as the *doldrums*.

Around 30° N. and S., the poleward flowing air begins to descend toward the surface in subtropical high-pressure belts. The sinking air is relatively dry because its moisture has already been released near the Equator above the tropical rain forests. Near the center of this high-pressure zone of descending air, called the "Horse Latitudes," the winds at the surface are weak and variable. The name for this area is believed to have been given by colonial sailors, who, becalmed sometimes at these latitudes while crossing the oceans with horses as cargo, were forced to throw a few horses overboard to conserve water.

The surface air that flows from these subtropical high-pressure belts toward the Equator is deflected toward the west in both hemispheres by the Coriolis Effect. Because winds are named for the direction from which the wind is blowing, these winds are called the northeast trade winds in the Northern Hemisphere and the southeast trade winds in the Southern Hemisphere. The trade winds meet at the doldrums. Surface winds known as "westerlies" flow from the Horse Latitudes toward the poles. The "westerlies" meet "easterlies" from the polar highs at about 50-60° N. and S.

Near the ground, wind direction is affected by friction and by changes in topography. Winds may be seasonal, sporadic, or daily. They range from gentle breezes to violent gusts at speeds greater than 300 kilometers/hour.

These dunes in the Algodones Sand Sea of southeastern California move as much as five meters per year. The dunes in this photograph, looking south, move toward the east (left).

Searles Lake, California.

Where Deserts Form

Dry areas created by global circulation patterns contain most of the deserts on the Earth. The deserts of our world are not restricted by latitude, longitude, or elevation. They occur from areas close to the poles down to areas near the Equator. The People's Republic of China has both the highest desert, the Qaidam Depression that is 2,600 meters above sea level, and one of the lowest deserts, the Turpan Depression that is 150 meters below sea level.

Deserts are not confined to Earth. The atmospheric circulation patterns of other terrestrial planets with gaseous envelopes also depend on the rotation of those planets, the tilts of their axes, their distances from the Sun, and the composition and density of their atmospheres. Except for the poles, the entire surface of Mars is a desert. Venus also may support deserts.

Types of Deserts

Deserts are classified by their geographical location and dominant weather pattern as *trade wind, midlatitude, rain shadow, coastal, monsoon,* or *polar* deserts. Former desert areas presently in nonarid environments are *paleodeserts,* and *extraterrestrial* deserts exist on other planets.

Trade wind deserts

The trade winds in two belts on the equatorial sides of the Horse Latitudes heat up as they move toward the Equator. These dry winds dissipate cloud cover, allowing more sunlight to heat the land. Most of the major deserts of the world lie in areas crossed by the trade winds. The world's largest desert, the Sahara of North Africa, which has experienced temperatures as high as 57° C, is a trade wind desert.

Midlatitude deserts

Midlatitude deserts occur between 30° and 50° N. and S., poleward of the subtropical high-pressure zones. These deserts are in interior drainage basins far from oceans and have a wide range of annual temperatures. The Sonoran Desert of southwestern North America is a typical midlatitude desert.

Rain shadow deserts

Rain shadow deserts are formed because tall mountain ranges prevent moisture-rich clouds from reaching areas on the lee, or protected side, of the range. As air rises over the mountain, water is precipitated and the air loses its moisture content. A desert is formed in the leeside "shadow" of the range.

Coastal deserts

Coastal deserts generally are found on the western edges of continents near the Tropics of Cancer and Capricorn. They are affected by cold ocean currents that parallel the coast. Because local wind systems dominate the trade winds, these deserts are less stable than other deserts. Winter fogs, produced by upwelling cold currents, frequently blanket coastal deserts and block solar radiation. Coastal deserts are relatively complex because they are at the juncture of terrestrial, oceanic, and atmospheric systems. A coastal desert, the Atacama of South America, is the Earth's driest desert. In the Atacama, measurable rainfall—1 millimeter or more of rain—may occur as infrequently as once every 5-20 years.

High dunes of the Namib Desert (coastal) near Sossus Vlei.

Monsoon deserts

"Monsoon," derived from an Arabic word for "season," refers to a wind system with pronounced seasonal reversal. Monsoons develop in response to temperature variations between continents and oceans. The southeast trade winds of the Indian Ocean, for example, provide heavy summer rains in India as they move onshore. As the monsoon crosses India, it loses moisture on the eastern slopes of the Aravalli Range. The Rajasthan Desert of India and the Thar Desert of Pakistan are parts of a monsoon desert region west of the range.

Polar deserts

Polar deserts are areas with annual precipitation less than 250 millimeters and a mean temperature during the warmest month of less than 10° C. Polar deserts on the Earth cover nearly 5 million square kilometers and are mostly bedrock or gravel plains. Sand dunes are not prominent features in these deserts, but snow dunes occur commonly in areas where precipitation is locally more abundant.

Temperature changes in polar deserts frequently cross the freezing point of water. This "freeze-thaw" alternation forms patterned textures on the ground, as much as 5 meters in diameter.

The Dry Valleys of Antarctica have been ice-free for thousands of years.

Paleodeserts

Data on ancient sand seas (vast regions of sand dunes), changing lake basins, archaeology, and vegetation analyses indicate that climatic conditions have changed considerably over vast areas of the Earth in the recent geologic past. During the last 12,500 years, for example, parts of the deserts were more arid than they are today. About 10 percent of the land between 30° N. and 30° S. is covered now by sand seas. Nearly 18,000 years ago, sand seas in two vast belts occupied almost 50 percent of this land area. As is the case today, tropical rain forests and savannahs were between the two belts.

Fossil desert sediments that are as much as 500 million years old have been found in many parts of the world. Sand dunelike patterns have been recognized in presently nonarid environments. Many such relict dunes now receive from 80 to 150 millimeters of rain each year. Some ancient dunes are in areas now occupied by tropical rain forests.

The Nebraska Sand Hills is an inactive 57,000 square kilometer dune field in central Nebraska. The largest sand sea in the Western Hemisphere, it is now stabilized by vegetation and receives about 500 millimeters of rain each year. Dunes in the Sand Hills are up to 120 meters high.

A dry community of vegetation grows among the dunes of the Nebraska Sand Hills.

Extraterrestrial deserts

Mars is the only other planet on which we have identified wind-shaped (*eolian*) features. Although its surface atmospheric pressure is only about one-hundredth that of Earth, global circulation patterns on Mars have formed a circumpolar sand sea of more than five million square kilometers, an area greater than the Empty Quarter of Saudi Arabia, the largest sand sea on our planet. Martian sand seas consist predominantly of crescent-shaped dunes on plains near the perennial ice cap of the north polar area. Smaller dune fields occupy the floors of many large craters in the polar regions.

The Viking spacecraft image of Mars shows alternating layers of ice and windblown dust near the north polar cap. Annual and other periodic climatic changes due to orbit fluctuations may occur on Mars.

Desert Features

Sand covers only about 20 percent of the Earth's deserts. Most of the sand is in sand sheets and sand seas—vast regions of undulating dunes resembling ocean waves "frozen" in an instant of time.

Nearly 50 percent of desert surfaces are plains where eolian deflation—removal of fine-grained material by the wind—has exposed loose gravel consisting predominantly of pebbles but with occasional cobbles.

The remaining surfaces of arid lands are composed of exposed bedrock outcrops, desert soils, and fluvial deposits including alluvial fans, playas, desert lakes, and oases. Bedrock outcrops commonly occur as small mountains surrounded by extensive erosional plains.

Oases are vegetated areas moistened by springs, wells, or by irrigation. Many are artificial. Oases are often the only places in deserts that support crops and permanent habitation.

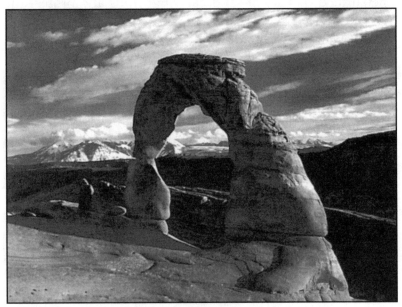

Delicate Arch, Arches National Park, Utah.

Soils

Soils that form in arid climates are predominantly mineral soils with low organic content. The repeated accumulation of water in some soils causes distinct salt layers to form. Calcium carbonate precipitated from solution may cement sand and gravel into hard layers called "calcrete" that form layers up to 50 meters thick.

Caliche is a reddish-brown to white layer found in many desert soils. Caliche commonly occurs as nodules or as coatings on mineral grains formed by the complicated interaction between water and carbon dioxide released by plant roots or by decaying organic material.

Plants

Most desert plants are drought- or salt-tolerant. Some store water in their leaves, roots, and stems. Other desert plants have long tap roots that penetrate the water table, anchor the soil, and control erosion. The stems and leaves of some plants lower the surface velocity of sand-carrying winds and protect the ground from erosion.

Deserts typically have a plant cover that is sparse but

Sparse, very dry, single species vegetation in Death Valley, California.

enormously diverse. The Sonoran Desert of the American Southwest has the most complex desert vegetation on Earth. The giant saguaro cacti provide nests for desert birds and serve as "trees" of the desert. Saguaro grow slowly but may live 200 years. When 9 years old, they are about 15 centimeters high. After about 75 years, the cacti are tall and develop their first branches. When fully grown, saguaro are 15 meters tall and weigh as much as 10 tons. They dot the Sonoran Desert and reinforce the general impression of deserts as cacti-rich land.

Although cacti are often thought of as characteristic desert plants, other types of plants have adapted well to the arid environment. They include the pea family and sunflower family. Cold deserts have grasses and shrubs as dominant vegetation.

Water

Rain does fall occasionally in deserts, and desert storms are often violent. A record 44 millimeters of rain once fell within 3 hours in the Sahara. Large Saharan storms may deliver up to 1 millimeter per minute. Normally dry stream channels, called arroyos or wadis, can quickly fill after heavy rains, and flash floods make these channels dangerous. More people drown in deserts than die of thirst.

Though little rain falls in deserts, deserts receive runoff from ephemeral, or short-lived, streams fed by rain and snow from adjacent highlands. These streams fill the channel with a slurry of mud and commonly transport considerable quantities of sediment for a day or two. Although most deserts are in basins with closed, or interior drainage, a few deserts are crossed by "exotic" rivers that derive their water from outside the desert. Such rivers infiltrate soils and evaporate large amounts of water on their journeys through the deserts, but their volumes are such that they maintain their continuity. The Nile, the Colorado, and the Yellow are exotic rivers that flow through deserts to deliver their sediments to the sea.

Lakes form where rainfall or meltwater in interior drainage basins is sufficient. Desert lakes are generally shallow,

The Wei River in the Loess Plateau, China.

temporary, and salty. Because these lakes are shallow and have a low bottom gradient, wind stress may cause the lake waters to move over many square kilometers. When small lakes dry up, they leave a salt crust or hardpan. The flat area of clay, silt, or sand encrusted with salt that forms is known as a playa. There are more than a hundred playas in North American deserts. Most are relics of large lakes that existed during the last Ice Age about 12,000 years ago. Lake Bonneville was a 52,000-square-kilometer lake almost 300 meters deep in Utah, Nevada, and Idaho during the Ice Age. Today the remnants of Lake Bonneville include Utah's Great Salt Lake, Utah Lake, and Sevier Lake. Because playas are arid land forms from a wetter past, they contain useful clues to climatic change.

The flat terrains of hardpans and playas make them excellent race tracks and natural runways for airplanes and spacecraft. Ground-vehicle speed records are commonly established on Bonneville Speedway, a race track on the Great Salt Lake hardpan. Space shuttles land on Rogers Lake Playa at Edwards Air Force Base, California.

Eolian Processes

Eolian processes pertain to the activity of the winds. Winds may erode, transport, and deposit materials, and are effective agents in regions with sparse vegetation and a large supply of unconsolidated sediments. Although water is much more powerful than wind, eolian processes are important in arid environments.

Eolian erosion

Wind erodes the Earth's surface by deflation, the removal of loose, fine-grained particles by the turbulent eddy action of the wind, and by abrasion, the wearing down of surfaces by the grinding action and sand blasting of windborne particles.

Most eolian deflation zones are composed of *desert pavement*, a sheetlike surface of rock fragments that remains after wind and water have removed the fine particles. Almost half of the Earth's desert surfaces are stony deflation zones. The rock mantle in desert pavements protects the underlying material from deflation.

A dark, shiny stain, called *desert varnish* or rock varnish, is often found on surfaces of some desert rocks that have been exposed at the surface for a long period of time. Manganese, iron oxides, hydroxides, and clay minerals form most varnishes and provide the shine.

Deflation basins, called *blowouts*, are hollows formed by the removal of particles by wind. Blowouts are generally small, but may be up to several kilometers in diameter.

Wind-driven grains abrade landforms. Grinding by particles carried in the wind creates grooves or small depressions. *Ventifacts* are rocks which have been cut, and sometimes polished, by the abrasive action of wind.

Sculpted landforms, called *yardangs*, are up to tens of meters high and kilometers long and are forms that have been streamlined by desert winds. The famous sphinx at Giza in Egypt may be a modified yardang.

Eolian transportation

Particles are transported by winds through suspension, saltation, and creep.

Small particles may be held in the atmosphere in *suspension*. Upward currents of air support the weight of suspended particles and hold them indefinitely in the surrounding air. Typical winds near the Earth's surface suspend particles less than 0.2 millimeters in diameter and scatter them aloft as dust or haze.

Saltation is downwind movement of particles in a series of jumps or skips. Saltation normally lifts sand-size particles no more than one centimeter above the ground, and proceeds at one-half to one-third the speed of the wind. A saltating grain may hit other grains that jump up to continue the saltation. The grain may also hit larger grains that are too heavy to hop, but that slowly creep forward as they are pushed by saltating grains. Surface creep accounts for as much as 25 percent of grain movement in a desert.

Eolian turbidity currents are better known as *dust storms*. Air over deserts is cooled significantly when rain passes through it. This cooler and denser air sinks toward the desert surface.

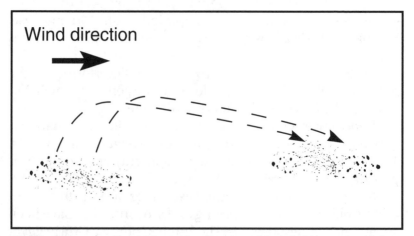

Saltation moves small particles in the direction of the wind in a series of short hops or skips.

When it reaches the ground, the air is deflected forward and sweeps up surface debris in its turbulence as a dust storm.

Crops, people, villages, and possibly even climates are affected by dust storms. Some dust storms are intercontinental, a few may circle the globe, and occasionally they may engulf entire planets. When the Mariner 9 spacecraft arrived at Mars in 1971, the entire planet was enshrouded in global dust.

Most of the dust carried by dust storms is in the form of silt-size particles. Deposits of this windblown silt are known as *loess.* The thickest known deposit of loess, 335 meters, is on the Loess Plateau in China. In Europe and in the Americas, accumulations of loess are generally from 20 to 30 meters thick.

Small whirlwinds, called *dust devils,* are common in arid lands and are thought to be related to very intense local heating of the air that results in instabilities of the air mass. Dust devils may be as much as one kilometer high.

Eolian deposition

Wind-deposited materials hold clues to past as well as to present wind directions and intensities. These features help us understand the present climate and the forces that molded it. Wind-deposited sand bodies occur as sand sheets, ripples, and dunes.

Sand sheets are flat, gently undulating sandy plots of sand surfaced by grains that may be too large for saltation. They form approximately 40 percent of eolian depositional surfaces. The Selima Sand Sheet, which occupies 60,000 square kilometers in southern Egypt and northern Sudan, is one of the Earth's largest sand sheets. The Selima is absolutely flat in some places; in others, active dunes move over its surface.

Wind blowing on a sand surface *ripples* the surface into crests and troughs whose long axes are perpendicular to the wind direction. The average length of jumps during saltation corresponds to the wavelength, or distance between adjacent crests, of the ripples. In ripples, the coarsest materials collect at the crests. This distinguishes small ripples from dunes, where

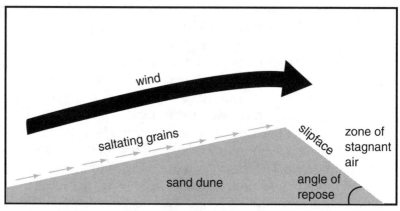

Windblown sand moves up the gentle upwind side of the dune by saltation or creep. Sand accumulates at the brink, the top of the slipface. When the buildup of sand at the brink exceeds the angle of repose, a small avalanche of grains slides down the slipface. Grain by grain, the dune moves downwind.

the coarsest materials are generally in the troughs.

Accumulations of sediment blown by the wind into a mound or ridge, dunes have gentle upwind slopes on the wind-facing side. The downwind portion of the dune, the lee slope, is commonly a steep avalanche slope referred to as a *slipface*. Dunes may have more than one slipface. The minimum height of a slipface is about 30 centimeters.

Sand grains move up the dune's gentle upwind slope by saltation and creep. When particles at the brink of the dune exceed the angle of repose, they spill over in a tiny landslide or avalanche that reforms the slipface. As the avalanching continues, the dune moves in the direction of the wind.

Some of the most significant experimental measurements on eolian sand movement were performed by Ralph Bagnold, a British engineer who worked in Egypt prior to World War II. Bagnold investigated the physics of particles moving through the atmosphere and deposited by wind. He recognized two basic dune types, the crescentic dune, which he called "barchan," and the linear dune, which he called longitudinal or "sief" (Arabic for "sword").

Types of Dunes

A worldwide inventory of deserts has been developed using images from the Landsat satellites and from space and aerial photography. It defines five basic types of dunes: *crescentic, linear, star, dome,* and *parabolic.*

The most common dune form on Earth and on Mars is the *crescentic.* Crescent-shaped mounds generally are wider than long. The slipface is on the dune's concave side. These dunes form under winds that blow from one direction, and they also are known as barchans, or transverse dunes. Some types of crescentic dunes move faster over desert surfaces than any other type of dune. A group of dunes moved more than 100 meters per year between 1954 and 1959 in China's Ningxia Province; similar rates have been recorded in the Western Desert of Egypt. The largest crescentic dunes on Earth, with mean crest-to-crest widths of more than 3 kilometers, are in China's Taklimakan Desert.

Straight or slightly sinuous sand ridges typically much longer than they are wide are known as *linear dunes.* They may be more than 160 kilometers long. Linear dunes may occur as isolated ridges, but they generally form sets of parallel ridges separated by miles of sand, gravel, or rocky interdune corridors. Some linear dunes merge to form Y-shaped compound dunes. Many form in bidirectional wind regimes. The long axes of these dunes extend in the resultant direction of sand movement.

Radially symmetrical, *star dunes* are pyramidal sand mounds with slipfaces on three or more arms that radiate from the high center of the mound. They tend to accumulate in areas with multidirectional wind regimes. Star dunes grow upward rather than laterally. They dominate the Grand Erg Oriental of the Sahara. In other deserts, they occur around the margins of the sand seas, particularly near topographic barriers. In the southeast Badain Jaran Desert of China, the star dunes are up to 500 meters tall and may be the tallest dunes on Earth.

Oval or circular mounds that generally lack a slipface, *dome dunes* are rare and occur at the far upwind margins of sand seas.

U-shaped mounds of sand with convex noses trailed by elongated arms are *parabolic dunes*. Sometimes these dunes are called U-shaped, blowout, or hairpin dunes, and they are well known in coastal deserts. Unlike crescentic dunes, their crests point upwind. The elongated arms of parabolic dunes follow rather than lead because they have been fixed by vegetation, while the bulk of the sand in the dune migrates forward. The longest known parabolic dune has a trailing arm 12 kilometers long.

Occurring wherever winds periodically reverse direction, *reversing dunes* are varieties of any of the above types. These dunes typically have major and minor slipfaces oriented in opposite directions.

All these dune types may occur in three forms: *simple, compound,* and *complex.* Simple dunes are basic forms with a minimum number of slipfaces that define the geometric type. Compound dunes are large dunes on which smaller dunes of similar type and slipface orientation are superimposed, and complex dunes are combinations of two or more dune types. A

Linear dunes in the Western Desert of Egypt.

crescentic dune with a star dune superimposed on its crest is the most common complex dune. Simple dunes represent a wind regime that has not changed in intensity or direction since the formation of the dune, while compound and complex dunes suggest that the intensity and direction of the wind has changed.

Remote Sensing of Arid Lands

The world's deserts are generally remote, inaccessible, and inhospitable. Hidden among them, however, are hydrocarbon reservoirs, evaporites, and other mineral deposits, as well as human artifacts preserved for centuries by the arid climate. In these harsh environments, the information and perspective required to increase our understanding of arid-land geology and resources often depends on remote-sensing methods. Remote sensing is the collection of information about an object without being in direct physical contact with it.

Remote-sensing instruments in Earth-orbit satellites measure radar, visible light, and infrared radiation. Radar imaging systems provide their own source of electromagnetic energy, so they can operate at any time of day or night. Additionally, clouds and all but the most severe storms are transparent to radar.

The first Shuttle Imaging Radar System (SIR-A), flown on the U.S. space shuttle *Columbia* in 1981, recorded images that show buried fluvial topography, faults, and intrusive bodies otherwise concealed beneath sand sheets and dunes of the Western Desert in Egypt and the Sudan. Most of these features are not visible from the ground. The radar signal penetrated loose dry sands and returned images of buried river channels not visible at the surface. These images helped find new archaeological sites and sources of potable water in the desert. These "radar rivers" are the remnants of a now vanished major river system that flowed across Africa some 20 million years before the development of the Nile River system. Radar imagery also is a powerful tool for exploring for placer mineral deposits in arid lands.

In 1972, the United States launched the first of a group of unmanned satellites collectively known as Landsat. Landsat satellites carry sensors that record "light," or portions of the electromagnetic spectrum, as it reflects off the Earth. Landsat acquires digital data that are converted into an image.

The scarcity of vegetation makes spectral remote sensing especially effective in arid lands. Rocks containing limonite, a hydrous iron oxide, may be identified readily from Landsat Multispectral Scanner data. The Landsat Thematic Mapper (TM) has increased our ability to detect and map the distribution of minerals in volcanic rocks and related mineral deposits in arid and semiarid lands.

More than a million images of Earth have been acquired by the Landsat satellites. A Landsat image may be viewed as a single band in black-and-white, or as a combination represented by three colors, called a color composite. The most widely used Landsat color image is called a false-color composite because it reproduces the infrared band (invisible to the naked eye) as red, the red band as green, and the green band as blue. Healthy vegetation in a false-color composite is red.

Desert studies still are hampered in many regions by lack of accurate climate data. Most desert weather stations are in oases surrounded by trees and buildings and have been subjected to many location and elevation changes throughout the life of the station. Data from oases do not reflect conditions from the surrounding desert. A wide variety of instruments has been used to record measurements over varying lengths of time and in different formats, making data difficult to interpret and compare.

To overcome some of these problems in deserts of the American Southwest, the U.S. Geological Survey (USGS) established its Desert Winds Project to measure in a standard format several key meteorologic characteristics of arid lands. Project scientists have successfully established instrument stations to measure windspeed, including peak gusts, which alter the landforms the most. A station recorded a windstorm near Vicksburg, Arizona, for example, with peak gusts of almost 150 kilometers per hour. Using low-maintenance, automatic, solar-powered sensors, the stations also measure wind direction, precipitation, humidity, soil and air temperatures, and barometric pressure at specific heights above the surface.

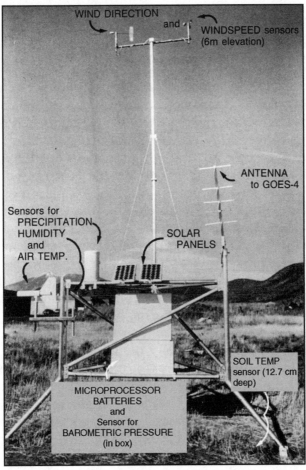

This geometeorologic station of the Desert Winds Project measures wind speed and direction, soil and air temperature, and precipitation and humidity in the Great Basin Desert.

Data are sampled at 6-minute intervals and transmitted every 30 minutes to a Geostationary Operational Environmental Satellite (GOES). From GOES, the data are transmitted to the USGS laboratory in Flagstaff, Arizona.

The Desert Winds Project's investigators combine analyses of data with detailed geologic field studies and repetitive remote-sensing coverage in order to investigate and understand the long-term changes produced by wind in deserts of differing geologic and climatic types.

Mineral Resources in Deserts

Some mineral deposits are formed, improved, or preserved by geologic processes that occur in arid lands as a consequence of climate. Ground water leaches ore minerals and redeposits them in zones near the water table. This leaching process concentrates these minerals as ore that can be mined. Of the 15 major types of mineral deposits in the Western Hemisphere formed by action of ground water, 13 occur in deserts.

Evaporation in arid lands enriches mineral accumulation in their lakes. Playas may be sources of mineral deposits formed by evaporation. Water evaporating in closed basins precipitates minerals such as gypsum, salts (including sodium nitrate and sodium chloride), and borates. The minerals formed in these evaporite deposits depend on the composition and temperature of the saline waters at the time of deposition.

Significant evaporite resources occur in the Great Basin Desert of the United States, mineral deposits made forever famous by the "20-mule teams" that once hauled borax-laden wagons from Death Valley to the railroad. Boron, from borax and borate evaporites, is an essential ingredient in the manufacture of glass, ceramics, enamel, agricultural chemicals, water softeners, and pharmaceuticals. Borates are mined from evaporite deposits at Searles Lake, California, and other desert locations. The total value of chemicals that have been produced from Searles Lake substantially exceeds $1 billion.

The Atacama Desert of South America is unique among the deserts of the world in its great abundance of saline minerals. Sodium nitrate has been mined for explosives and fertilizer in the Atacama since the middle of the 19th century. Nearly 3 million metric tons were mined during World War I.

Valuable minerals located in arid lands include copper in the United States, Chile, Peru, and Iran; iron and lead-zinc ore in Australia; chromite in Turkey; and gold, silver, and uranium deposits in Australia and the United States. Nonmetallic mineral resources and rocks such as beryllium, mica, lithium, clays, pumice, and scoria also occur in arid regions. Sodium carbonate, sulfate, borate, nitrate, lithium, bromine, iodine, calcium, and strontium compounds come from sediments and near-surface brines formed by evaporation of inland bodies of water, often during geologically recent times.

The Green River Formation of Colorado, Wyoming, and Utah contains alluvial fan deposits and playa evaporites created in a huge lake whose level fluctuated for millions of years. Economically significant deposits of trona, a major source of sodium compounds, and thick layers of oil shale were created in the arid environment.

This open-pit mine in the Sonoran Desert near Ajo, Arizona, has exposed an elliptical copper deposit about 1,000 meters long and 750 meters wide. The copper ore mined here is in a bed that averages 150 meters in thickness.

Some of the more productive petroleum areas on Earth are found in arid and semiarid regions of Africa and the Mideast, although the oil reservoirs were originally formed in shallow marine environments. Recent climate change has placed these reservoirs in an arid environment.

Other oil reservoirs, however, are presumed to be eolian in origin and are presently found in humid environments. The Rotliegendes, a hydrocarbon reservoir in the North Sea, is associated with extensive evaporite deposits. Many of the major U.S. hydrocarbon resources may come from eolian sands. Ancient alluvial fan sequences may also be hydrocarbon reservoirs.

Desertification

The world's great deserts were formed by natural processes interacting over long intervals of time. During most of these times, deserts have grown and shrunk independent of human activities. Paleodeserts, large sand seas now inactive because they are stabilized by vegetation, extend well beyond the present margins of core deserts, such as the Sahara. In some regions, deserts are separated sharply from surrounding, less arid areas by mountains and other contrasting landforms that reflect basic structural differences in the regional geology. In other areas, desert fringes form a gradual transition from a dry to a more humid environment, making it more difficult to define the desert border.

These transition zones have very fragile, delicately balanced ecosystems. Desert fringes often are a mosaic of microclimates. Small hollows support vegetation that picks up heat from the hot winds and protects the land from the prevailing winds. After rainfall the vegetated areas are distinctly cooler than the surroundings. In these marginal areas, human activity may stress the ecosystem beyond its tolerance limit, resulting in degradation of the land. By pounding the soil with their hooves, livestock compact the substrate, increase the proportion of fine material, and reduce the percolation rate of the soil, thus

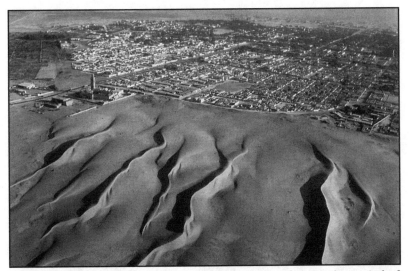

Linear dunes of the Sahara Desert encroach on Nouakchott, the capital of Mauritania. The dunes border a mosque at left.

encouraging erosion by wind and water. Grazing and the collection of firewood reduce or eliminate plants that help to bind the soil.

This degradation of formerly productive land—*desertification*—is a complex process. It involves multiple causes, and it proceeds at varying rates in different climates. Desertification may intensify a general climatic trend toward greater aridity, or it may initiate a change in local climate.

Desertification does not occur in linear, easily mappable patterns. Deserts advance erratically, forming patches on their borders. Areas far from natural deserts can degrade quickly to barren soil, rock, or sand through poor land management. The presence of a nearby desert has no direct relationship to desertification. Unfortunately, an area undergoing desertification is brought to public attention only after the process is well underway. Often little or no data are available to indicate the previous state of the ecosystem or the rate of degradation. Scientists still question whether desertification, as a process of global change, is permanent or how and when it can be halted or reversed.

Overgrazing has made the Rio Puerco Basin of central New Mexico one of the most eroded river basins of the American West and has increased the high sediment content of the river.

Problem

Desertification became well known in the 1930s, when parts of the Great Plains in the United States turned into the "Dust Bowl" as a result of drought and poor practices in farming, although the term itself was not used until almost 1950. During the dust bowl period, millions of people were forced to abandon their farms and livelihoods. Greatly improved methods of agriculture and land and water management in the Great Plains have prevented that disaster from recurring, but desertification presently affects millions of people on almost every continent.

Increased population and livestock pressure on marginal lands has accelerated desertification. In some areas, nomads moving to less arid areas disrupt the local ecosystem and increase the rate of erosion of the land. Nomads are trying to escape the desert, but because of their land-use practices, they are bringing the desert with them.

It is a misconception that droughts cause desertification. Droughts are common in arid and semiarid lands. Well-managed lands can recover from drought when the rains return. Continued land abuse during droughts, however, increases land degradation. By 1973, the drought that began in 1968 in the Sahel of West Africa and the land-use practices there had caused the deaths of more than 100,000 people and 12 million cattle, as well as the disruption of social organizations from villages to the national level.

While desertification has received tremendous publicity by the political and news media, there are still many things that we don't know about the degradation of productive lands and the expansion of deserts. In 1988, Ridley Nelson pointed out in an important scientific paper that the desertification problem and processes are not clearly defined. There is no consensus among researchers as to the specific causes, extent, or degree of desertification. Contrary to many popular reports, desertification is actually a subtle and complex process of deterioration that may often be reversible.

Global monitoring

In the last 27 years, satellites have begun to provide the global monitoring necessary for improving our understanding of desertification. Landsat images of the same area, taken several years apart but during the same point in the growing season, may indicate changes in the susceptibility of land to desertification. Studies using Landsat data help demonstrate the impact of people and animals on the Earth. However, other types of remote-sensing systems, land-monitoring networks, and global data bases of field observations are needed before the process and problems of desertification will be completely understood.

Local remedies

At the local level, individuals and governments can help to reclaim and protect their lands. In areas of sand dunes, covering the dunes with large boulders or petroleum will interrupt the wind regime near the face of the dunes and prevent the sand from moving. Sand fences are used throughout the Middle East and the United States, in the same way snow fences are used in the north. Placement of straw grids, each up to a square meter in area, will also decrease the surface wind velocity. Shrubs and trees planted within the grids are protected by the straw until they take root. In areas where some water is available for irrigation, shrubs planted on the lower one-third of a dune's windward side will stabilize the dune. This vegetation decreases the wind velocity near the base of the dune and prevents much of the sand from moving. Higher velocity winds at the top of the dune level it off and trees can be planted atop these flattened surfaces.

Oases and farmlands in windy regions can be protected by planting tree fences or grass belts. Sand that manages to pass through the grass belts can be caught in strips of trees planted as wind breaks 50 to 100 meters apart adjacent to the belts. Small plots of trees may also be scattered inside oases to stabilize the area. On a much larger scale, a "Green Wall," which will eventually stretch more than 5,700 kilometers in length, much longer than the famous Great Wall, is being planted in northeastern China to protect "sandy lands"—deserts believed to have been created by human activity.

More efficient use of existing water resources and control of salinization are other effective tools for improving arid lands. New ways are being sought to use surface-water resources such as rain water harvesting or irrigating with seasonal runoff from adjacent highlands. New ways are also being sought to find and tap groundwater resources and to develop more effective ways of irrigating arid and semiarid lands. Research on the reclamation of deserts also is focusing on discovering proper crop rotation to protect the fragile soil, on understanding how sand-fixing plants can be adapted to local environments, and on how grazing lands and water resources can be developed effectively without being overused.

If we are to stop and reverse the degradation of arid and semiarid lands, we must understand how and why the rates of climate change, population growth, and food production adversely affect these environments. The most effective intervention can come only from the wise use of the best earth-science information available.

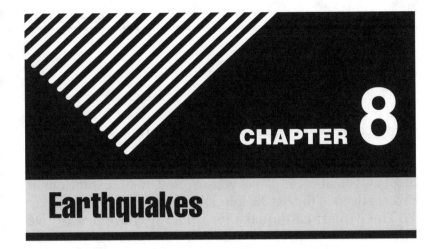

CHAPTER 8

Earthquakes

One of the most frightening and destructive phenomena of nature is a severe earthquake and its terrible aftereffects. An earthquake is a sudden movement of the earth, caused by the abrupt release of strain that has accumulated over a long time. For hundreds of millions of years, the forces of plate tectonics have shaped the Earth as the huge plates that form the planet's surface slowly move over, under, and past each other. Sometimes the movement is gradual. At other times, the plates are locked together, unable to release the accumulating energy. When the accumulated energy grows strong enough, the plates break free. If the earthquake occurs in a populated area, it may cause many deaths and injuries and extensive property damage.

Today we are challenging the assumption that earthquakes must present an uncontrollable and unpredictable hazard to life and property. Scientists have begun to estimate the locations and likelihoods of future damaging earthquakes. Sites of greatest hazard are being identified, and definite progress is being made in designing structures that will withstand the effects of earthquakes.

Earthquakes in History

The scientific study of earthquakes is comparatively new. Until the 18th century, few factual descriptions of earthquakes were recorded, and the natural cause of earthquakes was little understood. Those who did look for natural causes often reached conclusions that seem fanciful today; one popular theory was that earthquakes were caused by air rushing out of caverns deep in the Earth's interior.

The earliest earthquake for which we have descriptive information occurred in China in 1177 B.C. The Chinese earthquake catalog describes several dozen large earthquakes

The great 1906 San Francisco earthquake and fire destroyed most of the city and left 250,000 people homeless.

in China during the next few thousand years. Earthquakes in Europe are mentioned as early as 580 B.C., but the earliest for which we have some descriptive information occurred in the mid-16th century. The earliest known earthquakes in the Americas were in Mexico in the late 14th century and in Peru in 1471, but descriptions of the effects were not well documented. By the 17th century, descriptions of the effects of earthquakes were being published around the world—although these accounts were often exaggerated or distorted.

The most widely felt earthquakes in the recorded history of North America were a series that occurred in 1811-12 near New Madrid, Missouri. A great earthquake, whose magnitude is estimated to be about 8, occurred on the morning of December 16, 1811. Another great earthquake occurred on January 23, 1812, and a third, the strongest yet, on February 7, 1812. Aftershocks were nearly continuous between these great earthquakes and continued for months afterwards. These earthquakes were felt by people as far away as Boston and Denver. Because the most intense effects were in a sparsely populated region, the destruction of human life and property was slight. If just one of these enormous earthquakes occurred in the same area today, millions of people and buildings and other structures worth billions of dollars would be affected.

The San Francisco earthquake of 1906 was one of the most destructive in the recorded history of North America— the earthquake and the fire that followed killed nearly 700 people and left the city in ruins. The Alaska earthquake of March 27, 1964, was of greater magnitude than the San Francisco earthquake; it released perhaps twice as much energy and was felt over an area of almost 500,000 square miles. The ground motion near the epicenter was so violent that the tops of some trees were snapped off. One hundred and fourteen people (some as far away as California) died as a result of this earthquake, but loss of life and property would have been far greater had Alaska been more densely populated.

Where Earthquakes Occur

The Earth is formed of several layers that have very different physical and chemical properties. The outer layer, which averages about 70 kilometers in thickness, consists of about a dozen large, irregularly shaped plates that slide over, under, and past each other on top of the partly molten inner layer. Most earthquakes occur at the boundaries where the plates meet. In fact, the locations of earthquakes and the kinds of ruptures they produce help scientists define the plate boundaries.

There are three types of plate boundaries: spreading zones, transform faults, and subduction zones. At *spreading zones*, molten rock rises, pushing two plates apart and adding new material at their edges. Most spreading zones are found in oceans; for example, the North American and Eurasian plates are spreading apart along the mid-Atlantic ridge. Spreading zones usually have earthquakes at shallow depths (within 30 kilometers of the surface).

Transform faults are found where plates slide past one another. An example of a transform-fault plate boundary is the San Andreas fault, along the coast of California and northwestern Mexico. Earthquakes at transform faults tend to occur at shallow depths and form fairly straight linear patterns.

Subduction zones are found where one plate overrides, or subducts, another, pushing it downward into the mantle where it melts. An example of a subduction-zone plate boundary is found along the northwest coast of the United States, western Canada, and southern Alaska and the Aleutian Islands. Subduction zones are characterized by deep-ocean trenches, shallow to deep earthquakes, and mountain ranges containing active volcanoes.

Earthquakes can also occur within plates, although plate-boundary earthquakes are much more common. Less than 10 percent of all earthquakes occur within plate interiors. As plates continue to move and plate boundaries change over geologic time, weakened boundary regions become part of the interiors of the plates. These zones of weakness within the continents can cause earthquakes in response to stresses that originate at the edges of the plate or in the deeper crust. The New Madrid earthquakes of 1811-12 and the 1886 Charleston earthquake occurred within the North American plate.

How Earthquakes Happen

An earthquake is the vibration, sometimes violent, of the Earth's surface that follows a release of energy in the Earth's crust. This energy can be generated by a sudden dislocation of segments of the crust, by a volcanic eruption, or even by manmade explosions. Most destructive quakes, however, are

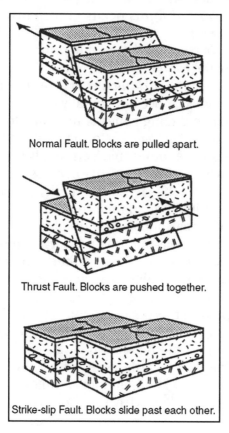

Normal Fault. Blocks are pulled apart.

Thrust Fault. Blocks are pushed together.

Strike-slip Fault. Blocks slide past each other.

caused by dislocations of the crust. The crust may first bend and then, when the stress exceeds the strength of the rocks, break and "snap" to a new position. In the process of breaking, vibrations called *seismic waves* are generated. These waves travel outward from the source of the earthquake along the surface and through the Earth at varying speeds depending on the material through which they move. Some of the vibrations are of high enough frequency to be audible, while others are of very low frequency. These vibrations cause the entire planet to quiver or ring like a bell or a tuning fork.

A *fault* is a fracture in the Earth's crust along which two blocks of the crust have slipped with respect to each other. Faults are divided into three main groups, depending on how they move. *Normal faults* occur in response to pulling or tension; the overlying block moves down the dip of the fault plane. *Thrust (reverse) faults* occur in response to squeezing or compression; the overlying block moves up the dip of the fault plane. *Strike-slip (lateral) faults* occur in response to either type of stress; the blocks move horizontally past one another. Most faulting along spreading zones is normal, along subduction zones is thrust, and along transform faults is strike-slip.

Geologists have found that earthquakes tend to reoccur along faults, which reflect zones of weakness in the Earth's crust. Even if a fault zone has recently experienced an earthquake, however, there is no guarantee that all the stress has been relieved. Another earthquake could still occur. In New Madrid, a great earthquake was followed by a large aftershock within 6 hours on December 16, 1811. Furthermore, relieving stress along one part of the fault may increase stress in another part; the New Madrid earthquakes in January and February 1812 may have resulted from this phenomenon.

The *focal depth* of an earthquake is the depth from the Earth's surface to the region where an earthquake's energy originates (the *focus*). Earthquakes with focal depths from the surface to about 70 kilometers (43.5 miles) are classified as shallow. Earthquakes with focal depths from 70 to 300

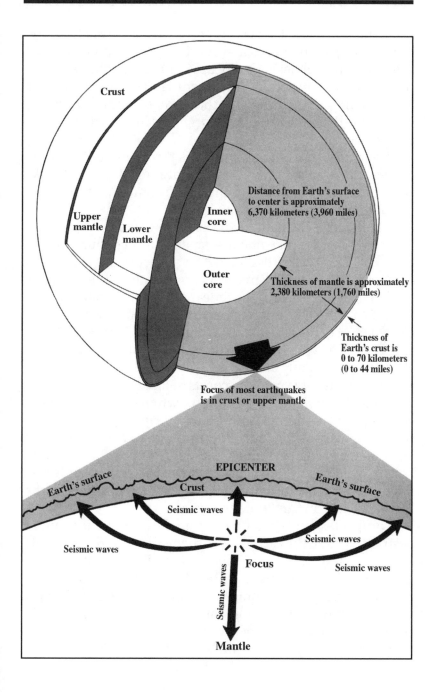

kilometers (43.5 to 186 miles) are classified as intermediate. The focus of deep earthquakes may reach depths of more than 700 kilometers (435 miles). The focuses of most earthquakes are concentrated in the crust and upper mantle. The depth to the center of the Earth's core is about 6,370 kilometers (3,960 miles), so even the deepest earthquakes originate in relatively shallow parts of the Earth's interior.

The *epicenter* of an earthquake is the point on the Earth's surface directly above the focus. The location of an earthquake is commonly described by the geographic position of its epicenter and by its focal depth.

Earthquakes beneath the ocean floor sometimes generate immense sea waves or tsunamis (Japan's dreaded "huge wave"). These waves travel across the ocean at speeds as great as 960 kilometers per hour (597 miles per hour) and may be 15 meters (49 feet) high or higher by the time they reach the shore. During the 1964 Alaska earthquake, tsunamis engulfing coastal areas caused most of the destruction at Kodiak, Cordova, and Seward and caused severe damage along the west coast of North America, particularly at Crescent City, California. Some waves raced across the ocean to the coasts of Japan.

Liquefaction, which happens when loosely packed, waterlogged sediments lose their strength in response to strong shaking, causes major damage during earthquakes. During the 1989 Loma Prieta earthquake, liquefaction of the soils and debris used to fill in a lagoon caused major subsidence, fracturing, and horizontal sliding of the ground surface in the Marina district in San Francisco.

Landslides triggered by earthquakes often cause more destruction than the earthquakes themselves. During the 1964 Alaska quake, shock-induced landslides devastated the Turnagain Heights residential development and many downtown areas in Anchorage. An observer gave a vivid report of the breakup of the unstable earth materials in the Turnagain Heights region: *I got out of my car, ran northward toward my driveway, and then saw that the bluff had broken back approximately 300 feet southward from its original edge. Additional slumping of the bluff caused me to return to my car and back southward approximately 180 feet to the corner of McCollie and Turnagain Parkway. The bluff slowly broke until the corner of Turnagain Parkway and McCollie had slumped northward.*

Liquefaction of sands and debris caused major damage throughout the Marina district in San Francisco during the Loma Prieta earthquake (1989).

Measuring Earthquakes

The vibrations produced by earthquakes are detected, recorded, and measured by instruments called seismographs. The zigzag line made by a seismograph, called a *seismogram,* reflects the changing intensity of the vibrations by responding to the motion of the ground surface beneath the instrument. From the data expressed in seismograms, scientists can determine the time, the epicenter, the focal depth, and the type of faulting of an earthquake and can estimate how much energy was released.

The two general types of vibrations produced by earthquakes are *surface waves,* which travel along the Earth's surface, and *body waves,* which travel through the Earth. Surface waves usually have the strongest vibrations and probably cause most of the damage done by earthquakes.

Body waves are of two types, *compressional* and *shear.* Both types pass through the Earth's interior from the focus of an earthquake to distant points on the surface, but only compressional waves travel through the Earth's molten core.

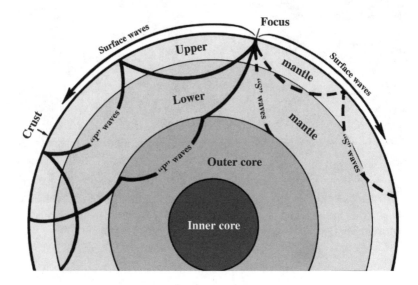

Because compressional waves travel at great speeds and ordinarily reach the surface first, they are often called "primary waves" or simply "P" waves. P waves push tiny particles of Earth material directly ahead of them or displace the particles directly behind their line of travel.

Shear waves do not travel as rapidly through the Earth's crust and mantle as do compressional waves, and because they ordinarily reach the surface later they are called "secondary" or "S" waves. Instead of affecting material directly behind or ahead of their line of travel, shear waves displace material at right angles to their path and are therefore sometimes called "transverse" waves.

The first indication of an earthquake is often a sharp thud, signaling the arrival of compressional waves. This is followed by the shear waves and then the "ground roll" caused by the surface waves. A geologist who was at Valdez, Alaska during the 1964 earthquake described this sequence: *The first tremors were hard enough to stop a moving person, and shock waves were immediately noticeable on the surface of the ground. These shock waves continued with a rather long frequency, which gave the observer an impression of a rolling feeling rather than abrupt hard jolts. After about 1 minute the amplitude or strength of the shock waves increased in intensity and failures in buildings as well as the frozen ground surface began to occur ... After about 3$^1/_2$ minutes the severe shock waves ended and people began to react as could be expected.*

The severity of an earthquake can be expressed in several ways. The *magnitude* of an earthquake, usually expressed by the *Richter Scale*, is a measure of the amplitude of the seismic waves. The *moment magnitude* of an earthquake is a measure of the amount of energy released—an amount that can be estimated from seismograph recordings. The *intensity*, as expressed by the *Modified Mercalli Scale*, is a subjective measure that describes how strong a shock was felt at a particular location.

The Richter Scale, named after Dr. Charles F. Richter of the

California Institute of Technology, is the best known scale for measuring the magnitude of earthquakes. The scale is logarithmic so that a recording of 7, for example, indicates a disturbance with ground motion 10 times as large as a recording of 6. A quake of magnitude 2 is the smallest quake normally felt by people. Earthquakes with a Richter value of 6 or more are commonly considered major; great earthquakes have magnitudes of 8 or more on the Richter scale.

The Modified Mercalli Scale expresses the intensity of an earthquake's effects in a given locality in values ranging from I to XII. The most commonly used adaptation covers the range of intensity from the condition of "I—Not felt except by a very few under especially favorable conditions," to "XII—Damage total. Lines of sight and level are distorted. Objects thrown upward into the air." Evaluation of earthquake intensity can be made only after eyewitness reports and results of field investigations are studied and interpreted. The maximum intensity experienced in the Alaska earthquake of 1964 was X; damage from the San Francisco and New Madrid earthquakes reached a maximum intensity of XI.

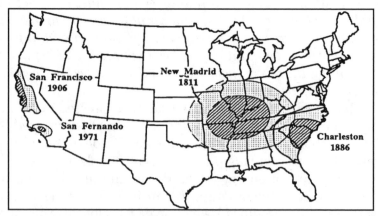

Large earthquakes cause more damage east of the Rocky Mountains; this map shows areas that suffered major architectural damage (striped areas) and minor damage (dotted areas) during the magnitude-8 earthquakes in New Madrid and San Francisco and the smaller but still damaging quakes in Charleston and San Fernando.

The January 17, 1994 earthquake at Northridge, California caused this collapse of a major highway interchange.

Earthquakes of large magnitude do not necessarily cause the most intense surface effects. The effect in a given region depends to a large degree on local surface and subsurface geologic conditions. An area underlain by unstable ground (sand, clay, or other unconsolidated materials), for example, is likely to experience much more noticeable effects than an area equally distant from an earthquake's epicenter but underlain by firm ground such as granite. In general, earthquakes east of the Rocky Mountains affect a much larger area than earthquakes west of the Rockies.

An earthquake's destructiveness depends on many factors. In addition to magnitude and the local geologic conditions, these factors include the focal depth, the distance from the epicenter, and the design of buildings and other structures. The extent of damage also depends on the density of population and construction in the area shaken by the quake.

The Loma Prieta earthquake of 1989 demonstrated a wide range of effects. The Santa Cruz mountains suffered little damage from the seismic waves, even though they were close to the epicenter. The central core of the city of Santa Cruz, about 24 kilometers (15 miles) away from the epicenter, was almost completely destroyed. More than 80 kilometers (50 miles) away, the cities of San Francisco and Oakland suffered selective but severe damage, including the loss of more than 40 lives. The greatest destruction occurred in areas where roads and elevated structures were built on unstable ground underlain by loose, unconsolidated soils.

The Northridge, California earthquake of 1994 also produced a wide variety of effects, even over distances of just a few hundred meters. Some buildings collapsed, while adjacent buildings of similar age and construction remained standing. Similarly, some highway spans collapsed, while others nearby did not.

Volcanoes and Earthquakes

Earthquakes are associated with volcanic eruptions. Abrupt increases in earthquake activity heralded eruptions at Mount St. Helens, Washington; Mount Spurr and Redoubt Volcano, Alaska; and Kilauea and Mauna Loa, Hawaii. The location and movement of swarms of tremors indicate the movement of magma through the volcano. Continuous records of seismic and tiltmeter (a device that measures ground tilting) data are maintained at U.S. Geological Survey volcano observatories in Hawaii, Alaska, California, and the Cascades, where study of these records enables specialists to make short-range predictions of volcanic eruptions. These warnings have been especially effective in Alaska, where the imminent eruption of a volcano requires the rerouting of international air traffic to enable airplanes to avoid volcanic clouds. Since 1982, at least seven jumbo jets, carrying more than 1,500 passengers, have lost power in the air after flying into clouds of volcanic ash. Though all flights were able to restart their engines eventually and no lives were lost, the aircraft suffered damages of tens of millions of dollars. As a result of these close calls, an international team of volcanologists, meteorologists, dispatchers, pilots, and controllers have begun to work together to alert each other to imminent volcanic eruptions and to detect and track volcanic ash clouds.

Predicting Earthquakes

The goal of earthquake prediction is to give warning of potentially damaging earthquakes early enough to allow appropriate response to the disaster, enabling people to minimize loss of life and property. The U.S. Geological Survey conducts and supports research on the likelihood of future earthquakes. This research includes field, laboratory, and theoretical investigations of earthquake mechanisms and fault zones. A primary goal of earthquake research is to increase the reliability of earthquake probability estimates. Ultimately, scientists would like to be able to specify a high probability for a specific earthquake on a particular fault within a particular year. Scientists estimate earthquake probabilities in two ways: by studying the history of large earthquakes in a specific area and the rate at which strain accumulates in the rock.

Scientists study the past frequency of large earthquakes in order to determine the future likelihood of similar large shocks. For example, if a region has experienced four magnitude 7 or larger earthquakes during 200 years of recorded history, and if these shocks occurred randomly in time, then scientists would assign a 50 percent probability (that is, just as likely to happen as not to happen) to the occurrence of another magnitude 7 or larger quake in the region during the next 50 years.

But in many places, the assumption of random occurrence with time may not be true, because when the strain is released along one part of the fault system, it may actually increase on another part. Four magnitude 6.8 or larger earthquakes and many magnitude 6-6.5 shocks occurred in the San Francisco Bay region during the 75 years between 1836 and 1911. For the next 68 years (until 1979), no earthquakes of magnitude 6 or larger occurred in the region. Beginning with a magnitude 6.0 shock in 1979, the earthquake activity in the region increased dramatically; between 1979 and 1989, there were four magnitude 6 or greater earthquakes, including the magnitude 7.1 Loma Prieta earthquake. This clustering of earthquakes leads scientists to estimate that the probability of a magnitude 6.8 or larger earthquake occurring during the next 30 years in the San Francisco Bay region is about 67 percent (twice as likely as not).

Another way to estimate the likelihood of future earthquakes is to study how fast strain accumulates. When plate movements build the strain in rocks to a critical level, like pulling a rubber band too tight, the rocks will suddenly break and slip to a new position. Scientists measure how much strain accumulates along a fault segment each year, how much time has passed since the last earthquake along the segment, and how much strain was released in the last earthquake. This information is then used to calculate the time required for the accumulating strain to build to the level that results in an earthquake. This simple model is complicated by the fact that such detailed information about faults is rare. In the United States, only the San Andreas fault system has adequate records for using this prediction method.

Both of these methods, and a wide array of monitoring techniques, are being tested along part of the San Andreas fault. For the past 150 years, earthquakes of about magnitude 6 have occurred an average of every 22 years on the San Andreas fault near Parkfield, California. The last shock was in 1966. Because of the consistency and similarity of these earthquakes, scientists

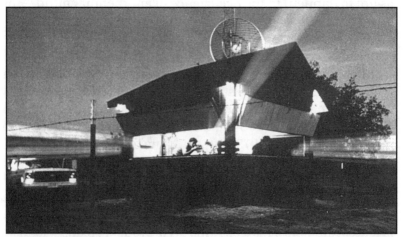

Using a two-color laser to detect movement along a fault near Parkfield, California.

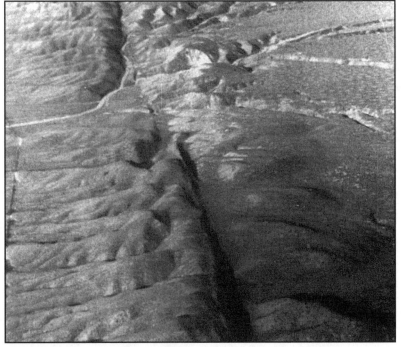

San Andreas fault in the Carrizo Plain, central California.

have started an experiment to "capture" the next Parkfield earthquake. A dense web of monitoring instruments was deployed in the region during the late 1980s. The main goals of the ongoing Parkfield Earthquake Prediction Experiment are to record the geophysical signals before and after the expected earthquake; to issue a short-term prediction; and to develop effective methods of communication between earthquake scientists and community officials responsible for disaster response and mitigation. This project has already made important contributions to both earth science and public policy.

Scientific understanding of earthquakes is of vital importance. As the population increases, expanding urban development and construction works encroach upon areas susceptible to earthquakes. With a greater understanding of the causes and effects of earthquakes, we may be able to reduce damage and loss of life from this destructive phenomenon.

CHAPTER 9

Historical Geology

Historical geology deals with the evolution of the life of the past, and with the development of the continents and oceans. It traces out, as accurately as our present knowledge will permit, the changes through which the earth has passed; it endeavors to gather from the available record the history of the life of geological times and the evolutional changes which the many classes of animals and plants have undergone and, as far as possible, to determine the cause or causes of these changes. This section of geology is concerned not only with the recording of facts, but is also, to an important degree, philosophical.

Human history is but a short chapter of geological history, the former being measured in thousands of years while the latter extends over millions of years. The immensity of geological time is beyond our comprehension, but some conception of it can be gained when it is remembered that the time necessary to excavate the Grand Canyon of the Colorado was, geologically, comparatively short; that a maximum thickness of sediments of not less than 40 miles has been laid down in the seas; that great mountain ranges have not only been raised but have been worn down to sea level during portions of the smaller divisions of geological history. Perhaps the most striking evidence of the length of geological time is to be seen in the evolution of life.

Fossils

A fossil is any remains or trace of an animal or plant preserved in the rocks of the earth. It may consist of the original substance of the animal, or it may be merely an impression, such as a footprint or a worm trail. Even the flint implements made by primitive man may be considered as fossils.

When a shell or other organic remain is buried in the mud or sand of an ocean or lake bottom, in the dune sand of a desert, in volcanic dust, in a peat bog, or in the flood plain of a river, the record of its existence may be preserved in a number of ways.

(1) *The original substance may be preserved*—In recent sediments the shells are often unchanged, even the nacreous luster being retained. In the ice of Siberia, mammoths have been found whose flesh had been so perfectly preserved that it was eaten by dogs and wolves and possibly by the natives themselves. Insects are found in amber—the fossil gum of cone-bearing trees—in which they were entrapped and covered.

(2) *Replacement*—The original substance may have been entirely replaced by some other mineral, and shells, corals, and bones are often found which, although bearing little external evidence of alteration, are composed entirely of silica or some other mineral.

Silica is not the only mineral which replaces the substance of shells, bones, and other hard parts. Pyrite, iron oxide, lime carbonate, and other minerals sometimes occur.

(3) *Casts and molds*—The original substance may be carried away in solution by underground water, leaving a cavity in which only the external form is preserved; in other words, a *mold* of the shell or bone is left. Often natural *casts* of these molds are formed by mineral matter carried into the mold or by the infiltration of mud. Molds of the interior and exterior are frequently encountered in porous rocks. Some fine opals in Nevada have the form of branches, but are in reality casts of the branches of trees, the cavities formed by the decay of the

wood having been filled with silica.

(4) *Footprints, trails, etc.* — Many animals are known from their footprints, trails, burrows, or the impressions (p. 201) made by their bodies in the soft mud.

Entombment of Plants and Animals—The most favorable conditions for the preservation of animal life are to be found on those portions of the ocean bottom which are not uncovered by tides and where sediments are accumulating. When, under such conditions, shellfish or other animals die, their bodies may be buried in the mud or sand and preserved. It is not unusual to find layers of rock made up largely of the remains of shells which were buried in this way. On the surface of some slabs of rock 250 or 300 fossils may sometimes be counted.

Animal and plant remains are often well preserved in lake deposits. In deposits of this class are found leaves, branches, and flowers which were carried from the surrounding land by the streams, insects which were beaten down by the wind to the surface of the lake, and vertebrates which were floated down the streams and found a burial on the lake bottom. Some of the most beautiful fossils were made in this way, but deposits of this class are much less important than those of marine origin, both because of their smaller extent and because the contained fossils seldom afford a means of exact correlation with those of other countries.

The fossils preserved in delta swamps and flood plains are often numerous, and during certain periods of the earth's history have afforded the chief record of the vertebrates of these periods.

Fossils are also preserved in wind-blown sand, in peat bogs, in caverns, and in travertine.

Imperfection of the Record—The record of ancient life must necessarily be imperfect for two reasons:

(1) Only a small percentage of the life of any one period is preserved. This can be seen best by observing the proportion

of the plant and animal life of today that will remain as a record of the life of the twentieth century. Of the life of the sea only the animals with shells or skeletons will be preserved in large numbers; the myriads of soft-bodied animals such as jellyfish and protozoans will not form recognizable fossils except under very exceptional conditions. The trees of the forest decay where they fall, and it is seldom that any are buried and leave a permanent record. The same fate awaits land animals, since upon their death their bones are soon disintegrated by the agents of the atmosphere and they crumble to dust. It is only the bones of the occasional carcass which floats downstream and is buried under favorable conditions that will form fossils.

(2) Even after being buried, the record is not always preserved. Thousands of square miles of sediments have been metamorphosed and the contained fossils destroyed. When marine sediments have been raised to form land, they are immediately attacked by the weather and erosion and are soon carried away. We consequently find that thousands of feet of rock have been removed and the record has been completely lost. Much of the fossiliferous strata is also either buried so far beneath younger rocks as to be inaccessible or is under the waters of the seas and so beyond the reach of the geologist.

Geological Chronology

Relative ages of strata are determined in two ways:

(1) *Order of Superposition*—If a series of strata or beds is in the order in which they were laid down, it is evident that the oldest will be at the bottom and the youngest at the top. It is for this reason that the strata of a geological section are always placed with the oldest at the bottom of the column. This order is conclusive proof of the relative age of rocks unless they have lost their original position by faulting or folding.

(2) *Chronology determined by fossils*—After the true order of a series of beds has been determined by their superposition, their contained fossils will usually make it possible to correlate

them with strata which may be hundreds or even thousands of miles distant. This is rendered possible by the fact that the inhabitants of the earth have undergone a progressive change which has, as a whole, been gradual, but which has taken place more rapidly at certain times than at others. Certain classes became dominant for a time, and then declined but seldom entirely disappeared. As a result of this change the assemblage of animals and plants of each division of geological history differs from that of every other. The fact that life has suffered such a progressive modification is of the greatest importance, since, as already indicated, it furnishes a means by which the relative age of the rocks in different parts of the world can be determined. Since certain species have a short geological life (their *vertical range* is short), when such are present the relative age of the rock is readily fixed.

Although fossils are the surest test of the relative age of widely separated strata it should not be concluded that they prove exact contemporaneity, since in favored regions an old fauna may live thousands of years after it has become extinct in others. An example is found in Australia today, where the indigenous fauna belongs to the early Tertiary.

Use of Fossils in Determining Physical Conditions—A study of the inclosed fossils usually tells definitely whether the rocks were laid down in the sea, in a lake, or on land. Fossils also give a clue to the depth of the water and the proximity of the shore. Corals show that the deposits containing them were laid down in warm seas some distance from land, or that the land was so low that little sediment was carried to the sea. Leaves and stems of plants as well as the fossils of land animals indicate nearness to shore.

The climate of the past is also told with considerable certainty by fossils. For example, relatively recent travertine deposits of northern France contain the canary laurel, a plant which blooms in winter and which now grows in the moist climate of the Canary Islands, where the temperature seldom falls below 59° F. It is evident, therefore, that when the canary laurel grew in northern France the climate of that region was probably warm and moist. The occurrence in the Pleistocene deposits of Denmark and England of Arctic willows which now grow only within the Arctic Circle is evidence of a cool climate in the past in those countries.

A typical example of the knowledge to be gained of the physical geography and climate of a region by a study of the fossils is illustrated by the limestones of Wisconsin. These strata are composed of practically pure limestone, being free from land sediments, and contain fossil corals, crinoids, brachiopods, and the remains of other marine animals. It is evident, therefore, that when the limestone was accumulating, a sea spread over a portion at least of Wisconsin, that the region in which the lime ooze was deposited was probably far from land, and that the climate, as shown by the corals, was probably warm.

Difficulties in Correlating Strata—(1) When rocks have been overturned or faulted, older beds are sometimes found to rest on younger ones. (2) In some regions a once widespread stratum may be represented now only by isolated patches which may be separated by distances of several miles. (3) Strata are sometimes separated by an unconformity which may represent a lost interval of many years. (4) The lithological character of a stratum may vary greatly even over short distances. In every case, however, fossils if present will usually give definite knowledge of the relative age of the rocks.

Since in no one region are the strata of even a majority of the systems of the earth represented, it is evident that one of the difficulties of geology is to bring together the data and place them in their true order so as to make a complete and accurate record. For example, unconformities representing a loss of two or three systems may occur in two sections, but when the two sections are compared it may be found that they complement each other, that which is lacking in the one being present in the other and vice versa. It is evident that when such sections exist, a complete record of a portion of geological time is available.

The difficulties may be seen by a study of the rocks upon which the city of Paris is situated. An examination of these strata shows that, at least ten times in the past, this region was covered by the sea and sediments accumulated on the sea floor, and as many times the sea bottom was raised above the water and was subjected to erosion. When the latter occurred, no sediments preserve the fossils of the periods during which land existed, and it is only by studying the fossils in strata of other regions that the whole history can be read and the age of the strata which are present be determined.

Divisions of Geological Time

The broad outlines of the earth's history have been learned as a result of such studies as those indicated above, and have been arranged in chronological order and separated into more or less clearly marked divisions which correspond to the chapters of human history. The divisions of time and corresponding divisions of the rocks have been given the following terms:

Time Scale	Rock Scale
Era	Group
Period	System
Epoch	Series
Age	Stage

An *era* consists of several *periods* during which a group composed of several *systems* of strata were accumulated. During an *epoch* a *series* composed of one or more *stages* was laid down. When one speaks of the *Cambrian System*, he means the succession of strata which were laid down in the *Cambrian Period*; when he speaks of the *Miocene Series*, he refers to strata deposited during the *Miocene Epoch*, i.e., during a definite portion of the *Tertiary Period*.

The following table includes the more important divisions of the geological record:

Cenozoic Era and Group	Quaternary	Period and System	Recent Pleistocene	Epochs and Series		
	Tertiary	Period and System	Pliocene Miocene Oligocene Eocene	Epochs and Series		
Mesozoic Era and Group	Cretaceous	Period and System				
	Jurassic	Period and System				
	Triassic	Period and System				
Paleozoic Era and Group	Carboniferous	Period and System				
	Permian					
	Pennsylvanian					
	Mississippian					
	Devonian	Period and System				
	Silurian	Period and System				
	Ordovician	Period and System				
	Cambrian	Period and System				
Pre-Cambrian Eras	Proterozoic Era and Group					
	Archæozoic Era and Group					

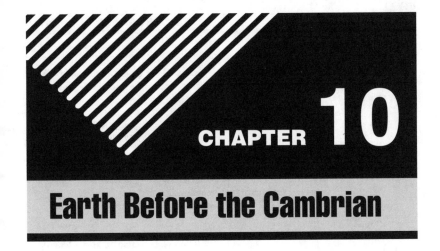

CHAPTER 10

Earth Before the Cambrian

Pre-Cambrian Eras

In no connection is the saying, "All beginnings are diffcult," more true than in the study of the earliest chapters of the earth's history. This is the case not only because the rocks which preserve the record in a given region have been subjected to all the foldings and metamorphisms which have affected all the subsequent rock formations of that region as well as earlier ones, but also because no fossils have been found except near the close of the Pre-Cambrian, and these are few and fragmentary.

A brief classification of the Pre-Cambrian of the Lake Superior region of North America appears on the following page:

| *Proterozoic* (Greek, *proteros*, early, and *zoe*, life) | One or more series separated by unconformities to which local names have been given since it has not been possible to determine their equivalents in distant regions. |

Upper Proterozoic (Keweenawan)
Unconformity
Middle Proterozoic (Upper Huronian)
Unconformity

Lower Proterozoic { Middle Huronian
Unconformity
Lower Huronian }

Great Unconformity

Proterozoic in the Lake Superior region.

Archæozoic (Greek, *arche*, beginning, and *zoe*, life)

(Laurentian) — Mainly light-colored (acid) granites, gneisses, and schists. These are largely intrusive, but some may represent the surface upon which the Keewatin was laid down.

(Keewatin) — Mainly dark-colored (basic) metamorphic rocks, composed largely of metamorphic lava flows and tuffs, with small amounts of metamorphic sediments.

The Archaeozoic Era

Distribution of the Archaeozoic Rocks—The rocks constituting the Archaeozoic system are the oldest of which we at present have any knowledge and, as far as known, underlie all the younger rocks of the earth's crust. In regions which have been repeatedly uplifted and eroded the Archaeozoic rocks are uncovered, and it is in such places that they have been studied. In North America the greatest area of Archaeozoic rocks lies in the eastern half of Canada, where they have an area of about 2,000,000 square miles, forming an irregular mass around Hudson Bay and extending south into Wisconsin and Minnesota. This is often designated as the "Laurentian shield." In the Adirondacks of New York, in New England, and in a belt stretching from Maryland south into Alabama (Piedmont

Plateau) are crystalline rocks which are partly of Archaeozoic age. In the cores of the mountains of the western half of the continent and in other isolated patches they also appear at the surface.

Our detailed knowledge of the Pre-Cambrian of North America is largely confined to the region about the Great Lakes and the St. Lawrence River. Here excellent and fresh exposures have been developed by glacial erosion, and the presence of valuable deposits of copper, iron, nickel, cobalt, and silver has led to a careful study of the region.

Archaeozoic rocks apparently corresponding to the Archaeozoic of North America occur in Scandinavia and other parts of Europe, over a large area in Brazil, in central Africa, in China, in India, and elsewhere, but the determination of the age of the crystalline rocks of many regions is yet in doubt. It has been roughly estimated that the Pre-Cambrian rocks appear at the surface over one fifth of the land area. The term "surface" is used to mean that the formation is not covered by younger rock formations, although it may be hidden in many places by soil or glacial deposits.

The difficulty of any attempt to correlate the Archaeozoic rocks of distant or isolated regions is obvious, since fossils are absent, and this exact method of determining the age of rocks is consequently unavailable. Moreover the fact that the lithological character of the rocks of a formation may vary greatly, even in short distances, makes such characters of a formation an extremely uncertain criterion upon which to base a correlation. However, since fossils are lacking, the lithological character, superposition, and the degree of metamorphism and deformation must be taken advantage of in making provisional correlations.

Characteristics of Archaeozoic Rocks—No rocks are more complex than those of this system. In fact, their very complexity is a character which aids in their determination. In the Lake Superior region the system is, in general, composed of a great series (Keewatin) made up predominantly of dark-colored (basic) schists and great masses of granitoid gneisses and light-colored (acid) schists (Laurentian) which have apparently for the most part been intruded into the Keewatin schists. The Keewatin schists are, therefore, as far as present investigation shows, the oldest rocks of the Earth's crust (unless some of the gneisses prove to be of even greater age). They are composed largely of lava flows and tuffs, with occasional conglomerates, shales, and beds of iron ore, which have been folded, contorted, and so metamorphosed that their former condition is with difficulty recognized. They have, moreover, been broken by faults and by massive intrusions.

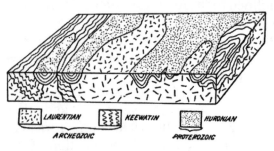

LAURENTIAN   KEEWATIN   HURONIAN

ARCHEOZOIC   PROTEROZOIC

Block diagram showing the occurrence and complicated structure of Pre-Cambrian rocks in the Lake Superior region.

Dikes through which was forced the lava that flowed over surfaces that have since been worn away now cut both the schists and the Laurentian granites and gneisses. Great batholiths of granite (Laurentian) occur so frequently as to make them almost characteristic of the Archaeozoic systems, and in certain regions they constitute the larger part of the surface rock. These batholiths have, in turn, been broken, faulted, and intruded by lavas of later age, and these by even younger intrusions. Formerly, before they were recognized as intrusive masses, the granites and gneisses of the Archaeozoic systems were considered to be portions of the original crust of the Earth. That surfaces must have existed upon which the lava flows and ash deposits spread and from which the material was derived to form the sedimentary beds is obvious. Nevertheless, no such surface has yet been recognized with certainty, either because it is still buried beneath the overlying rocks or because it has been so welded into them by heat and pressure that it cannot be determined.

Thickness—The rocks referred to the Archaeozoic systems are of great but unknown thickness. The lower limits of the system, as stated, have never been observed, even where they have been cut down many hundreds of feet in mountain ranges or in the great "Pre-Cambrian shield" of Canada, which has apparently been repeatedly subjected to prolonged and profound erosion such perhaps as few other regions of the world have experienced.

Causes of Metamorphism and Deformation—The cause of the metamorphic character of the Archaeozoic rocks is readily understood when the disturbances which have affected them are considered. The Archaeozoic was a time (1) of unusual volcanic activity, as well as (2) of great deformations. Moreover (3) the rock that now appears at the surface was probably, for the most part, deeply buried beneath younger formations. These three factors alone would produce metamorphic changes of the first order. Deformation was produced in a number of ways: (1) The great masses of lavas which intruded into the rocks caused the rocks to fold and crumple. Moreover (2) as the lava was withdrawn from below the surface and poured out upon it, settling resulted which caused a further folding. These elements, taken in connection with (3) the lateral pressure resulting from the contraction of the interior of the earth, must have altered profoundly the original structure and composition of the rocks, changing the lavas and tuffs and sedimentary rocks to schists of various kinds, and the granites to gneisses and even schists.

Conditions during the Archaeozoic Era—A few deductions can be made concerning the conditions which prevailed in this earliest era. The presence of successive lava flows and of volcanic ash and cinders shows that volcanoes were abundant and active, at least locally. The conglomerates and shales prove that the surfaces of the land were worn down by running water and that the rocks were weathered, since the clays from which shales are formed were produced by the weathering of igneous or other rocks. The presence of limestone suggests the possibility that shell-bearing animals were in existence, but since limestone is known to be formed by chemical precipitation as well as by organic remains, the evidence is not conclusive. The Grenville series of the St. Lawrence valley, estimated to be 50,000 feet thick, is distinctly stratified and is one of the greatest limestone series in the earth's crust, a part if not all of which is believed to be of Archaeozoic age. Graphite, which may be metamorphic organic matter, indicates the presence of plants. Graphite, however, may be of inorganic origin, derived perhaps from petroleum. No remains that can be positively identified as fossils have been found in the Archaeozoic rocks.

It has been suggested (Daly) that the Pre-Cambrian limestones were entirely products of chemical precipitation. This is based on the assumption that the land areas were at first relatively small, and that the abundance of decaying, soft-bodied organisms on the sea floor produced ammonium carbonate, which led to a continuous precipitation of such lime as was available. Hence the ocean was limeless, and it was not until the lands became more extended that a sufficient quantity of lime salts was brought in by rivers to counterbalance that thrown down by the ammonium carbonate and sodium carbonate on the sea floor. If this be true, the earlier organisms could not form calcareous shells or skeletons. The fact that Pre-Cambrian and Cambrian limestones, even when unaltered, show no signs of having originated from shell remains is offered in proof.

Duration—An immense but unknown duration is assigned to the Archaeozoic era, an era so vast that even if it were possible to state the duration in terms of years the number would be so large as to convey little meaning to the human mind. If millions of years were consumed by the later eras, tens of millions must be ascribed to this era. In fact, it is possible that the Archaeozoic may have been longer than all the subsequent eras taken together.

Bearing Upon the Theories of the Earth's Origin—(1) According to the theory that the earth was originally a molten globe (nebular hypothesis) which upon cooling first formed a crust, we should expect to find the earliest sedimentary rocks underlain by an igneous or metamorphic-igneous floor, provided that igneous activity was slight after the crust became cool enough to permit the operation of the agencies of erosion and the weather. It seems more probable, however, that in these early stages igneous activity would be unusually prevalent, with the result that lava flows, volcanic products of enormous thickness, as well as great intrusions might completely hide the original crust, if indeed it were not remelted.

(2) According to the planetesimal theory, the matter gathered in from space became so hot at the (*a*) center that it recrystallized to form an essentially igneous core. (*b*) A thick zone outside of the central core, made up largely of planetesimal matter, partly of igneous rock erupted from below, theoretically underlies the (*c*) next, and relatively thin zone which is composed largely of extrusive igneous rock, with smaller amounts of sediments and of planetesimal matter gathered from space. This is the zone which, according to the planetesimal theory, appears at the surface and is known as the Archaeozoic.

It will be seen from the above that according to either the planetesimal or the modified nebular hypothesis, the "crust" of the earth would have practically the same characters and that, therefore, even though fundamentally different, no means is afforded of testing the two theories.

The Proterozoic Era

Archaeozoic and Proterozoic Contrasted—The Archaeozoic systems are separated from the overlying Proterozoic by a great and widespread unconformity upon which rests a series of rocks of enormous thickness, which extend to the fossiliferous Cambrian. The two groups differ in a number of particulars. "The Archaean (Archaeozoic) is a group dominantly composed of igneous rocks, largely volcanic, and for extensive areas submarine. Sediments are subordinate. The Algonkian (Proterozoic) is a series of rocks which is mainly sedimentary. Volcanic rocks are subordinate. The Algonkian (Proterozoic) sediments, where not too greatly metamorphosed, are similar in all essential respects to those which occur in the Paleozoic and later periods. When the Algonkian (Proterozoic) rocks were laid down essentially the present conditions prevailed on earth. The Archaean (Archaeozoic) rocks, on the other hand, indicate that during this era the dominant agencies were igneous. On the whole, the deformation and metamorphism of the Archaean (Archaeozoic) are much farther advanced than the Algonkian (Proterozoic). The two groups are commonly separated by an unconformity which at many localities is of a kind indicating that the physical break was of the first order of importance." (Van Hise)

The Proterozoic in Different Regions—*Lake Superior Region.* (Table, p. 174.) South of the "Pre-Cambrian shield" the lowest member of the great series which constitutes the

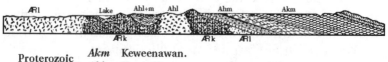

Proterozoic	*Akm*	Keweenawan.
	Ahl m	Huronian.
Archaeozoic	Æ*l*	Granites and gneisses of the Laurentian.
	Æ*k*	Schists and iron-bearing formations of the Keewatin.

Section through a portion of northern Minnesota, showing the relation of the Pre-Cambrian rocks.

Proterozoic, is the Lower Proterozoic (Huronian, named from the fine development north of Lake Huron), and is composed of quartzites, slate, schists, interbedded lava flows, and igneous intrusions, together with limestone and beds of iron ore. The rocks are usually much folded and occur in the form of long, narrow belts, separated by the Archaeozoic schists and gneisses, being small remnants of a once extensive system. Locally, at least, the Lower Proterozoic (Huronian) is divided into two systems (Lower and Middle) by an unconformity.

Resting unconformably upon the Lower Proterozoic (Middle Huronian) is the Middle Proterozoic (Upper Huronian), which resembles the lower system lithologically in that it is composed of similar sedimentary rocks and lava flows, but is somewhat less metamorphic. In this system occur the largest and richest deposits of iron in North America. The unconformity which separates the Lower Proterozoic (Lower and Middle Huronian), and Middle Proterozoic (Upper Huronian), is considered by some geologists to be of an importance almost equal to that between the Archaeozoic and Proterozoic systems.

The closing system of the Proterozoic (Keweenawan), separated from the Middle Proterozoic (Upper Huronian) by an unconformity, differs from the preceding Proterozoic systems in the presence of numerous, and in the aggregate enormously thick lava beds, which apparently welled up through fissures (much as in Iceland today) and did not flow from distinct volcanoes. The total thickness of these lava flows is estimated at nearly six miles, making this the most notable time of local volcanism in geological history. In northwestern Minnesota and contiguous portions of Wisconsin there are sixty-five distinct lava flows and five conglomerate beds, none of the former being more than 100 feet thick. In the section cited neither the upper nor the lower limits are known. The maximum thickness of the Keweenawan is estimated at 50,000 feet, of which sedimentary beds constitute about 15,000 feet. Towards the close of the period the igneous outbursts became less frequent, with a corresponding increase in the proportion of sedimentary deposits. The great Lake Superior copper deposits, which have up to this time yielded many millions of dollars in profits to their owners, occur in the lavas and conglomerates of this system. (The Keweenawan is by some writers considered to be Cambrian.)

The unconformities which separate the various systems of the Proterozoic in the Lake Superior region are well marked. They are evidenced (1) by basal conglomerates that represent the shores of an encroaching sea, (2) by the irregular erosion surfaces of the underlying rocks, (3) by differences in the amount of volcanism, and (4) by the differences in the metamorphism of the sediments of the overlying and underlying formations.

In the Grand Canyon of the Colorado the Pre-Cambrian formations are more than 10,000 feet thick and differ in many respects from those of the Lake Superior region. The lower portion of the gorge is sunk into the Archaeozoic gneisses. These are overlain unconformably by a strongly dipping series of sedimentary (Proterozoic) strata separated by minor unconformities, and they, in turn, underlie unconformably the Cambrian strata. Some measure of the length of time represented by the unconformities is shown by the flatness of the floor upon which the Proterozoic rests, and also of that above the tilted Proterozoic sediments upon which the Cambrian lies.

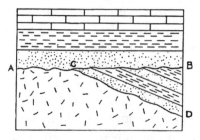

Section of the Grand Canyon of the Colorado River, Arizona. The lower portion shows the complex schists of the Archaeozoic. Upon them, separated by an unconformity CD, rest a series of Proterozoic strata. The Proterozoic strata are separated from the overlying Cambrian by the unconformity.

In the Black Hills of South Dakota, in the cores of many of the mountain ranges of the west, as well as in the Adirondacks and the Piedmont Plateau of eastern North America, Proterozoic rocks have been identified with some certainty.

Rocks of this age are believed to occur on other continents, but their correlation has not yet been definitely determined. In China, for example, the Pre-Cambrian rocks have a threefold division, the upper two of which are believed to be Proterozoic.

Iron and Copper Deposits—A discussion of the Proterozoic

A generalized section through the Black Hills, South Dakota, showing the basal Archaeozoic rocks underlying the Cambrian and younger strata.

would be incomplete without mention of the valuable deposits of iron which they contain. In the five years previous to 1914, 216,981,280 long tons of iron ore were mined from the Proterozoic rocks of the Lake Superior region alone, making this the most important iron-ore center in the world. The ore, chiefly as hematite (Fe_2O_3), occurs in the form of thick beds in the sedimentary strata. Originally, some of the formations contained large quantities of iron minerals intermingled with silica and other non-metallic minerals, and if it had remained in this state it would probably not have been of commercial value. The iron ore was later concentrated through the agency of underground waters which dissolved out and carried away the silica and other impurities, leaving pure, or nearly pure, iron ore. Some deposits were further enriched by "replacement," ore being deposited as the non-metallic minerals were removed.

One of the greatest known deposits of native copper occurs in the rocks of the Keweenawan system of the Lake Superior region. The copper occurs in the cracks of igneous rocks, in the pores of some of the lava flows, and in the spaces between the pebbles and grains of sand of the conglomerates and sandstones. The copper was originally diffused in small quantities through the lava, but was partly dissolved out by underground water, carried into porous layers, and there deposited, in some cases in such quantities as to constitute a cementing material.

Life of the Proterozoic Era—The indirect evidences of life in the Proterozoic are more abundant than in the Archaeozoic, although of much the same character. Limestones imply but do not prove the existence of shell-bearing animals, such as are now forming the calcareous ooze and shell deposits of the ocean bottom. Graphite and black shales are suggestive of plant remains. The great deposits of iron ore are thought to indicate the existence of life, since organic matter seems necessary to have furnished the carbon dioxide by means of which the insoluble iron minerals were decomposed, and as soluble iron carbonates were carried away and redeposited where the further movement of the underground water was prevented. It is possible, however, that decomposing organic matter may not have been essential to this process.

Direct evidence is furnished by a few fossils that have been found in the Proterozoic rocks of the Grand Canyon of the Colorado in Arizona, and in rocks of this age in Montana and Ontario. The known animal life consists of several species of worms, a large crustacean, a sponge-like fossil (Atikokania), some of which are 15 inches in diameter, and a brachiopod. Abundant fossils of a calcareous alga, individuals of which are more than two feet in diameter, form layers of limestone three feet thick. It is probable that when all parts of the world become geologically better known, fossils will be discovered in Proterozoic formations as distinctive in character as those of the Cambrian and overlying systems.

Duration—The fossils of the Proterozoic, though few and fragmentary, show that some forms of life were well up in the scale of life. Crustaceans, worms, and brachiopods (p. 204) are so high in the scale as to force the conclusion that life had been in existence many millions of years prior to this time. Moreover, judging from the extreme slowness with which evolutionary changes take place, the great differentiation in the life proves a great antiquity. When this evidence, even though theoretical, is taken in connection with the great thickness of the sediments and lava flows, as well as the long periods represented by the unconformities, it seems probable that the Proterozoic was very much longer than all of Paleozoic time. In fact, if the degree of life development is taken as a basis by which to measure time, it is thought that the appearance of the Cambrian fauna, although many millions of years ago, was a comparatively recent event.

Climate—Little can be said of the climatic conditions of this remote age. The presence of fossils in Montana, Arizona, and Ontario indicates a climate that was certainly not frigid. The presence of scratched boulders in formations believed to be Proterozoic in Norway, China, and Australia, and perhaps in southern Africa, sometimes resting upon a striated rock pavement possessing such characters as to make the glacial origin of the deposits undoubted, leads to the surprising conclusion that even at this time the earth was visited by periods of glaciation such as that of the Great Ice Age. There is some question as to the age of these glacial formations, some investigators believing that they belong to the Lower Cambrian. According to the theory of a cooling earth, with an atmosphere that was at first heavy, it is difficult to explain the presence of continental ice sheets in this early era.

Life Before Fossils—The earliest rocks in which an abundance of fossils of which any records have been found occur in the Cambrian. These fossils are highly organized and are not the simple, unspecialized ancestors of modern animals that the theory of evolution demands. They are of a degree of specialization which indicates a long period of preceding life. Can the life which antedates the first known fossils be inferred?

In the seas of today the number and aggregate bulk of

minute and microscopic soft-bodied animals and plants which live near or at the surface of the ocean is astonishing. Small jellyfish sometimes cover the ocean for many miles, tiny crustaceans live in myriads and microscopic animals in countless numbers. The reasons for the abundance of microscopic life near the surface (i.e., within a few hundred feet of the surface) are evidently to be found in the abundance and uniform distribution of mineral food in solution, in the presence of sunlight, and in the uniformity of temperature. Practically all of the life of the ocean depends upon these simple forms, either directly or indirectly. Yet, with the conditions as they are today, almost nothing of this profuse life would be preserved in a fossil state as a record of their existence, since, with few exceptions, fossils are confined to such forms as possessed some hard parts, such, for example, as shells or skeletons. The study of embryology teaches that all classes of life were descended from minute, possibly swimming creatures. The starfish, coral, shellfish, and other marine animals all began life as minute, free-swimming forms. It seems probable that preceding the Cambrian the oceans were tenanted by such small, soft-bodied animals as those which populate the surface today, and that they were in equal abundance.

The question next to be answered is: Why did any of these animals seek the bottom of the ocean and become stationary forms ? The first settlers on the bottom probably did not secure more or better food than their swimming relatives, but they had one advantage: they were able to devote their superfluous energies to growth and multiplication and thus to become larger and to increase in numbers faster than the swimming forms. Consequently those which first acquired the habit of resting on the bottom soon began to multiply faster than their swimming relatives. But this rapid increase must soon have given rise to crowding and competition which led to a struggle for existence. Thus the stronger forms increased at the expense of the weaker.

The development of hard coverings, such as the shell of the mollusk (the clam is an example, p. 204) and the crustacean (the crawfish is an example, p. 200), may have been due largely to such competition, since the animal which was protected in some way would have a better chance to escape being devoured. Or the development of hard protective coverings may have been due to the appearance of some especially voracious creature, and the trilobite (p. 200), the largest and most active of the inhabitants of the early ocean bottom, has been suggested as the aggressive animal. Later, however, some animal arose more formidable and active than the trilobite, such perhaps as the ancestor of the fish, and may have caused the development of still heavier armor.

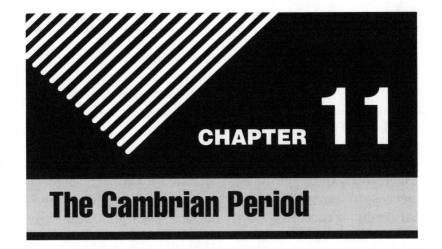

CHAPTER 11

The Cambrian Period

The Paleozoic Era—Lying above the Pre-Cambrian formations is the Paleozoic group, which includes the following systems:

Younger Paleozoic Characterized by the presence of verte- brates, both fishes and amphibians	7. Permian ⎫ 6. Pennsylvanian ⎬ Carboniferous 5. Mississippian ⎭ 4. Devonian
Older Paleozoic Characterized by the scarcity of verte- brates	3. Silurian 2. Ordovician 1. Cambrian

The first three of these systems (the Cambrian, Ordovician, and Silurian) are sometimes grouped together as the older Paleozoic, since they are characterized by invertebrate life, vertebrate remains being absent in the first half and rare in the second half. The younger Paleozoic (Devonian, Mississippian, Pennsylvanian, and Permian) is characterized by the higher forms of life, such as fishes, amphibians, and reptiles, although invertebrates were as abundant as in the older Paleozoic. These seven systems are not of equal length, nor are they of equal importance in the evolution of life, but may be recognized in any portion of the world in which they occur by their peculiar fauna. Locally, the systems are often clearly marked by unconformities, but it is upon differences in the faunas that the separations are ultimately based.

The maximum thickness of the Paleozoic rocks in Europe is estimated at 100,000 feet, and in the Appalachian region of this country a maximum up to 40,000 feet is exposed. The duration of the Paleozoic era was immense, exceeding that of all subsequent time.

Divisions of the Cambrian—The first great period of the Paleozoic is the Cambrian (Latin name for Wales), so called because of its development in Wales, where it was first studied with care. This is the oldest fossiliferous system known at present (although a new series of fossils may yet be discovered in the youngest Proterozoic rocks), if one excepts the few fossils found in the Proterozoic; and upon it we must depend to a large extent for our knowledge of the early life of the world.

The Cambrian is usually separated into three subdivisions: the Lower (Waucobian), the Middle (Acadian), and the Upper (Croxian). These divisions are based upon differences in the character of the sediments in certain regions, but chiefly upon the differences in the faunas. A study of the fossils of the Cambrian formations has shown (as is true of all later systems) that the fossils of the earliest and latest formations of the system differ markedly, although some of them are the same. This is due to the gradual disappearance of some species and the introduction of others. Among the trilobites in the Lower Cambrian is a world-wide genus (Olenellus, pg. 203, *A*) which is not found in the Middle Cambrian, while in the Middle Cambrian a trilobite appears (Paradoxides, pg. 203, *B*) at about the time that the Olenellus drops out. The Upper Cambrian likewise is distinguished by the presence of another genus (Dicellocephalus, pg. 203, *C*). The fact that these trilobites are practically restricted to one series each has given rise to the use of their names in indicating the divisions of the system. Thus the life of the Lower Cambrian is spoken of as the *Olenellus* fauna, that of the Middle Cambrian as the *Paradoxides* fauna, and that of the Upper Cambrian as the *Dicellocephalus* fauna. Not only are certain trilobites characteristic of these three epochs of the Cambrian, but other forms of life as well, so that even though trilobites are absent, the age of the rocks can be determined by other genera and species. Although certain genera and species are practically confined to one formation, others have a wide vertical range, i.e., are found in several formations. Such fossils, while showing that the rocks are of Cambrian age, do not, without the presence of those of more restricted vertical range, tell to which series they belong.

Location of Cambrian Rocks—The Cambrian formations outcrop around the borders of the Pre-Cambrian rocks; as, for example, on the border of the Pre-Cambrian shield and the Pre-Cambrian mass of the Adirondacks, and in regions where the Cambrian has been exposed by the deep erosion of regions which have been raised and folded, as in the folded Appalachians, from the St. Lawrence to Alabama, and in portions of the West. For the most part, however, Cambrian rocks in North America, although of wide extent, are not exposed at the surface over large areas, being deeply buried under younger strata.

Physical Geography of Ancient Periods—The determination of the distribution of land and water in such remote periods as the Cambrian is very difficult, and at best the outlines of the continents, oceans, and seas are only approximately known. Maps of the kind shown here (pgs. 192-3) are based upon several lines of evidence.

(1) When the fossils of a formation of known age are found to be of practically the same species in outcrops that are widely separated, it is assumed that the waters in which they lived were either connected by broad straits, or, if nothing points to a different conclusion, that they inhabited the same seas. If, however, they are found to differ widely in species in regions which may, for example, be less than fifty miles apart, although the conditions under which they lived were apparently the same, it is assumed that the seas which they inhabited were

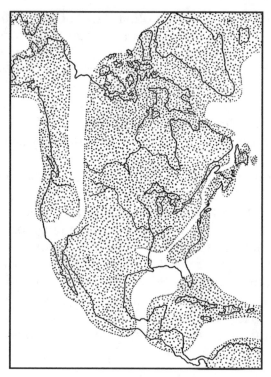

Map showing the probable distribution of land and water in North America during Lower Cambrian times. The shaded portion is land. The Lower Cambrian sediments were laid down in long, narrow straits.

separated by dry land or other barrier to their spread. Here, however, is an opportunity for error, since currents of cold water are favorable for one fauna, while in the warm waters of the same sea, a short distance away, a very different assemblage of animals may flourish. Such a distribution has often been reported from the seas of today. This objection is not as serious as at first appears, since during much of the geologic past climatic zones were probably not as well established as now.

(2) The character of the deposits furnishes aid in determining ancient shore lines. If a certain formation is a conglomerate, it is evident that it was laid down at or near the shore, since only strong waves and currents, such as are effective

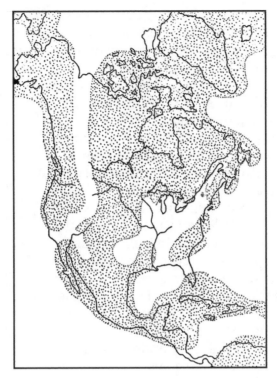

Map showing the probable distribution of land and water in the Upper Cambrian. The shaded portions are land.

in shallow waters, are able to move coarse gravel. Sandstones are also good indicators of shores, or at least nearness of land. Muds point to shallow seas, while limestones are indicative of seas of wider extent, with more distant shores in which the accumulation of lime carbonate from the remains of shell-bearing and coral-secreting animals, and that chemically precipitated, was built up with little intermixture of muds and sands.

(3) The above, as well as other evidences of which space will not permit mention, taken in connection with the distribution of the formations as shown on geological maps, gives a clue to the extent of the continents and the positions of the shallow

seas (epicontinental; Greek, *epi*, upon) which at various times in the past covered large areas of what is now land. These maps, showing the distribution of the land and water in ancient periods, must be considered as mere approximations, since (1) the absence of strata does not always prove the absence of seas in the past in any particular region because if the strata had been laid down they might have been subsequently carried away by erosion. For example, 80 miles from the nearest rocks of a certain age (Devonian) in Illinois, fossils of this age were found in a fissure in rocks of older age (Silurian), the strata of the former having been entirely eroded away. If this accidental discovery had not been made, there would have been doubt as to the extension of the seas in Devonian times. Also, in the buried extensions of strata there may be many interruptions where islands and peninsulas formerly existed. (2) Much of the strata is often buried deeply under younger formations, and its distribution in such regions is uncertain.

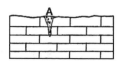

Devonian sediments, A, found in fissures of Silurian limestone. This is the principal evidence that Devonian strata at one time covered an area in Illinois.

Basal Unconformity—The lower layers of the Cambrian formations usually rest upon the eroded surface of older rocks, showing that at the close of the Pre-Cambrian the continent of North America was probably even larger than at present. The comparative levelness of the Pre-Cambrian surface, except where it has been deformed by later movements, indicates that erosion had been active and that the land had been reduced to a comparatively level plain. Upon such a surface the sea appears to have gradually encroached. The reason for the spread of the water may be found either (1) in the actual sinking of the land or (2) in the raising of the sea level in an amount equal to the volume of the sediments which were being carried into the sea, displacing the water and causing it to overflow the land. As the sea encroached upon the land, it left upon its ancient surfaces the coarse gravels and sands composed of fragments of the older rocks, which occur at the base of the Cambrian system and constitute the "basal conglomerate."

Physical Geography of the Cambrian—On evidence such as that already mentioned, it has been found that at the beginning of the Cambrian (pg. 192) the continent of North America was much expanded, the Atlantic shore being farther east than now. On the east a narrow sea stretched from Alabama northeast to Labrador, separated from the ocean by a land of unknown eastern extent called Appalachia, but whose western shore line was drawn near the site of the present Blue Ridge. In the west a similar sea existed which, at its greatest extent, reached from California to the Arctic Ocean.

The submergence of the continent continued in the Middle Cambrian, at which time a portion of the central United States was covered by seas whose shallowness is shown by ripple marks in the sandstone, and even by sun cracks made by the drying out of sediments exposed to the sun's heat. In the Upper Cambrian (pg. 193) the seas spread over portions of the continent which were land in the Middle Cambrian and were withdrawn from others which had been covered by the Middle Cambrian seas, and in still other portions the sedimentation continued, showing that the seas remained as before. Consequently near the close of the Cambrian the physical geography was very different from that in the early epoch, the water covering a much larger area than in the latter.

Character of the Cambrian Rocks—The Cambrian formations are composed of sedimentary rocks which vary in character from place to place. Where the sea advanced over a low shore in which there was an abundance of soil or other loose material, the waves and currents worked them over and spread them upon the sea bottom. Such was doubtless the origin of the Middle and Upper Cambrian sandstones which are so widespread in the interior of the United States. The occurrence of limestones and shales in the West and in the Appalachian Mountains indicates either that the shores were distant in these regions, or were so low that the gradients of the streams were insufficient to permit the latter to move any but fine material and such salts as were in solution.

The thickness of the formations of the period varies from a few hundred to twelve thousand feet. This variation is due to the fact (1) that in some places deposition took place longer than in others, and (2) that in other places where erosion was rapid and the conditions favorable to sedimentation, the ocean bottom was built up rapidly by the sand and gravel brought in by the streams and waves. (3) In other regions, where the land was low, or (4) in portions of the seas distant from the shore, the sedimentation may have taken place with extreme slowness, so that in thousands of years the thickness of sediment accumulated was a small fraction of that laid down in an equal time in more favorable locations. This should be kept in mind in future discussions, since too often the student forgets that a comparatively thin formation may have required in its upbuilding as long or a longer time than a much thicker one of different material.

Present Condition of the Sediments—Some of the Cambrian sediments have undergone important changes since their deposition. The gravels have been changed to hard conglomerates; the sands to sandstones, and where the quartz grains have been cemented by quartz, into flint-like quartzites; the calcareous ooze of the clear seas into limestone. When metamorphism has been intense, shales have been converted

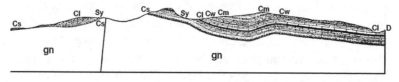

Section showing the relation of the Cambrian, Cs, and overlying strata to the Pre-Cambrian gneiss, *gn*. Crested Butte, Colorado.

into slates and schists, sandstones into schists, and limestones into marble. All of the Cambrian formations, however, have not been metamorphosed, some having been little changed. In many places the Cambrian strata have been intensely folded, tilted, and faulted (pg. 196). Some of the mountain ridges of the Appalachians are formed of the hard, upturned edges of the quartzites of this age. In other regions, as in Wisconsin and northern Minnesota (below), where the comparatively thin beds are not folded, the formations spread over a wide extent of territory.

A section in northern Minnesota showing the relation of the Cambrian to the Pre-Cambrian strata.

Volcanism—The Cambrian seems to have been a time of little volcanic activity over the greater part of the world. In North America, scarcely a trace of volcanic material has been discovered. Scotland and Wales, however, were the scenes of intense volcanic activity.

Close of the Cambrian—The Cambrian is not separated from the rocks of the overlying system (Ordovician) by great unconformities, although local ones exist, but so gradual was the change that it is often difficult to draw a line between them. In fact, it has been suggested that the former dividing line be disregarded and the Upper Cambrian and a portion of the Lower Ordovician constitute a separate system called the Ozarkian.

Other Continents—The Cambrian system is represented in Wales (20,000 feet) and Brittany by formations of great thickness. It also occurs in Scandinavia, Russia, Siberia, China, India, Australia, Argentina, and other parts of the world.

There appears to have been a land connection between

North America and Europe, or at least a chain of islands separated by shallow water, in the Cambrian. This is shown by the strong resemblance of the fossils of this age in eastern North America and Europe, a similarity which would not have been possible had the animals inhabiting the shallow waters of the shores been unable to migrate from one continent to the other.

Life of the Cambrian

The richness of the life of the Cambrian is in marked contrast to that of the Pre-Cambrian, although the presence of worm trails and a highly developed crustacean in the latter indicates that the life of that ancient time comprised many forms of invertebrates. However, so few specimens have been found and so obscure is the evidence that little more can be said at present than that the facts indicate that life was well developed before Cambrian times began.

The apparently abrupt appearance of the earliest known Cambrian fauna is probably to be explained by the absence on our present land areas of the sediments and fossils of the period between the Proterozoic and the Cambrian. This resulted from the continental areas being above sea level during the development of the unknown ancestry of the Cambrian fauna, and consequently the sediments of that time are now covered by the sea and cannot be studied.

The indirect evidence of the existence of life long antedating the Cambrian is even stronger than the direct. A comparison of the life of the Cambrian with that of today shows that of the eight branches of the animal kingdom all except the vertebrates have representatives in the former. If, as is generally believed, this differentiation was the result of slow evolutional changes, it is probable that a greater length of time was required to produce such a divergence than for all the changes in life that have taken place since the Cambrian.

Another indirect evidence is found in embryology. Each animal in its development from the egg to the adult condition passes through a series of stages which resemble those through which the race passed in its evolution, many embryonic stages representing those of mature but remote ancestors. It is evident, therefore, that the embryonic and larval stages of the individual furnish somewhat of a basis upon which to estimate the length of the evolutional history of the race to which the individual belonged. Some of the larval stages of the trilobites are preserved and give firm ground for the belief that this class had a long line of ancestors previous to the Cambrian.

Plants

Since all animals depend directly or indirectly upon vegetation for their food, it is evident that plants must have been in existence in large numbers in the Cambrian in order to supply with food the abundant marine animal life of that time. When, however, a search for plant fossils is made, none are found that can with certainty be recognized as plant remains. The inference is forced upon us that Cambrian plants were not highly organized, and that they possessed little or no woody tissue and were, consequently, incapable of fossilization. Some poorly defined, stemlike impressions found in the Cambrian strata at Burlington, Vermont, and elsewhere strongly suggest the stems of seaweeds, but some of these may be worm tracks; some, rill marks; and some, trailings made by animals. The difficulty in determining such "fossils" is well shown in the controversy over the determination of certain specimens (Oldhamia) found in the Cambrian rock of Ireland, which, as the illustration shows (right), have the appearance of vegetable growth. Some investigators have classed them as the remains of animals, some as plants, and some as inorganic markings.

A problematic fossil, *Oldhamia antiqua*, has sometimes been thought to be a plant.

The scarcity of plant fossils may

perhaps be attributed to the fact that only Cambrian rocks of marine origin have been studied, since it is seldom, even in the sediments that are being deposited today, that plants are embedded in marine sediments.

Small calcareous algae have been found in the Cambrian rock of the Antarctic Continent and elsewhere, and it appears probable that plants which secreted lime played an important part in the formation of the Cambrian limestone.

Animals

The oldest known fossiliferous rocks of the Cambrian contain a varied fauna. Corals, sponges, worms (in the form of trails and borings and impressions), brachiopods, pteropods, and crustaceans have been identified, and there is no doubt that the discoveries have brought to light only a fraction of the life of the time. Doubtless the lowly protozoans were in existence at that time, as well as other classes which have not been found. This diverse life, as has been shown, did not arise suddenly, but was derived from a long line of ancestors which lived during Proterozoic times.

Crustacea

Trilobites—The highest and most striking form of life of the period was the trilobites (Greek, *tri-*, three, and *lobos*, a lobe or rounded projection), a group of animals belonging to the same phylum as the crabs and lobsters. The name trilobite is a very descriptive one, since the animal was marked by two grooves running lengthwise of the body, which divided it into three, usually well-marked lobes. Transversely, there were also three divisions: the head shield (cephalon); the body, composed of jointed segments (thorax); and the tail shield (pygidium). Trilobites were marine animals and had delicate antennae,

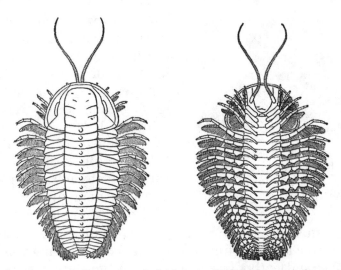

Dorsal and ventral views of an Ordovician trilobite, showing the restored appendages.

doubtless for touch, and numerous legs and breathing organs (above). They were probably able both to walk and swim. Their trails and burrows show that they burrowed and pushed their way through the muds and soft sands. A series of tracks probably made by a trilobite is shown at the right.

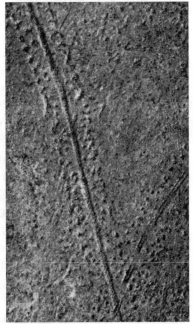

Tracks supposed to have been made by a trilobite.

Eye of a Devonian trilobite, much enlarged, showing the many eyelets forming the compound eye.

The eyes of trilobites were usually raised, crescent-shaped elevations and were compound like those of an insect (left), the number of lenses in each eye varying in different species from 14,000 to 15,000. A few species were eyeless. Cambrian trilobites (pg. 203, *A, B, C, D*) varied greatly in size, in form, and in ornamentation: some were a fraction of an inch long (Agnostus, pg. 203, *D*), while others (Paradoxides, pg. 203, *B*) attained a length of from one to two feet; some had a smooth surface and few body segments (two in Agnostus), while others were ornamented with spines and had a large number of body segments (Paradoxides had from 17 to 20); some had large eyes, while others were eyeless. These features show that the trilobite race must have extended far back into Pre-Cambrian times, since such a great diversity of form and structure could only be developed as a result of long evolution. Although trilobites varied greatly, it should be distinctly understood that in nearly every particular they were very primitive or simple in structure and closely agree with a theoretical crustacean ancestor.

Since trilobites moulted their shells at certain times and the great majority of their fossils consist of these fragments, a complete specimen usually indicates the death of an individual.

Not only were trilobites the most conspicuous animals of the period, but since new species and genera appeared, while the older became extinct, they furnish the best means of correlating the formations of different continents and of widely separated portions of the same continent. The three divisions of the system are consequently named for the three dominant genera of trilobites: the Lower or Olenellus zone, the Middle or Paradoxides zone, and the Upper or Dicellocephalus zone.

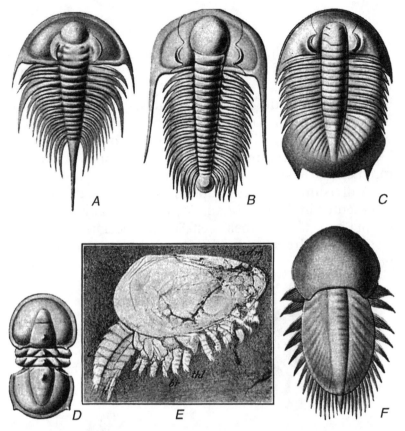

Cambrian crustaceans. Trilobites: *A, Olenellus thompsoni; B, Paradoxides harlani; C, dicellocephalus minnesotensis; D, Agnostus interstrictus.* Other crustaceans: *E, Hymenocaris perfecta; F, Naraoia compacta.*

Other Crustaceans—In addition to trilobites a number of crustaceans (above, *E, F*) of a different group, representatives of which are living today, have been found in the Middle Cambrian of British Columbia. That a large and varied crustacean fauna preceded these is certain.

Mollusca

Gastropods (Univalves)—This class is now represented by snails, conchs, and wrinkles. The most conspicuous feature of the shelled forms is the single, usually spiral shell. Gastropods lived throughout the period but were seldom abundant. The earliest forms were chiefly simple, conical shells (below, *C, E*), while later in the period coiled and spiral forms (below, *B, D*) became more common. Some of the spiral forms bear a close resemblance to some modern gastropods.

A division of the gastropods, the pteropods, was well represented in the Cambrian. The fossils usually consisted of simple, conical shells (below, *A*). Several specimens have been discovered with distinct impressions of the characteristic fleshy portions.

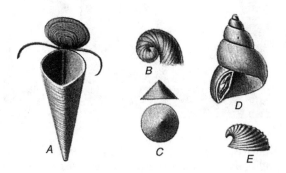

Cambrian gastropods: *A, Hyolithes carinatus; B, Pelagiella (Platyceras) primoevum; C, Scenella varians; D, Trochus saratogensis; E, Stenotheca rugosa.*

Molluscoidea

Brachiopods—This great class was especially important in the Paleozoic, not only because of the abundance of individuals, but also because certain species, though prolific, were short-lived, being abundant in one period or a subdivision of one period and becoming extinct at its close. As a result, when the

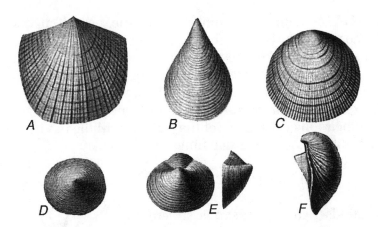

Cambrian brachiopods: *A, Billingsella coloradoensis; B, Lingulepis acuminata; C, Obelella atlantica; D, Acrothele subsidua; E, Micromitra bella; F, Kutorgina cingulata.*

fossil remains of such species are found in a stratum, proof is offered of the age of the formation. Brachiopods or Lamp Shells (so-called because of the resemblance of some of them [above, *A, F*] to Roman lamps) are enclosed by two shells or *valves* and can usually be readily distinguished from other shellfish by two characteristic features: (1) the bilateral symmetry of their shells, i.e., a line drawn from the beak to the front divides them into equal parts; and (2), in most cases, by the dissimilarity and unequal size of the two valves. The name brachiopod (Greek, *brachion*, arm, and *pous*, foot) refers to the two long spiral "arms" enclosed between the valves by means of which food is obtained and respiration carried on. These "arms" are attached to a shelly apparatus, sometimes in the form of loops and sometimes in spirals (pg. 226 *M*).

Brachiopods are divided into two great subdivisions: the hingeless (above, *B, D, E*) or inarticulate, with phosphate of lime shells the two valves of which were held together only by the muscles of the animal; and the hinged (above, *A, C*) or articulate, with calcareous shells and well-developed hinges and dissimilar valves. Of these two subdivisions the first and most primitive was more abundant in the Cambrian, and the second, later in the Paleozoic. In the Lower Cambrian, 22 genera of brachiopods have been found in Europe and North America, showing that the class was probably well developed in the preceding era (Proterozoic). The two subdivisions of brachiopods are living in the seas of the present, having undergone many changes during their long existence; yet the class as a whole has been little modified since Cambrian times.

Echinodermata

Cystoids (right) of very simple structure lived in the Cambrian, and *sea cucumbers* have been discovered in Middle Cambrian strata in British Columbia.

Worms

Perhaps the fossils next in abundance to the trilobites are the trails and borings of worms. In certain Lower Cambrian beds, vertical tubes (Scolithus) are so common as

Cambrian cystoid: *Eocystites longidactylus.*

to give a striped appearance to the rock. Many fossils which were at one time thought to be fossil marine plants are now known to be the trails of worms, or borings which have been filled with sand or clay. Since worms are, as a rule, destitute of hard parts, it is seldom that any traces of the actual animals have been preserved. In some fine shales of the Middle Cambrian in

Cambrian worms: A, Ottoia prolifica; B, Worthenella cambria. These fossils are represented by a thin film which is darker than the shale containing them, the contents of the animal being preserved as a glistening surface.

British Columbia, however, the fleshy parts of the animal are sometimes preserved as a glistening surface, even to the fine details of the structure (pg. 206 *A, B*).

Coelenterata

Corals—Cambrian corals were so simple in structure that some of them have been called sponges by certain writers and corals by others. This group was abundant locally, but in general was rare throughout the period.

Graptolites—This group will be discussed more fully in the next chapter, since it reached its greatest abundance and development in the Ordovician. The word graptolite (Greek, *graptos*, written, and *lithos*, a stone) is descriptive, since, when preserved in shale, as is usually the case, graptolites have the appearance of lead-pencil marks (pg. 221) with saw teeth on one or both sides. Graptolites were slender organisms, plant-like in appearance, usually resembling the modern hydroids. As is true of hydroids, they were composite animals in which the individuals lived in cells strung on one or both sides of a slender, horny axis which united the "colony." The form of these colonies varied greatly, as will be shown later. Graptolites appear for the first time in the Upper Cambrian.

Jellyfish—The discovery of fossil jellyfish in Cambrian rocks is most surprising, since these animals have no bony skeletons or shells. Specimens preserving the external form as well as something of the interior structure have been found. They must have been buried in mud soon after they died, for otherwise they would have been destroyed by the worms and predatory crustaceans associated with them.

Sponges lived in some abundance in portions of the Cambrian and were represented by several genera. They are known by their siliceous spicules, which were either embedded in horny fibers or interlaced into a supporting framework, and which were preserved because of their resistant character.

Protozoa

Theoretically there is every reason to believe that the simple unicellular protozoa were as abundant in the Cambrian seas as in the present oceans, but no fossils have been discovered which are known with certainty to belong to this group.

Summary

Evolution during the Cambrian—It has been seen that at the beginning of the Cambrian many classes of animals were already in existence, and that the advanced stage of development of some of them, notably the trilobites, taken in connection with the traces of Pre-Cambrian life, indicates that life was well-advanced before Cambrian time began. Whether or not the evolution of this Pre-Cambrian life was rapid is not known. Evolution during the Cambrian was continuous, but more rapid at certain times than at others, the fauna at the close of the period being distinctly more advanced than that at the beginning. The evolution of life was profoundly influenced by environment, this perhaps more than any other cause being responsible for the marked difference between the faunas of the Lower and Upper Cambrian.

Since the life of the Cambrian changed from time to time during the period, a study of the fossils of any stratum, as has been said, gives definite information as to the relative age of the beds containing them. This change in the fauna was brought about (1) by the slow evolution of species when conditions were somewhat uniform; (2) by rapid evolution due to changes in environment, such as occurred when seas were enlarged, shore lines shifted, and new conditions of food and temperature imposed; (3) by competition resulting from the immigration of large numbers of new species from other regions, which caused the extinction of many species and the modification of others.

Climate and Duration—The widespread occurrence of coral-like organisms in the Lower Cambrian and the vast numbers of individuals of various species of trilobites and other classes indicate a warm and more or less uniform climate. In fact, throughout at least the greater part of the period the character and distribution of the fossils imply nearly uniform climatic conditions over the entire world.

The duration of the Cambrian is to be expressed in terms of hundreds of thousands of years. The time required to remove and deposit thousands of feet of rock must have been enormous. If limestone is deposited on an average of one foot a century, it would require 600,000 years for the accumulation of the 6000 feet of Cambrian limestone of some portions of the West, omitting the time necessary for the formation of the thick sandstones of the same regions. Perhaps 1,000,000 years may be placed as the minimum duration of the period and 4,000,000 years as the maximum.

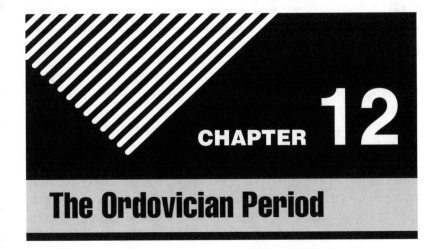

CHAPTER 12

The Ordovician Period

The system next younger than the Cambrian is the Ordovician. The name Lower Silurian has been replaced by the above, although still occasionally used by writers.

Ordovician Physical Geography—In North America during this period, the epicontinental seas varied greatly in size and position in the different stages (pg. 213), shifting so often and to such an extent that an attempt to define their borders would require a more extended description than seems advisable.

In general, it can be said that the lands about the epicontinental seas were low and that the seas were shallow, as is shown by the character of the sediments, perhaps not exceeding 200 to 300 feet in depth. It is impossible to characterize the rocks of a system in a few words, since at all times in the Earth's history sedimentary deposits of every description were being laid down in some portion of the world. This is also true of the Ordovician, during which gravels and sands were deposited in certain places, but limestones and shales form a much larger proportion of the deposits than perhaps in any other Paleozoic period. One of the physical conditions which brought this about was the limited area and probably slight elevation of the land, which consequently yielded little sediment, leaving extensive areas of the seas free from sands and muds. However, although built up largely of the remains of lime-secreting animals, such as brachiopods, corals, and bryozoans, the lime was ultimately derived from the land by solution.

In the Appalachian trough (pg. 214), which was first formed in the Cambrian, sands, muds, and limestones were laid down. The Appalachian trough was separated from the Atlantic by the great island or continent of Appalachia, whose eastern extent was unknown. Its western border was a shifting shore which during certain times was west of the present Mississippi River. Sandstones occur in the West, where islands formerly existed, but are seldom extensive. In Newfoundland, Ottawa, and west of the Adirondacks, limestone is the prevalent rock of the system, although shales are abundant, showing either that the land was low or covered by vegetation which prevented rapid erosion, or that the drainage of the land was not discharged in the direction of these regions.

The Ordovician was a period of quiet, during which the epicontinental seas gradually increased until the middle of the period (pg. 213, top-right), at which time a larger portion of the continent was under water than at any stage since the Pre-Cambrian, more than half of the continent being at this time submerged, the epicontinental seas, broken by peninsulas and large and small islands, extending at certain times from ocean to ocean. The seas were for the most part much less extensive in the latter part of the period (pg. 213, bottom-left), at which time deposits of mud were laid down over extensive areas. A submergence almost equal to that earlier in the period (pg. 213, top-right), however, again occurred (pg. 213, bottom-right), and epicontinental seas spread widely over the continent. At the close of the period these seas were again drained, and the outlines of the continent were probably not unlike those of today.

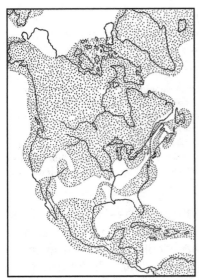

Probable distrubution of land and water in the Lower Ordovician.

Probable distribution of land and water in the Middle Ordovician. The continent has probably never been so completely submerged since that time.

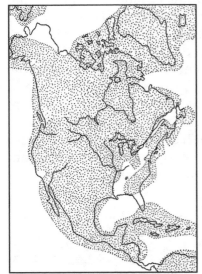

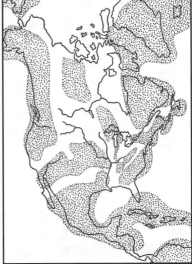

A later stage of the Ordovician, when the land was greatly extended.

Later than at right, when the continent was again greatly submerged.

The classic section of the Ordovician in the United States is in New York, where it was first extensively studied.

Ordovician system
{
Pulaski stage (shale)
Frankfort stage (shale)
Utica stage (shale)
Trenton stage (limestone)
Black River stage including Lowville limestone
Chazy stage (limestone)
Beekmantown stage (limestone)
Tribes Hill stage (limestone)
}

In this region limestone deposits prevailed during the Lower (Canadian) and Middle (Mohawkian) Ordovician, but shales in the Upper (Cincinnatian) Ordovician.

Close of the Ordovician

Close of the Ordovician—The close of the Ordovician was marked by horizontal and vertical movements of considerable importance in eastern North America (Taconic deformation) and Great Britain. During the Cambrian and Ordovician in North America sediments had been accumulating in a subsiding trough lying between the Adirondack land mass and a land mass in New England, and which stretched from the St. Lawrence River to the City of New York and to the south (pg. 193). These sediments, after having been accumulated to a thickness of more than a mile, were subjected to great lateral pressure which folded them and brought them above sea level to form a mountain range of which the present Taconic Mountains of western New England are perhaps rather insignificant remnants. The folding was so intense that

A section through the Appalachian Mountains in Alabama: Sc, Sr, Silurian; Dc, Devonian; Cb, Cl, Cp, Cw, Carboniferous.

An east-west section through the Appalachian Mountains near Mercersburg, Pennsylvania, showing the relation of the Ordovician, *Osr, Oc, Om, Oj*, and the Silurian, *Sc* and *St.*

limestones were recrystallized to marbles, of which the most famous are those of Vermont and Massachusetts; the sandstones were changed to quartzites and schists, and the muds and shales to slates and schists. These disturbances affected the region east of the Taconics, but were comparatively local, as is shown by the slightly folded rocks of this period in New York, New Jersey, and Canada, only short distances from the scene of maximum deformation. The region north of the St. Lawrence seems to have been little affected, since the sedimentation continued from the Ordovician to the Silurian with slight interruption, almost the entire record being preserved in the strata of Anticosti Island in the Gulf of St. Lawrence. The date at which the Taconic deformation occurred is known, because the Silurian rocks rest upon the eroded and upturned edges of the Ordovician (below), showing that after the deformation, the strata were elevated and eroded for many years, and were again covered by the sea and Silurian deposits laid down on them. These disturbances, which culminated in the Taconic deformation and in the draining of the interior seas, were of long duration.

An east-west section in eastern New York showing the Silurian resting unconformably upon the upturned edges of the Ordovician. We have here the proof that this portion of the continent was raised above the sea during the Ordovician, and that after the land was eroded, it sank, and upon this old land surface, Silurian sediments were deposited.

Cincinnati Anticline—The first evidence of a deformative movement in the Middle States is found in the formation of the Cincinnati and other anticlines, which appeared as low folds in the Middle Ordovician (Trenton). The Cincinnati arch, though later submerged, was again elevated and greatly enlarged at the close of the period. This fold extends over an oval area in Ohio, Indiana, and Kentucky, with the longer axis in a north and south direction.

The withdrawal of the epicontinental seas at this time may have been due to the sinking of the ocean bottoms (oceanic segments) or to the raising of the land.

Volcanism—There is little evidence of igneous activity in North America during the period, although in England and Wales, great masses of lava and volcanic ash form thick strata. Indeed, the Ordovician volcanism of Great Britain was one of the most extensive in Europe since Pre-Cambrian times.

Ordovician of Other Continents—Ordovician rocks occur in Great Britain, where a thickness of 24,000 feet has been measured. A deformation comparable to that in North America folded, crumpled, and metamorphosed the Ordovician strata at the close of the period. In Scotland, the folding was exceptionally severe, producing overturned folds and faults, in one locality thrusting strata ten miles along a fault plane.

In Europe, although the Ordovician often underlies the Silurian unconformably, the disturbances which ushered in the latter appear to have been slight. On both continents the important disturbances took place where thick beds of sediment had accumulated.

Petroleum and Natural Gas

Conditions Favoring the Accumulation of Oil and Gas— The importance of the oil and gas industry is such that the essential features of their geological occurrence demand attention.

Petroleum and natural gas occur in varying quantities in all of the fossiliferous rocks, from the Ordovician through the Tertiary, but oil never occurs in paying quantities unless there is a porous stratum overlain by an impervious one, in this respect resembling artesian wells. In an artesian well, however, it is essential that the porous stratum be open to the surface in order that the supply of water may be replenished. In an oil well, on the contrary, if the porous stratum reaches the surface, the oil is lost by evaporation, since the supply of oil comes from below.

Oil and gas usually occur at or near the crest of broad anticlines or other "reservoirs," where their further movement upward is prevented. The oil moves up the porous stratum through the water which permeates the bed, since oil and gas are lighter than the former. If, however, water is absent from the porous stratum, the oil will be at the bottom of the syncline and the gas in the anticline.

One of the modes of occurrence common in the eastern United States and Canada is shown at right, A, in which the oil and gas gradually move up the bed until (1) the

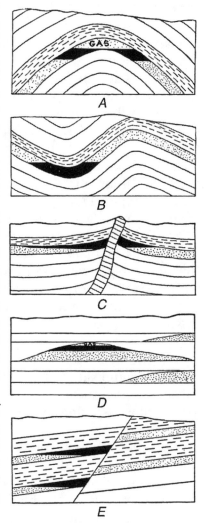

Diagrams showing the more important modes of the occurrence of oil and gas. Oil is represented in solid black. A, the oil-bearing stratum (dotted) contains water, and the oil and gas are consequently near the crest of the anticline (Pennsylvania and Illinois). B, the oil-bearing stratum is devoid of water. The oil is in the syncline. C, the strata are bent upward around a volcanic neck, and the oil has accumulated around the latter (Mexico). D, oil and gas occur in lenses of sandstone (Oklahoma). E, oil is accumulated as a result of faulting (California).

anticline is reached. If the stratum is saturated with water, the oil and gas will accumulate under the anticline, but (2) if water is absent, the oil will accumulate in the syncline (pg. 217, *B*), while the gas passes on to the highest attainable point.

(3) Oil is accumulated also when the strata are domed up, as in Texas. The principle of the accumulation of the oil is the same as in the anticline. (4) In Mexico, there are numerous volcanic necks which have burst through the strata (pg. 217, *C*), raising them and producing a dome-like structure which thus brings about favorable conditions for the concentration of oil. (5) Oil sometimes accumulates also when porous strata are faulted (pg. 217, *E*) (California). The oil ascending along an inclined, porous layer is prevented from escaping to the surface by a fault which has displaced the strata to such an extent that the porous layer is sealed by an impervious one. (6) In Oklahoma, big lenses of sandstone covered over by impervious beds (pg. 217, *D*) sometimes yield large quantities of oil.

The size of oil-producing regions is usually comparatively small.

Origin of Oil and Gas—Petroleum has not a definite chemical composition, but is made up of a large number of substances (hydrocarbons), ranging from gases to solids, the gas in oil wells being merely a liquid that is volatile at low temperatures.

Since oil and gas are probably rarely indigenous to the rock containing them, their origin has given rise to much speculation. There are two principal theories of the origin of oil, (1) the organic and (2) the inorganic, of which the former is more generally held. The inorganic theory is based upon laboratory experiments with metallic carbides, and holds that when water percolating downward through the earth's crust reaches heated rocks it becomes converted into steam which attacks the iron carbides, believed to exist there, generating hydrocarbons (oil). According to the organic theory, petroleum and its products are derived from animal or plant remains or both, which were embedded in the sediments and were later decomposed to oil. It is often stated that oil and gas were derived from beds of shale, either underlying or overlying the oil-bearing rock.

Life of Oil Wells and Fields—The amount of oil yielded by single wells in various parts of the world in one year has exceeded 100,000 tons, but such an enormous production lasts but a few weeks at the most. The oil wells of Pennsylvania have an average life of about seven years, those of Texas about four years, and those of California about six years. The average production of the wells of the Appalachian region was less than two barrels in 1907, and that of the California field was forty-two and a half barrels. Since the discovery of oil in the United States, the production has increased from decade to decade, but this increased yield has been the result of the sinking of new wells and the discovery of new fields. The reason for the short life of oil and gas wells is that, unlike water, there is no perennial supply. The great spouting wells, or "gushers" are the fortunate tappings of the accumulations of ages which, though enormously productive when first opened, are also in about the same proportion rapidly exhausted.

Oil is more commonly found in the younger rocks than in the older, although some of the richest "pools" are in the Ordovician and Devonian. The reason for this seems to be that the older a formation is, the more opportunity there has been for the escape of the oil and gas (1) by faulting which permits the escape of the oil and gas from the porous rock, and (2) by the erosion of the edges of the oil-bearing strata when it is lost by evaporation. Only those Paleozoic strata which have been deeply buried and sealed by newer formations and have remained practically undisturbed are likely to yield large quantities of petroleum. The Ordovician limestones of Ohio have yielded large quantities of high-grade oil and gas; the Devonian sandstones of New York, Pennsylvania, and West Virginia, however, have furnished the richest oil-bearing strata of the eastern United States.

Life of the Ordovician

The life of the Ordovician differed from that of the Cambrian in the abundance of certain classes which were rare in the latter, and in the higher level of development in many cases. Graptolites, although rare in the Cambrian, attained their greatest abundance in the Ordovician. The primitive corals of the Cambrian were followed by well-developed forms; the cephalopods became the largest animals of the period; gastropods were much more modern in appearance; and brachiopods show a great increase in variety and abundance.

Protozoa

Siliceous protozoa (Radiolaria) are found in the Ordovician strata of some regions in sufficient numbers to show that they were abundant in the seas of that period.

Coelenterata

Sponges are well represented by forms that secrete siliceous skeletons (below, *A, B*), and some of them attained a diameter of a foot or more. Certain beds of the Ordovician (Chazy), of New York, are composed almost entirely of sponges.

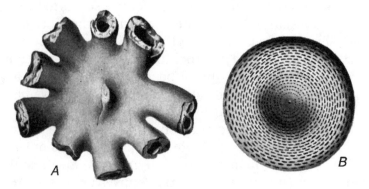

A *B*

Ordovician sponges: *A, Brachiospongia digitata; B, Receptaculites ohioensis.*

Graptolites—This class (below, *A-K*) can be traced from its earliest appearance to its final extinction, through all the stages of development, and is consequently well-adapted to illustrate some principles of evolution.

Graptolites began in the Cambrian as small, bushy forms (below, *A*) which, as a rule, lived throughout life attached to the sea bottom. Before the close of this period however, a change in the mode of life occurred which was to give the class entirely different habits and as a result bring about important

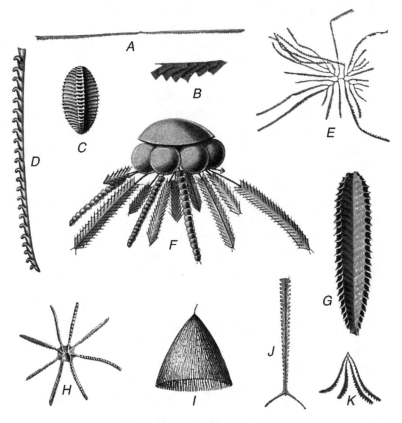

Graptolites: *A* and *B, Didymograptus nitidus; C, Phyllograptus typus; D, Monograptus cintonensis; E, Goniograptus postremus; F, iplograptus pristis; G, Phyllograptus angustifolius; H, Dichograptus octobrachiatus; I, Dictyonema flabelliforme; J, Climacograptus bicornis; K, Tetragraptus fruticosus.*

modifications in the structure. For some unknown reason, perhaps to avoid a new creeping enemy, the colonies left the sea bottom. At first the branches of the bush-like colonies hung suspended, later they became horizontal, and still later the branches were turned upward. This change was accompanied by a reduction in the number of branches. The irregular many-branched early forms gave way to regular, many-branched colonies (Bryograptus, below, *B*), then to eight-branched (Dichograptus, below, *C*, and pg. 221, *H*), these, in turn, to four-branched forms (Tetragraptus, below, *D*, and pg. 221, *K*), and these to two-branched forms (Didymograptus, pg. 221, *A*, and below, *E*). In addition to the changes in the main line of descent, many aberrant forms came into existence, but were not long-lived.

Another important change in the race was brought about when the graptolites became detached from seaweeds and led an independent floating existence, being buoyed up by "floats" (pg. 221, *F*, and below, *C, D, E*) to which they were attached by threads.

Towards the end of the race, numerous spines appeared on some species, a network of protecting fibers was developed on others, and the colonies became small. They were on the defensive and soon disappeared. The forms which survived the longest were those inconspicuous ones which had remained attached to the sea bottom from the beginning of the race.

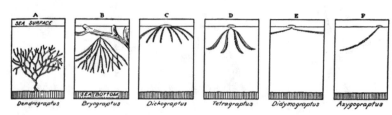

Diagram showing the evolution of one branch of the Graptolitoidea. At first attached to the sea floor, they later became attached to floating seaweeds and finally acquired floats. This change in their mode of life induced important changes in structure, one of which resulted in a reduction in the number of branches.

Because of the many progressive changes which the race underwent, and also because the colonies were carried about by currents over the seas of the world, graptolites are excellent fossils for correlating (determining identity of age) widely separated beds. The simple forms are especially characteristic of the Upper Cambrian and Lower Ordovician. The group in the main became extinct in the Silurian, but a few species lived even into the Carboniferous.

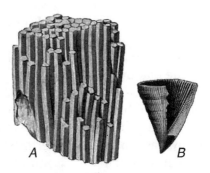

Ordovician corals: *A, Columnaria halli; B, Streptelasma profundum.* (A portion has been removed to show the interior.)

Stromatopora—An extinct order of organisms known as stromatoporoids were especially abundant as reef builders in the Ordovician, Silurian, and Devonian. They were allied to the corals and consisted of colonies of minute polyps which secreted concentric layers of thin calcareous plates connected by vertical rods. Limestone masses, sometimes five by ten feet in horizontal extent and several inches thick, were built by them. The aggregate amount of limestone built by the stromatoporoids was very large.

Corals—This class was present in the Ordovician and was represented by several types, among which were the simple, horn-shaped cup corals (above, *B*) and those living in colonies (above, *A*). The description of these types will be taken up under the Silurian.

Echinodermata

Cystoids (Greek, *custis*, a bladder) were so named because of the bladder-like shape of the body. Essentially, the animals had a sack-like or bladder-like body made up of calcareous plates, on the upper side of which two arms were sometimes attached, while some species were armless (below, *A-C*). The body was rooted by a tapering stem to the sea bottom. Cystoids first appeared in the Cambrian and reached their climax in the Ordovician and Silurian, after which they suddenly diminished in the number of species, although locally a few forms lived in considerable abundance. They are characteristic of the Ordovician and probably became extinct early in the Carboniferous.

A   B   C

Ordovician cystoids: *A, Lepidodiscus (Agelacrinus) cincinnatiensis; B, Pleurocystis filitextus; C, Amygdalocystites florealis.*

Crinoids (Greek, *crinon*, a lily) are living in the present seas and still constitute a vigorous stock, even though the race began in the Cambrian. The name "sea lily" was given to this class of animals because of their flower-like appearance. The animal (pg. 225, top, *A-D*) consists of a body composed of plates, as in cystoids, and is attached to the sea bottom by a jointed stem. From the upper margin of the body (calyx) spring the arms, which are short and simple in some species and long and many-branched in others. Within the arms is the mouth, to which food particles are carried by the currents set in motion by the arms.

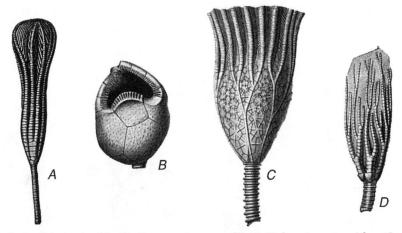

Ordovician crinoids: *A, Ectenocrinus gradis; B, Hybocrinus tumidus; C, Glyptocrinus decadactylus; D, Heterocrinus (Iocrinus) subcrassus.*

Blastoids, Starfish (right), **Brittle Stars, and Sea Urchins** lived in this period, but as they were rare they will be discussed in later chapters. The origin of the starfish probably goes back to the Proterozoic, as may be inferred from the complex metamorphism of the starfish larva. The absence of fossil starfish in the Cambrian sediments may mean that a preservable starfish skeleton was not evolved until the Ordovician.

Ordovician starfish: *Palaeaster simplex.*

Molluscoidea

Brachiopods—The preponderance of hinged (articulate) (pg. 226, except *C* and *L*) species of brachiopods over hingeless (inarticulate) (pg. 226, *C* and *L*) is very marked in the Ordovician. A conspicuous feature of many of the species was the greater thickness of the shells and the ribbing (pg. 226, *F, G, H*) of the exterior by means of which the shell was strengthened. Brachiopods were very abundant and are important in

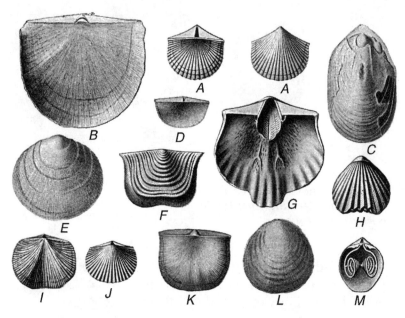

Ordovician brachipods: *A, Orthis tricenaria; B, Raphinesquina alternata; C, Lingula rectilateralis; D, Plectambonites sericeus; E, Trematis ottowaensis; F, Leptaena rhomboidalis; G, Platystrophia lynx; H, Rhynchotrema capax; I, Dalmanella testudinaria; J, Herbertella borealis; K, Strophomena rugosa; L, Schizocrania filosa; M, Zygospira recurvirostris.*

determining the subdivisions of the series.

Bryozoa—Fossil bryozoans (Greek, *bruon*, moss, and *zoon*, animal) consist of small branching stems and lacelike mats (pg. 227, top, *A-E*), the skeletons of minute animals that lived in colonies. They resemble certain corals in external appearance, but are related to the brachiopods. They can, as a rule, easily be distinguished from corals by the smaller size of the colonies in which the polyps lived. Bryozoan fossils are very common in limestones of Ordovician age and were important limestone makers. They are valuable "index fossils" in determining the age of Ordovician strata, since they were abundant not only in

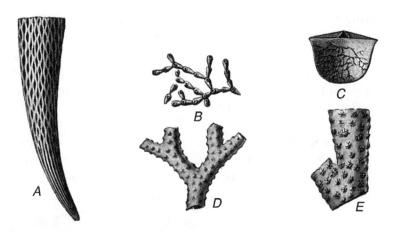

Ordovician bryozoans: A, *Escharopara subrecta;* B, *Corynotrypa inflata* (enlarged); and the same, *C,* in its natural position and size on a brachiopod; D, *Callopora pulchella;* E, *Constellaria florida.*

individuals but also in species.

Pelecypods are abundant in the salt and fresh waters of the present, being represented by the clam, pecten, oyster, and many others. They have bivalve shells in which the two valves are usually nearly alike (below, *A-E*). In external form they differ from brachiopods, which they resemble, in the lack of bilateral

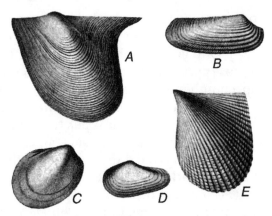

Ordovician pelecypods: *A, Pterinea demissa; B, Rhytimya radiata; C, Cyrtodonta billingsi; D, Ctenodonta nasuta; E, Byssonychia radiata.*

symmetry. Aside from fossils whose relationships are doubtful (Fordilla and Modioloides) this great class is almost unknown previous to the Ordovician. As a rule, pelecypods are rather rare fossils in the Ordovician rocks, being more abundant in sandstone and shales than in limestones, thus showing that they lived best on sandy and muddy bottoms.

Gastropods—This class was more abundant than the pelecypods, and even in the early Ordovician was represented by a considerable variety of forms (below, *A-G*) which closely

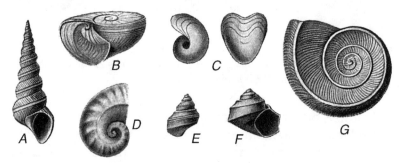

Ordovician gastropods: *A, Hormotoma gracilis; B, Maclurea logani; C, Protowarthia cancellata; D, Cyrtolites ornatus; E, Lophospira bicincta; F, Trochonema umbilicatum; G, Ophileta compacta.*

resemble modern relatives in external appearance.

Cephalopods—This is the most highly developed class of the mollusks. All Ordovician cephalopods (pg. 229, *I-D*) have shells such as those possessed by the nautilus of today. The shell is divided into a number of chambers by transverse partitions, called *septa*, through which a tube, the *siphuncle* (pg. 229, *A* and *D*), extends from one end to the other, the animal living in the body chamber (pg. 229, *A*) at the larger end. The juncture of the septa with the shell is called the *suture*, and the shape of this line is of great importance in determining the evolution of many genera. The shape and size of Ordovician cephalopods varied greatly, some being straight (pg. 229, *D*), some curved (pg. 229 *B, C*), and some tightly coiled (pg. 229, *A*). The straight

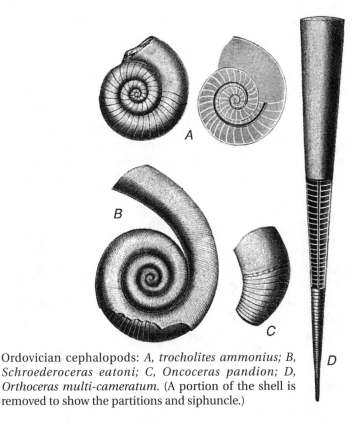

Ordovician cephalopods: *A, trocholites ammonius; B, Schroederoceras eatoni; C, Oncoceras pandion; D, Orthoceras multi-cameratum.* (A portion of the shell is removed to show the partitions and siphuncle.)

forms, represented by Orthoceras (Greek, *orthos*, straight, and *ceras*, a horn), were most characteristic of the period, some (Endoceras) attaining a length of ten or more feet and a maximum diameter of about one foot. At the other extreme were some less than an inch in length and one eighth of an inch in diameter. Cephalopods have been called the scavengers of the Ordovician, and they were probably the most powerful animals then living. The great diversity of the Ordovician cephalopods is evidence that the group began in the early Cambrian.

Crustacea

Trilobites—The rapid evolution of the trilobites noted in the discussion of the Cambrian was continued in the Ordovician, during which period the class attained its greatest development (below, *A-I*), more than half of all the known genera of trilobites being represented at that time. During the remainder of the Paleozoic they gradually declined until their extinction was reached in the closing stages. When Cambrian and Ordovician

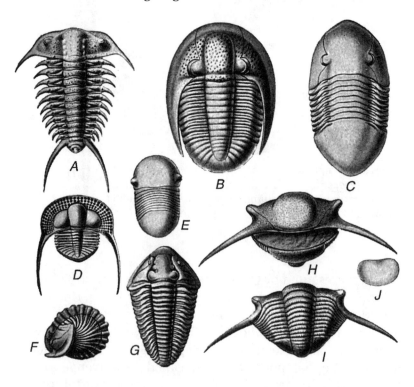

Ordovician crustaceans. Trilobites: *A, Ceraurus pleurexanthemus; B, Bathyurus longispinus; C, Isotelus gigas* (grealy reduced); *D, Trinucleus (Cryptolithus) tessellatus; E, Bumastus trentonensis; F, G,* rolled and straight specimens of *Calymene callicephala; H, I,* two views of a rolled specimen of *Thaleops ovata.* Ostracod: *J, Leperditia inflata.*

trilobites are compared, it is seen that the latter have rounder eyes, that the tail shield (pygidium) is larger, and that they have acquired the ability to roll themselves up (pg. 230, *F, H, I*) and thus protect the lower portions of the body, many being found in this position which was apparently taken at the approach of death. The Ordovician trilobites were not as large as some Cambrian species, the maximum length being about 18 inches as compared with about two feet for one Cambrian species.

Other Arthropods continued from the Cambrian. *Ostracods* (pg. 230, *J*), small bivalve crustaceans, flourished during portions of the period, and also *eurypterids.*

Fishes

An important addition to the fauna of the Ordovician, and one which had a profound effect on the evolution of life in subsequent ages, was the fishes, remains of which have been found in Colorado and Wyoming.

Plants

Seaweeds—Our knowledge of the plants of the Ordovician is almost as incomplete as that of the Cambrian. No land plants have been found and no marine plants higher than seaweeds (right) and calcareous algae. The absence of land plants in Ordovician strata, however, does not prove that a land vegetation was lacking, since the known plant-bearing strata are all of marine origin and consequently the absence of land plant fossils would not be remarkable, even though land plants had been abundant.

Ordovician plants (S p h e n o p h y c u s latifolius). They are probably appendages of floating algae.

Summary

Progress and Character of Ordovician Life—The life of the period as shown by the fossils was fuller, more varied, and of a higher grade than that of the Cambrian. Trilobites, cephalopods, gastropods, pelecypods, cystoids, graptolites, and corals became diversified and of higher types; and bryozoans, crinoids, and fishes are known for the first time. During this period, graptolites and cystoids attained their climax and were never again important. In this period, too, the straight cephalopods rapidly developed from small to gigantic forms and into many species, but occupied a subordinate place after the Silurian. Before the close of the period all of the great types of life and most of the important subdivisions were present.

When the faunas of the Ordovician stages of North America are compared with those of Europe, it is found that, although the genera are usually identical, the species are different though similar.

Adaptation to environment was almost as well established then as now. Certain species lived almost exclusively on muddy bottoms, certain ones on sandy, and still others on calcareous bottoms. There was also adaptation to shallow and deep water.

The effect of isolation is noticeable when, for example, a portion of an epicontinental sea was cut off by some barrier, such as a gentle upfolding of a portion of the sea bottom, or a bar, or when ocean currents, because of their lower or higher temperature, prevented the life of different portions of the sea from mingling, the isolation of the fauna permitting an independent development without interference from outside. It consequently sometimes happened that the faunas of adjoining seas differed considerably. When the barriers were removed, a rapid and marked change in the fauna was often quickly brought about.

The evolution of the life of the period gave rise to many new species, with the result that when the fauna of the earliest and latest Ordovician are compared, they are found to differ widely.

It is because of the appearance of new species that the Ordovician series of strata have been divided into several stages, which are usually easily distinguished by their contained fossils.

Climate and Duration of the Ordovician—Fossils found in Ordovician strata of Arctic lands show that the climate there was not unlike that of the temperate and tropical regions of the same time. During the Middle Ordovician, and again later in the period, reef corals were common from Alaska to Texas. The conclusion is that climatic zones did not exist, but that the climate of the world was uniformly equable and less diversified than now.

The duration of the period was about the same as that of the Cambrian, perhaps 4,000,00 years.

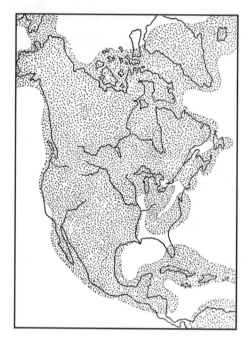

Map showing the probable outline of North America during a portion of the Lower Silurian.

Map showing the probable outline of North America during a portion of the Middle Silurian.

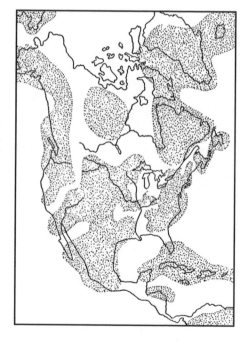

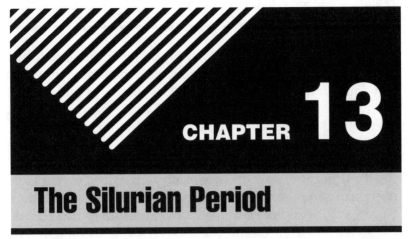

CHAPTER 13

The Silurian Period

This system has been divided into a number of subdivisions which in New York are as follows:

Rondout water lime
Cobbleskill limestone
Salina shales, salt, water lime
Lockport and Guelph dolomites
Rochester shale
Clinton shale, sandstone, limestone, and iron ore
Medina and Oneida sandstone and conglomerate

In eastern North America, the Silurian strata, for the most part, rest unconformably upon the deformed and eroded Ordovician rocks. In the Middle States, the Lower Silurian is usually absent, and the Middle Silurian strata rest unconformably upon the Ordovician or upon older rocks, showing that during early Silurian times the central portion of the continent was land. In Montana and Utah the strata of the Ordovician, Silurian, and Devonian are apparently conformable, and their separation is more or less arbitrary because of the scarcity of fossils.

Geography of the Silurian—The period can, for convenience, be divided into three epochs. (1) During the first (early lower) the epicontinental seas were apparently restricted to three principal bays (pg. 234, top): one stretching up the Mississippi Valley to northern Illinois; a second extending across Newfoundland and northern New Brunswick; and a third occupying the Appalachian trough, and stretching east and west over central New York and Ontario. Later in the Lower Silurian, the seas were, for a time, withdrawn from the Appalachian trough and New York. (2) The middle (later lower) of the period (pg. 234, bottom) saw an expansion of the seas over a large portion of Canada to the Arctic Ocean and over the United States east of the Mississippi River, and an extension of two seas on the west, one from California through Idaho to Canada and another from Mexico into Arizona and New Mexico. It was during this time that the great limestone strata (Niagaran) were deposited. (3) The epicontinental seas were again restricted in the Upper Silurian (pg. 237), the most important of them extending from Wisconsin and Illinois through New York and over the Appalachian trough. The three subdivisions of the period are therefore characterized (1) by constricted seas, (2) by expanded seas, and (3) by a later shrinking and shifting of the seas. It should be pointed out in this connection that Silurian strata do not cover all of the areas shown in the maps (p. 193), but often lie only in widely separated patches which appear to be remnants of a once continuous formation. The age and correlations of these patches are determined by their contained fossils.

The sediments which were deposited in the Appalachian trough were derived from the broad island or continent of Appalachia.

Character and Thickness of the Sediments—Limestones are the common strata of the Silurian, but in eastern North America conglomerates, sandstones, and shales predominate. These latter were deposited in shallow seas, as the ripple marks and cross-bedding show. The formation and distribution of these coarse sediments teach an important lesson. Along the western boundaries of the eastern lands, gravel and sand were carried by rapid streams flowing from the areas newly raised by the Taconic deformation and were spread along the shores, forming wide beaches, the gravel (forming the Oneida conglomerate) rapidly thinning toward the west. Sand (Medina) was carried farther out into the sea by the currents and formed extensive sandstone extensive sandstone strata. When in the course of time the lands were worn down, the streams were unable to carry such large quantities of coarse sediment as formerly, and the belt of gravel along the shores was consequently narrowed and sand was deposited nearer shore and upon the earlier gravels. It is evident from the above that the conglomerates and sandstones were contemporaneous. A still greater lowering of the land, either by erosion or by subsistence, further reduced the capacity of the streams for cutting, and during one epoch lime ooze (Niagaran) and during another mud (Salina) accumulated on the sandstones and conglomerates. Where the conglomerates have been tilted by later folding and cut by erosion, their upturned edges form mountain ridges.

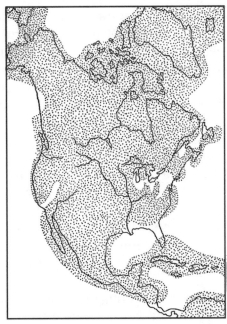

The Silurian formations west of New York are largely limestone, portions of which well-developed coral reefs are to be distinguished. The falls of Niagara are due to the presence of a massive layer of limestone of Silurian Period, from which an impor-

Map showing the probable outline of North America during a portion of the Upper Silurian.

tant stage, the Niagaran, received its name. The limestone of this stage is thickest in the Mississippi Valley, where it probably had been accumulating for a longer time than in New York. Such a great thickness of limestone indicates a long period during which few oscillations in level occurred, and when the lands were so low that little sediment was carried to the sea.

The thickness of the Silurian in Maryland is about 3000 feet; in western Tennessee, about 1500 feet; in Alabama, about 500 feet; in central Tennessee, 300 feet; in Wisconsin, about 800 feet; and in Nevada, about 1000 feet.

Clinton Iron Ore —One of the most widespread iron ore deposits known was accumulated during the Clinton stage of the Silurian Period. It outcrops in one or more broken belts from New York, through Pennsylvania to Alabama, and occurs in beds at different horizons in the formation, sometimes as many as four beds being present in one locality. The thickness of the ore beds varies from 40 feet to a fraction of an inch, but a bed 10 feet thick is unusual. The ore is called "fossil" and "pea" ore because fossil fragments are commonly found in it with the shell substance entirely replaced with hematite, while some beds are made up of rounded grains of a concretionary character. The ore was deposited close to shore, probably in lagoons and marshes, and was probably a chemical precipitate, the iron having been brought to the sea by streams which had leached it from the igneous rocks over which they flowed.

The presence of iron ore, limestone, and coal within short distances of each other near Birmingham, Alabama, has made that city a great center for iron and steel industries. Coal is necessary to reduce the iron, and limestone is used as a flux to carry away the siliceous impurities.

Deserts—During a portion of the Silurian (Salina), in eastern North America the climate was arid and desert conditions prevailed. This is shown by the beds of salt and gypsum, and by the red color of the shales. In New York state 325 feet of solid salt have been penetrated by wells. These salt beds are lens-shaped, and the conditions under which they were deposited may not have been unlike those today in the region of the Caspian Sea, the Dead Sea, and Great Salt Lake, or back of bars as described below. Such an arid climate may have been produced by high lands to the south and east, which shut off the moist winds from the Atlantic and the Gulf of Mexico.

Origin of Rock Salt—Salt is primarily formed by the evaporation of the salt water of lakes or the ocean, and is accumulating today in certain salt lakes which have been greatly concentrated. The evaporation of inland salt lakes does not, however, seem adequate to produce thick beds of pure salt such as occur in certain regions.

The theory which best explains the origin of massive salt deposits assumes that a body of ocean water had been shut off partly or completely by a low bar. If the region in which this occurred was arid, the evaporation of the water back of the bar would exceed that carried in by the rivers and that derived from the ocean. The lowering of the water of the bay by evaporation would permit the ocean water to flow in if the bar were incomplete; if, however, the bar were complete and the bay entirely shut off from the ocean, forming a lake, ocean water would enter only during storms or at high tide. In time, the concentration of the water would be so great that common salt and other salts would be precipitated. Under conditions such as those outlined above, pure salt might accumulate to a considerable thickness without the admixture of mud. Occasionally, the purity of the salt might be broken by sheets of mud brought in by streams swollen by the torrential showers of desert regions.

The Silurian salt of New York seems to have been deposited either in extensive salt lakes or in an arm of the sea which was partially shut off from the sea by a bar.

Igneous Rocks—The Silurian was a period of quiet as far as volcanism was concerned. In North America some igneous intrusions of this age are known, but they are not extensive.

Other Continents—The Silurian epicontinental seas of Europe were extensive and had much the same position as in the Ordovician. Two distinct seas, one in the north and the other in the southern part of the continent, were separated by a land ridge. The life of the two seas was unlike in many particulars, that of the northern sea being typical of the period in other continents, while that of the southern had many peculiarities which indicate that it was partly inclosed.

Life of the Silurian

Aside from a notable increase in the number and variety of corals, crinoids, brachiopods (spire bearers), and eurypterids, and a decrease in the graptolites and straight cephalopods, the life of the Silurian did not differ greatly in general aspect from that of the Ordovician. When, however, one looks for identical species and genera, one sees a marked change.

Coelenterata

Corals forged ahead and became important in the Silurian. Instead of the few, usually simple forms of the Ordovician, a varied and abundant coral fauna, many of the corals compound, throve in the clear seas of the time. Four well-marked types were abundant: (1) *chain corals* (Halysites, pg. 241, *A*), made up of vertical tubes joined together in such a way as to give them the appearance of a linked chain. Since chain corals began in the Ordovician and became extinct in the basal Devonian, their presence in a formation shows it to be either Ordovician or Silurian. (2) *Honeycomb corals* (Favosites) were composed of six-sided parallel columns, like a honeycomb, which were

Silurian corals: *A*, chain coral, *Halysites catenulatus; B* and *C*, cup corals, *Streptelasma (Enterolasma) calicula; D, Syringopora retiformis; E, Goniophyllum pyramidale.*

divided by horizontal partitions. Honeycomb corals were rare in the Ordovician, but built coral reefs in the Silurian and Devonian. (3) *Cup corals* (Enterolasma, above, *B, C, E*) were horn-shaped, with a depression in the top. A peculiar cup coral of the period (Goniophyllum, above, *E*) was provided with a cover which consisted of four triangular plates, hinged to the margins of the cup. The covering was evidently for protection against enemies, but since the genera which possessed it have no living representatives, it is probable that the device was not successful. Cup corals occurred in the Ordovician and continued until the close of the Paleozoic; many of these were separate individuals, while some were in hemispherical colonies (see pg. 285). (4) *Organ-pipe corals* (Syringopora, above, *D*) were similar to chain corals, but their cylindrical columns were not attached along their entire length.

Coral reefs date from the later Ordovician. Before this time simple corals predominated, and even these were rare. When, however, compound corals such as the honeycomb became abundant, the limestone secreted by many generations, together with that of associated animals such as brachiopods, gradually built up reefs at short distances from the shores. Silurian coral reefs were seldom of great thickness.

Other Coelenterates—*Stromatopora* were important reef builders, but *graptolites* no longer played an important role in America, and by the end of the period were practically extinct. *Sponges* (pg. 242, top) are common in certain beds, the peculiar family *Receptaculites* (below, *B*) which began in the Ordovician being not uncommon in some localities.

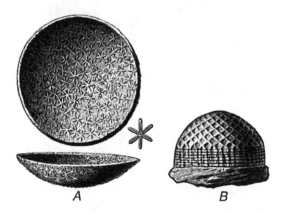

Silurian sponges: *A, Astraeosponia meniscus,* two views; *B, Receptaculites oweni.*

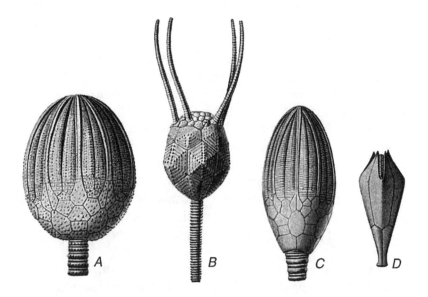

Silurian crinoids: *A, Eucalyptocrinus elrodi; C, Eucalyptocrinus crassus* (closed). Cystoid: *B, Caryocrinus ornatus.* Blastoid: *D, Troostocrinus reinwardi.*

Echinodermata

Crinoids (pg. 242, bottom, *A, C*) became so numerous in the Silurian that their "joints" constitute an important part of the beds of certain limestones. Where these flowerlike animals were abundant on the sea bottom, they must have presented an appearance not unlike that of a field of lilies. Not only did they live in great numbers, but the variety of forms which were developed was large.

Cystoids (pg. 242, bottom, *B*) continued to be abundant where conditions were favorable for their growth, but at the close of the period they were no longer an important element of the fauna.

Molluscoidea

Brachiopods—Although the Silurian brachiopods (below, *A-D*) differed almost entirely from those of the Ordovician in species, the importance of the race did not diminish. Some improvements in structure were accomplished, and new genera

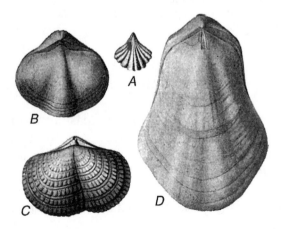

Silurian brachiopods: *A, Rhynochotreta cuneata americana; B, Spirifer radiatus; C, Streptis grayi; D, Pentamerus oblongus.*

which later became important were evolved. The evolutionary changes were doubtless directly or indirectly the result of the changes in environment, which consisted in shiftings of the epicontinental seas and the consequent frequent migrations of faunas and struggles between them.

Bryozoa—The coral-like bryozoans (below) were less important reef-builders in the Silurian than in the Ordovician.

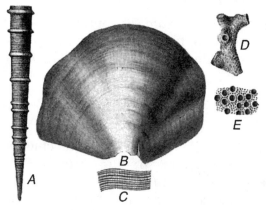

Silurian bryozoans (*B-E*), and pteropod (*A*): *A, Tentaculites gyracanthus; B, Lichenalia concentrica; C,* a portion of *B* enlarged; *D, Callopora elegantula; E,* a portion of *D* enlarged.

Mollusca

Gastropods—Aside from an increase in the number and variety of species with elevated spines and in a somewhat greater abundance, no important changes in the *gastropods* (pg. 245, top, *B-E*) are shown.

Pelecypods (pg. 245, top, *A*) also continued much as in the Ordovician.

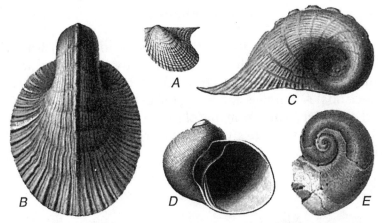

Silurian pelecypod: *A, Pterinea emacerata.* Silurian gastropods: *B* and *C,* two views of *Trematonotus alpheus; D, Strophostylus cyclostomus; E, Platyostoma (Diaphorostoma) niagarense.*

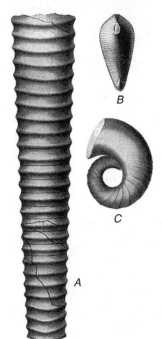

Cephalopods—Curved and coiled *cephalopods* (left, *B, C*) were more numerous than straight forms (left, *A*), while the latter were smaller and were commonly ornamented by rings and low, transverse ridges. The body chamber of many Silurian cephalopods was constricted (left, *B*), the constriction being apparently for the purpose of protecting the soft parts of the animal from enemies. As in the Ordovician, this class was the most powerful of the time.

Silurian cephalopods: *A, Dawsonoceras americanum; B, Phragmoceras parvum; C, Trochoceras desplainense.*

Arthropoda

Trilobites—This interesting class (below, *A-D*) was still important, but the decline had already begun and it was numerically decidedly less prominent than in the Ordovician. Since no new families appeared, the general aspect of the class did not differ greatly from that of the preceding period. The most significant change from the Ordovician was in the disappearance of Ordovician genera.

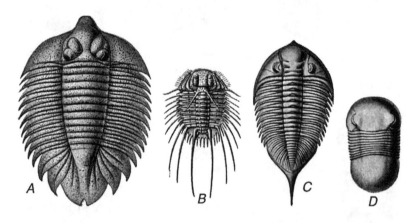

Silurian trilobites: *A, Arctinurus (Lichas) bigsbyi (boltoni); B, Ceratocephala dufrenoyi; C, Dalmanites limulurus; D, Bumastus (Illaenus) ioxus.*

Eurypterids—The arthropods (Greek, *arthron*, joint, and *pous*, foot), the branch of which the crustaceans and insects are members, reached their greatest size in the eurypterids (pg. 247, top). Some of the Silurian forms attained a length of one and a half feet, while in the Devonian there were giants eight feet long. They, together with the giant cephalopods, were probably the terrors of the sea until the fish obtained the mastery. They had elongated bodies covered with a leathery or horny integument. On the under side were six pairs of legs, of which the first had large or small pincers. Eurypterids are related to the horseshoe crabs (Limulus). The presence of gills and their association with cephalopods and trilobites in the Ordovician show that they lived in water and were for the most part mud crawlers, although some were good swimmers.

Silurian eurypterid:
*Dolichopterus
macrocheirus.*

They were at first marine animals, but late in the Paleozoic became adapted to brackish and possibly to fresh-water conditions, and there is evidence for the belief that some even lived in lagoons where the water was more salty than that of the sea.

Scorpions (right) first appear in the Silurian, but probably lived in water and got their oxygen there, not on land as do their modern relatives, the ability to breathe in air having been acquired later in the Paleozoic.

Restoration of a
Silurian scorpion.

Fishes

Fragmentary fish remains have been found in Silurian rocks (below, *A, B*) of both Europe and America. The fact that fishes were abundant and of considerable variety in the Devonian is presumptive evidence that a somewhat varied fish fauna existed during the closing days of the Silurian.

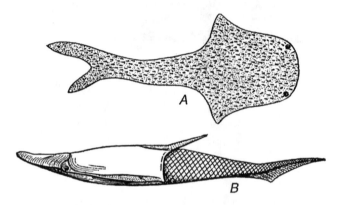

Restorations of Silurian fishes (ostracoderms): *A, Thelodus; B, Pteraspis.*

Summary

Life on the Land—It is probable that the lands of the period were clothed with plants, but if so, little evidence is afforded either from the remains of plants or of land animals. The highly developed land plants of the Devonian (p. 269), however, are indirect evidence of the existence of land plants in the preceding period. Nevertheless, the absence of the sediments of fresh-water lakes in America, where land fossils are likely to be preserved, leaves us without evidence, although it does not prove that there were no land plants or animals.

Migration—The presence of thirty or more identical species in the Silurian strata of Iowa and northwestern Europe indicates that migration between the two continents took place. This presumption is strengthened by the discovery of a peculiar genus of coral (pg. 241, *E*,) whose quadrangular opening was protected by a calcareous covering. Since the interior seas of North America had no free communication on the east, it is thought that the migration took place along a belt of shallow water which extended through Canada, Alaska, and perhaps the Arctic region.

Climate and Duration—The climate of this period seems to have been uniform over the entire world, as during the preceding periods, there being no positive evidence of the existence of climatic zones. The presence of salt and gypsum beds, locally 40 to 80 feet in thickness, in the Silurian strata (Salina) of New York and Ohio is evidence that desert conditions prevailed during a portion of the period, probably over a considerable area.

The Silurian Period was probably not more than one half as long as the Ordovician.

Close of the Silurian—The change from the Silurian to the Devonian in eastern North America is even less clearly marked than that between the Ordovician and the Silurian, the formations of one often passing into the other without an unconformity. In portions of Great Britain, an unconformity separates the two systems, but in other parts of Europe there is no break in the sedimentation. The separation in such cases is based upon the fauna.

CHAPTER 14

The Devonian Period

The passage from the Silurian to the Devonian in eastern North America was without any physical break, the transition from one to the other being marked by no great unconformity. In fact, so gradual was the change that much controversy has arisen as to the exact limits of the two systems. Not only was the change in the lithological character and structure of the strata slight, but the life at the close of the Silurian and the beginning of the Devonian was very similar. Wherever the Devonian seas spread over a wider or different area from that of the Silurian, the sediments were laid down unconformably on older rocks. Unconformities of this sort occur in Iowa and elsewhere, but they are inconspicuous and are sometimes with difficulty recognized as unconformities, since the underlying rocks are horizontal and their surfaces had not been greatly roughened by erosion. The lack of great relief as shown in these unconformities affords an evident explanation for the fineness of the sediments in the Western Interior in the closing stages of the Silurian, as well as of those of the early days of the Devonian.

Subdivisions of the Devonian—The subdivisions of this period in New York state are given below, both because it was in this state that they were first studied with care in North America, and also because the system is best developed there.

> Catskill and Chemung sandstones
> Portage shale and sandstones
> Genesee shale
> Tully limestone
> Hamilton shale
> Marcellus shale
> Onondaga limestone
> Oriskany sandstone
> Helderberg limestone

Geography—The close of the Silurian found few epicontinental seas in North America. In the east (below) the

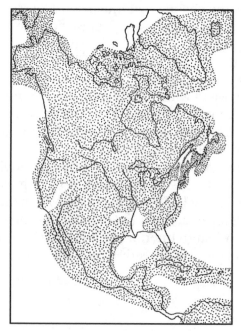

Map showing the probable outline of North America during a portion of the Lower Devonian.

Appalachian trough, portions of New York, and certain areas in the Maritime Provinces of Canada were covered with seas, and a bay extended from the Gulf of Mexico toward the north along the valley of the Mississippi. In the west, an arm of the sea extended from the Pacific across the site of the Sierra Nevada into Utah. The outlines of these seas were not constant but changed from stage to stage. Later in the period (pg. 254) the seas spread widely over the continent, calling to mind the submergent condition of the Middle Ordovician and Middle Silurian.

In New York state, the formations of the first half of the Devonian are for the most part limestones with occasional shales and sandstones, but in the later half of the period shales and sandstones predominate. The shales and sandstones were brought into the sea by streams from the Taconics of Massachusetts, and probably from land areas which existed to the north in Canada. The Devonian strata cover a greater area at the surface in New York than any other rocks, and their combined thickness is more than 4000 feet. They are much thicker in Pennsylvania, but thinner in the Mississippi Valley, and are said to be 8000 feet thick in portions of Nevada.

The Devonian in New York—There are three Devonian formations in New York which deserve special mention: the Oriskany, the Onondaga, and the Catskill. The Oriskany is a sandstone formation made, for the most part, of clean beach sands, which in New York is from a foot to several feet thick and in the Middle Atlantic States is several hundred feet thick. The formation indicates a raising of the land or an increase in the rainfall, or a combination of the two, since stronger currents are necessary to supply coarse waste.

The Onondaga limestone, with its wealth of corals and brachiopods, indicates warm, clear seas of long duration surrounded by low lands.

The Catskill formation, thousands of feet thick, extends from Virginia to the Catskill Mountains in New York and is a great delta deposit, made of alternate layers of sandstone and shale, sometimes the one and sometimes the other predominating. Upon this assumption the whole constitutes a delta. The form of the plain was probably somewhat similar to that of the high plains region of the western interior of North America. While the Catskill delta was being built up, muds were accumulating in the shallow seas.

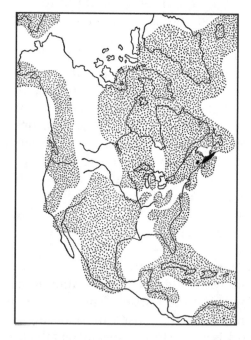

Map showing the probable outline of North America during a portion of the Middle Devonian. Solid black shows continental deposits.

Continent of Appalachia—The continent of Appalachia, situated east of the present Appalachian Mountains, which during the preceding periods of the Paleozoic was supplying the streams with sediment for the Appalachian geosyncline, was extensive at this time and was probably a broad, mountainous upland whose eastern boundary may have been beyond the present eastern limit of the continental shelf. This conclusion is justified when the volume of sediments laid down in the Appalachian trough is computed. Such a computation shows that the crest of Appalachia would have had to be lowered from five to seven miles to supply the Upper Devonian sediments, if it had not extended beyond the continental shelf. It seems likely, therefore, that Appalachia extended from the edge of the Appalachian trough eastward over the present site of the continental shelf and probably fifty miles beyond. The broad Appalachian continent probably never reached Alpine heights, but was rather slowly raised as the Appalachian trough sank. The sediments of the trough are those formed from igneous rocks of the land which had been subjected to chemical decay, and are not such as would have resulted from the mechanical disintegration of frost or changes in temperature. The sediments, moreover, are seldom coarse, showing that the streams did not flow from a high, mountainous region in proximity to the sea.

Igneous Rocks—In Maine, New Brunswick, and Nova Scotia, granite intrusions and volcanic extrusions took place during the Devonian. The city of Montreal lies at the foot of a volcano, and there are other volcanoes to the southeast. This was the first premonitory indication of the movements which were later to form the great Appalachian Mountains. When North America as a whole is considered, the Devonian Period closed with almost no deformation.

Devonian Oil and Gas—A discussion of the Devonian would be incomplete without mention of the important oil and gas-bearing strata of West Virginia, Pennsylvania, and southwestern New York. The oil and gas are more likely to be found at or near the crests of low anticlines than in any other situation.

Devonian of Other Continents—Epicontinental seas were widespread in Europe and Asia during the Devonian, and smaller seas covered portions of Africa, South America, and Australia. The Devonian of England is of unusual interest because of the development of a continental deposit of red sandstone, called the "Old Red Sandstone." It appears to have been laid down under desert conditions, although no gypsum or salt beds prove this contention. In addition to the Old Red Sandstone, there are marine deposits containing abundant fossils. Volcanic action in Europe during the Devonian is proved by thick volcanic accumulations in Great Britain and west central Europe.

Life of the Devonian

The invertebrate life of this period was, in general aspect, like that of the Silurian, but there were many changes in general and an almost total change in species. The contrast between the invertebrate life of the Silurian and the Devonian was about as marked as that between the Ordovician and the Silurian. As in the foregoing periods, certain species were characteristic not only of the period as a whole but of each of its stages.

Coelenterata

Corals (below, *A-C*) were present in great numbers and species, being almost or quite as abundant as in the Silurian. *Cup corals* (Tetracoralla, below, *A,C*), *honeycomb corals* (Favosites), and *organ-pipe corals* (Syringopora) flourished

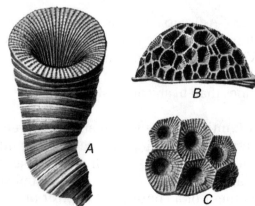

Devonian cup and compound corals: *A, Heliophyllum halli; B, Pleurodictyum stylopora; C, Acervularia davisoni.*

when conditions were favorable, but *chain corals* (Halysites) had become extinct in the beginning of the period.

Corals are not equally abundant in all Devonian formations; they are rare in shales and sandstones, but are usually common in limestones. This is not remarkable, since corals do not thrive in muddy waters.

Echinodermata

Crinoids (below, *B*) and *starfish* (below, *A*) were much more abundant than in previous periods, but *cystoids* were rarer than in the Silurian.

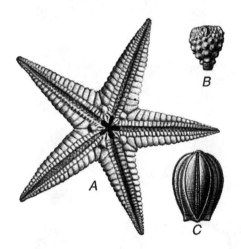

Devonian starfish, crinoid, and blastoid:
A, Palaeaster eucharis;
B, Melocrinus milwaukeensis;
C, Nucleocrinus verneuili.

Blastoids (Greek, *blastos*, bud) were locally abundant. These echinoderms (above, *C*), as the name implies, were oval, with five petal-like divisions resembling a flower bud. They were armless and were attached to the sea bottom by a jointed stem. Beginning in the Ordovician, they culminated in the Mississippian, after which they occurred sparsely and disappeared with the Paleozoic.

The *starfish* of Devonian times had already acquired the

habits of feeding which they possess today.

Molluscoidea and Mollusca

Brachiopods (below, *A-P*) were never more abundant in individuals and species than during portions of this period, and many characteristic species were present. Long-hinged *spirifers* were especially abundant and highly developed throughout the Devonian.

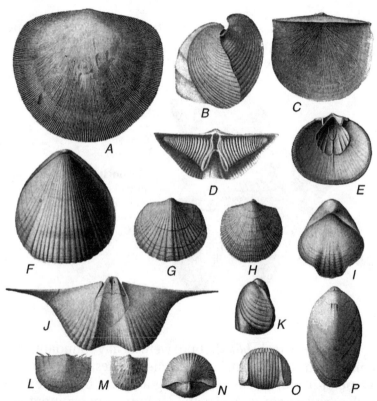

Devonian brachiopods: *A, Hipparionyx proximus; B, Spirifer acuminatus; C, Stropheodonta demissa; D, Spirifer mucronatus; E, Rhipidomella oblata; F, Camarotoechia endlichi; G, Tropidoleptus carinatus; H, Atrypa reticularis; I, Gypidula galeata; J, Spirifer disjunctus; K, Eatonia medialis; L, Chonetes coronatus; M, Productella spinulicosta; N and 0, two views of Hypothyris cuboides; P, Rensseloeria ovoides.*

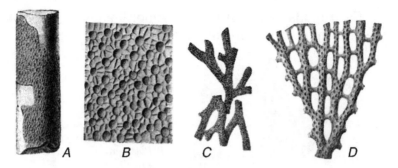

Devonian bryozoans: *A, Fistulipora micropora* surrounding a crinoid stem; *B*, a portion of the same greatly enlarged to show the arrangement of the cells; *C*, branches of *Cystodictya hamiltonensis*; *D, Polypora liloea.*

Bryozoans (above, *A-D*) were locally abundant.

Pelecypods (below, *A-E*) flourished where the bottoms were muddy and other conditions favorable.

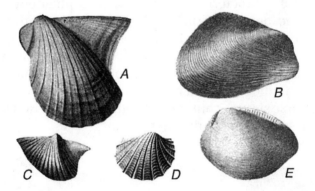

Devonian pelecypods: *A, Pterinea flabellum; B, Modiomorpha concentrica; C, Conocardium ohioense; D, buchiola retrostriata; E, Palaeoneilo constricta.*

Gastropods (below, *A-C*) were subordinate in numbers to the pelecypods but were not uncommon.

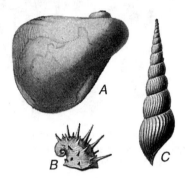

Devonian gastropods:
A, *Strophostylus expansus;*
B, *Platyceras dumosum;*
C, *Loxonema noe.*

Cephalopods—A rather inconspicuous member of this class, the goniatite (Greek, *gonia*, angle), but one whose modified descendants were to become the most prominent invertebrates of the Mesozoic, began in the Devonian. The important characteristic of this coiled cephalopod was the angled and lobed suture line (below, *A, B*), i.e., instead of smooth partitions (septa, p. 228) which joined the outer shell in straight lines or in simple curves, the septa were crumpled at the edges at the juncture with the outer shell, forming *angled sutures.* The straight (Orthoceras) and coiled (Gomphoceras) cephalopods with simple sutures continued throughout the period but were much less common in the later portion.

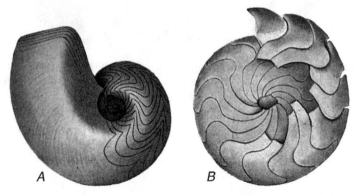

Devonian cephalopods: A, Manticoceras oxy; B, Tornoceras mithras.

Arthropoda

Trilobites—During the earlier stages of the Devonian more than 50 species of trilobites (below, *A-C*) are known to have existed, but the numbers rapidly decreased during the later stages. In the earlier portions of the period especially, a number of highly ornamented, spinous forms (below, *A*) lived, but later these extravagant species largely disappeared, and those of simpler outlines remained. The decline of the trilobites during the Devonian was very marked, and at its close they were on the verge of extinction, although a few survived until the close of the Paleozoic. Other crustaceans (bottom right) also lived in considerable abundance.

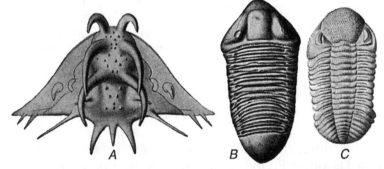

A *B* *C*

Devonian trilobites: *A, Lichas (Gaspelichas) forillonia (cephalon); B, Dipleura dekayi; C, Phacops rana.*

Barnacles, which are retrograde crustaceans that have given up a free-moving existence for a stationary one in which they are protected by a calcareous covering, began in the Ordovician, but the common acorn barnacle began in this period.

Echinocaris punctata, a Devonian Crustacean.

Eurypterids attained their greatest size during the Devonian, one species reaching a length of almost eight feet (right).

Insects — No undoubted remains of insects have been found in strata of this or earlier periods, although their discovery may be expected at any time.

Stylonurus, a gigantic Devonian eurypterid, some of which were eight feet long.

Fishes

The appropriateness of the term "Age of Fishes" as applied to the Devonian Period is evident when the importance of this great class, not only in the life of that time but as the probable progenitors of all subsequent vertebrate life, is considered. Fish were the rulers of the Devonian seas and rivers, perhaps even to a greater degree than were the trilobites in the Cambrian and the cephalopods in the Ordovician. The fact that fish were abundant during the period does not imply that other forms of life were less abundant than in previous periods. For example, brachiopods are exceedingly common fossils in almost all Devonian strata, while in many of the rocks of this age fish fossils are extremely rare and a search of many days may not be rewarded by even a fragment.

Ostracoderms—One of the strangest classes of Devonian

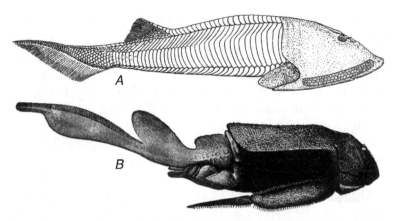

Devonian ostracoderms: *A, Cephalaspis,* about six inches long; *B, Bothriolepis,* about seven inches long.

animals was the ostracoderm (Greek, *ostrakon,* shell, and *derma,* skin). These were fishlike in shape but were probably not even closely related to fishes. The description of one well-known member of the class (Cephalaspis, Greek, *cephale,* head, and *aspis,* shield) gives a general notion of this group (above, *A*). The most striking feature was the crescent-shaped plate which covered the head and fore part of the body. Besides the protection afforded by this head shield, the tail was covered with rhomboidal scales. The eyes were situated close together on the top of the head. The lower jaw, if it ever existed, has not been found. Although many hundreds of the bony parts have been found, no internal skeleton has been discovered, and it is therefore probable either that none existed or that it was cartilagenous and was consequently incapable of fossilization. In another genus (Bothriolepis) a pair of appendages encased in bony plates, somewhat as are the appendages of a lobster, extended from the sides of the head. Ostracoderms seldom attained a size greater than six or seven inches.

Certain inferences as to the habits of ostracoderms can be drawn from their structure. They probably lived on the sea bottom as did the trilobites, either burrowing in the mud above which only their eyes and their dorsal shield showed, or because of their dull coloring crawling over it inconspicuously. The fact that they were protected implies that they had to contend with enemies more powerful than themselves. They lived in large numbers in certain localities, as is shown by the great abundance of their shields which form thin beds in some places and are said to be hardened by the oil from their remains.

Ostracoderms began in the Ordovician, reached their climax in the Devonian, and became extinct at its close.

Sharks—Sharks lived in the Devonian in considerable numbers, but since their skeletons were cartilagenous, the fossil evidence of their existence consists largely of teeth, spines (which probably stood in front of the dorsal fin), and small bony denticles which were doubtless embedded in the skin. The best known and most simple in structure of these ancient sharks (Cladoselache, below, *A*) varied from two to six feet in length. It had a short, blunt snout with the mouth situated on the lower side but farther front than in the modern shark. The teeth

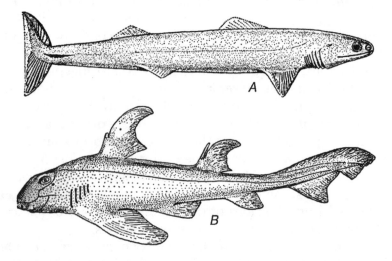

A, shark, *Cladoselache,* which sometimes reached a length of six feet (see next page); B, the Port Jackson shark, *Cestracion,* a modern shark of ancient type.

occurred in clusters (right) and were arranged in six or seven rows one behind the other. The fins were very simple, consisting of a flap of skin strengthened by straight rods of cartilage. It

Teeth of *Cladoselache.*

was very unlike modern sharks in the contour of its body.

Other sharks, now represented by the Port Jackson shark of Australian waters (Cestracion, pg. 264, *B*), were abundant in the later Paleozoic, judging from the number of their spines and pavement teeth. The teeth of these sharks have been called "cobblestone" pavement teeth because of their resemblance in shape and arrangement in the jaw to a pavement. Such teeth would be of use in crushing thin-shelled crustaceans and shellfish, but could not have been used for rending. It is evident, therefore, that their possessors probably lived on muddy bottoms and fed on brachiopods, pelecypods, or crustaceans. The structure of the tooth plate of sharks is very much like that of a shark's skin, and it is the teeth of these and other sharks that best illustrate the fact that teeth are really modifications of the skin and do not belong ill the same category as bones.

Lungfish—(1) *Armored Lungfish*—The most formidable and remarkable fish of the Devonian, as far as appearance and size is concerned, were related to the rare lungfish of today, although they probably did not possess lungs. One of the most remarkable of the Devonian lungfish was the Dinichthys (Greek, *deinos*, terrible, and *ichthus*, a fish) (pg. 266, top, *A*) which grew, in one species, to be 25 feet long, and resembled an overgrown catfish, in external form. The head, which in one species was six feet broad, and the front of the body were protected by thick bony plates, although the posterior portions seem to have been quite naked unless covered by a leather-like skin, as is perhaps indicated by certain marks upon the exterior of the bony plates. Their powerful jaws were adapted for tearing and cutting, and

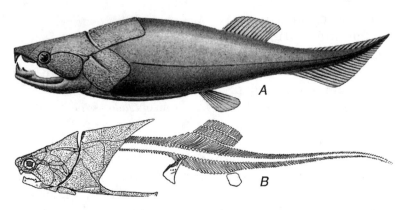

Armored lungfishes: *A, Dinichthys,* the giant fish of the Devonian, some of which attained a total length of more than ten feet, with head three feet long; *B, Coccosteus,* a fish of much smaller size.

their shape formerly led to the belief that these fish were fierce, predaceous creatures, but it is more probable that they lived on the ocean bottom and subsisted largely on shellfish, using their powerful jaws for crushing. With their heavy armor and clumsy shape they were probably sluggish in their movements.

(2) *Unarmored Lungfish*—Another abundant group of lungfishes whose descendants have succeeded in living to the present were unhampered by armor but were covered with thin scales (below). A modern representative (Ceratodus) lives in Australian waters, and two other genera are known, one in Australia and one in South America.

Ganoids—(1) *Fringed-finned Ganoids* (*Crossopterygians*)—

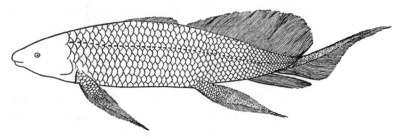

An unarmored lungfish, *Scaumenacia.*

Evolutionarily, this is the most important of the Devonian fishes, since it possessed so many characters in common with early amphibians that it is probable that the latter arose from this order. The fringe-finned ganoids had *conical teeth generally fluted*, were covered with scales which were rhomboidal in some species and rounded in others, and had *limblike fins* (below) which were jointed to the skeleton within the body.

(2) Another order of ganoids (*Actinopteri*) may, for

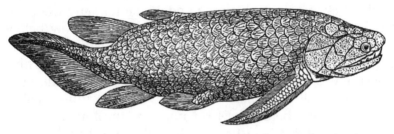

A fringe-finned ganoid, *Holoptychius*. Some of these were four feet long.

convenience, be called *typical ganoids* to distinguish them from the fringe-finned ganoids. These fishes, for the most part, had thick, rhomboidal scales, such as those of their modern representative, the gar pike (below). Typical ganoids were the most abundant and characteristic fish of the Triassic and Jurassic.

Teleosts or Bony Fish—The typical fish of today, such as the

A typical ganoid, the modern gar pike.

trout, perch, cod, and mackerel, were absent and do not appear until the Mesozoic.

Comparison of Devonian and Modern Fish—The teeth of Devonian and Carboniferous fishes were adapted for crushing and few had the sharp, rending teeth possessed by fish today. The fishes of the Devonian and Carboniferous were, as a class, massive and clumsy as compared with those of the present, and their bodies were probably less flexible. In Devonian fishes the backbone runs through to the end of the tail, and the fin is formed by vertical rays extending from above and below (below, *D*). In some, the resulting fin is symmetrical, but in others, as in the modern shark, it is unsymmetrical, the backbone turning upwards with an unequal lobe formed of rays on the under side. The tails of all Devonian fishes were either symmetrical or unsymmetrical (below, *C, D*), but none had the homocercal tail (below, *E*) of modern bony fish (Teleosts), in which the backbone ends in a broad plate from which diverging rays

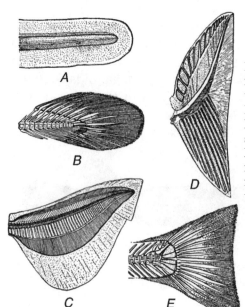

Evolution of the tail fin of fishes. *A*, embryonic tail fin; *C*, uneven-lobed tail fin of the Port Jackson shark; *D*, uneven-lobed tail fin of a Devonian shark, *Cladoselache*; *B*, even-lobed tail fin of the fringe-finned ganoid; and *E*, the tail fin of a modern *Teleost*, such as the trout.

spread to form a symmetrical tail of a type very different from that of Devonian fish.

It is interesting in this connection to note that the heavily armored fish were the first to become extinct. They were admirably suited for certain conditions, being protected from their enemies by heavy armor, but when the environment and food changed, their very weight and size were of disadvantage, and they failed to survive.

Why the Vertebrate Type was "Fit"—The reason for the establishment of the vertebrate type of animals and their rapid rise when once they appeared is evident when their structure is considered. An internal skeleton offers an excellent attachment for muscles, and at the same time permits a great flexibility of the body. As flexibility is necessary for rapid movement through the water, such animals as possessed it were better able both to escape their enemies and to secure their prey. Moreover, it permitted of greater size than as a whole appears to have been possible in other classes. The position and arrangement of the nervous system appears also to have been especially advantageous.

Plants

In the Devonian, for the first time in the history of the world, land plants are known to have been abundant. Discoveries of Silurian ferns and club mosses have been announced, but they are still open to doubt. The plants of the Devonian were of about the same general level of organization as some of those of the present day, although very different in appearance; changes have occurred, as will be pointed out from time to time, but the plants of this early period were, nevertheless, so highly developed as to prove an enormous antiquity.

At this time horsetails, ferns, club mosses, gymnosperms (of which the cypress, yow, and pines are members), and the extinct sphenophylls and seed ferns (pteridosperms) are known to have existed. The discussion of the Devonian flora will be taken up in the description of the Carboniferous, since in this later period it reached the climax of its development.

Summary

Migration and Evolution—The faunas of the various stages of the Devonian often differ so widely from one another as to suggest that at certain times evolution proceeded at an unusually rapid rate. The difference in the faunas of succeeding stages is due to several causes. At the beginning of the period there were a number of embayments so isolated that the evolution of the faunas of each proceeded independently, until each possessed many characteristic and peculiar species. As the seas spread over the land later in the period these embayments were, one after another, joined together, and as quickly as a waterway opened species from each embayment spread to the others, and a struggle for existence resulted which produced rapid and marked changes in the life, exterminating many species. The conflict thus brought about also caused the rapid rise of new forms not found in any of the original faunas.

The changes in the physical conditions were another cause of rapid evolution. As a result of the extension of the epicontinental seas, new food was doubtless introduced and currents were developed which may have brought about changes in temperature.

It should not be forgotten in this connection, however, that the differences in the faunas of beds of nearly the same age may be due entirely to the fact that one bed was deposited, for example, in shallow water and consequently had a shallow water fauna, and another in deep water and had a deep water fauna. The life of two such beds may, consequently, differ more widely than those of very different ages which were deposited under similar conditions.

Climate and Duration—Little more can be said of the climate of the Devonian than of the Cambrian and Ordovician, and the evidence, as in the latter, points to a uniformly warm climate over the entire world. In certain places, deserts existed as now, while in others extensive swamps were present.

The period was probably little more than half as long as the Ordovician.

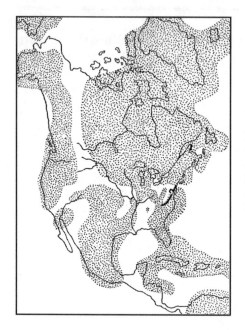

Map showing the probable outline of North America during a portion of the Lower Mississippian. Continental deposits are shown in solid black.

Map showing the probable outline of North America during a portion of the Upper Mississippian. The areas in black show where continental deposits were laid down.

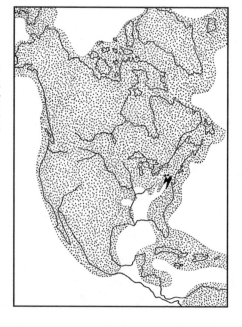

CHAPTER 15

The Carboniferous Periods

The Carboniferous formerly included the Lower Carboniferous (Mississippian), the Upper Carboniferous (Pennsylvanian), and the Permian. American geologists have been led to the conclusion that each of these three subdivisions is of a rank equal to that of the Ordovician, Silurian, or Devonian, and should be called a period. In this study, it seems advisable to discuss the life of the three periods together (the Lower and Upper Carboniferous, and Permian) since by so doing the sequence of life changes can best be followed.

Mississippian or Lower Carboniferous

The epicontinental seas (pg. 272, top) of the early portion of this period were about as extensive as in the Devonian and occupied much the same regions. As a result, over large areas the transition between the Devonian and Mississippian is not indicated by abrupt changes. Towards the close of the period (pg. 272, bottom), the seas again became constricted.

The sediments brought into the Appalachian trough from the continent of Appalachia were for the most part coarse sands and muds. Sun cracks, ripple marks, the footprints of amphibians, and other evidences indicate an arid or semi-arid climate, and that the sediments (Pocono and Mauch Chunk) were portions of a great delta or alluvial plain built by shifting streams which flowed over it. The Mississippian conglomerates are important mountain makers in the Appalachians.

In the central and western states the Mississippian sediments are finer than in the East, and limestone becomes increasingly abundant until, west of Ohio, it constitutes the greater mass of the sediments.

The presence of gypsum and salt in portions of the strata of this age in Michigan shows that the climate, for a time at least, was dry, and that the sea or seas of this region were isolated. The Mississippian gypsum of Nova Scotia implies a similar condition for that region, and, as has been seen, an arid or semiarid climate was present over the lands contiguous to the Appalachian trough.

The Mississippian seas spread over a large area of the Cordilleras of the West, from Mexico to the Arctic, the strata of this period being several thousand feet thick in certain places.

Close of the Mississippian—The extensive seas of the early Mississippian were gradually drained, so that before the close of the period the eastern portion of North America was land. In the west the seas also seem to have been withdrawn.

Other Continents—Mississippian seas spread over a large area in England, Ireland, and Europe, and the sediments which had been accumulating in them were locally folded at the close of the period. The term Paleozoic Alps which has been applied to this folded region assumes that the present folded areas are the "roots of former mountain ranges." Whether erosion kept pace with elevation or not cannot be stated. Strata which are believed to be of this age occur in northern and southern Africa, in western and central Asia and China, in Australia and New Zealand, and in Argentina and Chile. Coal occurs in the strata of this age in China, Russia, England, and elsewhere.

Pennsylvanian or Upper Carboniferous

The Pennsylvanian system is generally separated from the Mississippian by an unconformity, the Mississippian strata in some parts of the central United States having been gently folded, faulted, and eroded before the deposition of the Pennsylvanian sediments. A few seas persisted in Utah and Arizona in which sedimentation continued throughout the period without interruption, but such areas are rare. During the emergent condition of the continent the surface rocks were weathered, leaving a residual layer of insoluble quartz and clay. In the east this easily removable material was carried into a long, narrow sea, formed by the down-warping of the eastern part of the old Appalachian trough, where it was worked over and sorted by the seas to form the conglomerate (Pottsville) of the basal Pennsylvanian. As this trough was weighted down by the sediments carried into it by streams, it sank intermittently. For long intervals this area was slightly above the sea level, and the

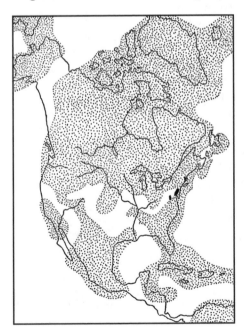

Map showing the probable outline of North America during a portion of the Upper Pennsylvanian. Continental deposits are shown in solid black.

sediments which then accumulated were continental and not marine; at other times the sea encroached on the land and formed immense, shallow seas which, upon being further shallowed by sediment from the land, formed vast, fresh, and brackish water swamps in which were accumulated the great coal beds of the Carboniferous. When the sinking kept pace with the accumulation of the vegetable matter, for many years, deposits of peat 100 or more feet in thickness were sometimes accumulated. These, when compressed to coal, formed workable coal beds. The accumulation of coal began in the Lower Pennsylvanian (Pottsville), but it was in the upper half of the period that its formation took place on a large scale.

While coal was accumulating in large quantities (never perhaps more than two percent of the total thickness of the deposit [right] in any one place) in Pennsylvania, West Virginia, Ohio, Tennessee, Illinois, and Iowa, marine conditions prevailed in the west and southwest, and limestones and shales were deposited with no coal. The red Pennsylvanian sandstone of South Dakota and the red conglomerate of Colorado were probably deposited on land by streams in an arid climate. Marine sediments to a depth of several thousand feet were deposited over the site of the Sierra Nevada Mountains.

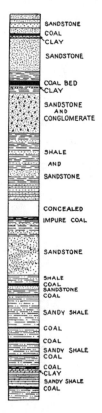

Section of coal-bearing strata in Pennsylvania, showing the relative amount of coal and barren rock in a rich field.

The thickness of the Pennsylvanian system varies from 4000 to 5000 feet in the Appalachians, to 1000 feet in Kansas and Nebraska. In Texas it is said to be 5000 feet thick and in Nevada about 10,000 feet thick. The probable distribution of the seas and swamps of portions of the Pennsylvanian is shown in the accompanying map (pg. 275).

Coal Fields of North America

Productive Coal Fields—(1) *Eastern Canadian and New England Fields*—Important coal deposits occur in Nova Scotia and New Brunswick on both sides of the Bay of Fundy. Metamorphic coal is also found in Rhode Island, but it is so graphitic as to be of little value at present.

(2) *Appalachian Field*—The great coal field (below) of the world is that which underlies an area of about 50,000 square miles in central and western Pennsylvania, western Maryland and Virginia, West Virginia, and eastern Ohio, Kentucky, and Tennessee. In this should be included the anthracite field,

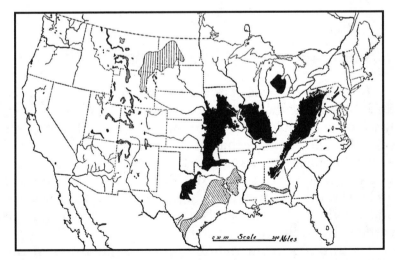

Map showing the distribution and extent of the Carboniferous coal fields (black), and more recent coal fields (lines).

confined to an area of 484 square miles in eastern Pennsylvania.

(3) *Michigan Coal Field*—This field covers an area of only 11,000 square miles and was probably formed in an isolated basin. It is not of great value as compared with the Appalachian field, since it is deeply buried and the coal beds are usually comparatively thin.

(4) *The Indiana-Illinois Field* covers an area of about 58,000 square miles, of which 30,000 square miles are underlain by workable coal.

(5) *The Iowa-Missouri-Texas Field* extends from northern Iowa to central Texas and covers an area of about 94,000 square miles, being about 800 miles from north to south. The Indiana-Illinois and Iowa-Texas fields were probably once continuous, but are now separated by the Mississippi Valley.

The Pennsylvanian strata dip beneath much younger strata when traced westward. In the mountains of the Great Basin region, they are found to consist of marine deposits and contain no coal. The coal fields of Wyoming and Colorado are of a later date.

Summary of the Pennsylvanian

Iron and Oil—Beds of iron ore occur associated with coal. Such beds are sometimes continuous, but the ore is often in the form of nodules. The origin of these beds of iron ore is probably the same as that of the "bog iron ore" which is accumulating in the swamps and lakes of the present. Surface waters containing carbon dioxide dissolved iron from the soil and rocks; the dissolved mineral was then carried to swamps and lakes by the streams, and there precipitated, either as iron carbonate or iron hydroxide.

Oil and gas occur in some of the sandstones of the Pennsylvanian system in Illinois, Kansas, and Oklahoma.

Duration—The exact length of the Pennsylvanian Period is as doubtful as that of the preceding periods, but is usually stated as being about 2,000,000 years. In estimating the duration of former periods it has been necessary to depend upon the rate of sedimentation, but in the Pennsylvanian an additional basis is afforded by the coal. This measure is, however, inaccurate, since the rate of accumulation is not definitely known. The aggregate thickness of the coal in a single section of the Carboniferous is often 150 feet, and sections are known where the total thickness of the coal beds is 250 feet. If the vigorous vegetation of a fertile region in North America today were accumulated for 1000 years without loss and compressed to the density of coal, it would form a layer only seven inches thick, but since in the making of coal it is probable that four fifths of the vegetation disappears as carbon dioxide (CO_2), methane (CH_4), and other gases, the rate of accumulation would be only one and one half inches in 1000 years. It is readily seen, therefore, that 1,000,000 or 2,000,000 years may have been required for the accumulation of the Pennsylvanian coal.

Other Continents—The Pennsylvanian was the greatest coal-producing period of the world. Workable beds occur in Great Britain and Ireland and in all of the principal countries of Europe except Norway, Sweden, Denmark, and Italy. The Pennsylvanian strata in China, Asia Minor, and eastern Siberia contain coal beds, many of which are of great value; in China, especially, coal beds of great thickness and excellent quality have been reported. The Carboniferous strata of Africa, Australia, and South America are seldom coal bearing, but in some areas valuable deposits occur and much coal is mined.

Permian

In North America, the Permian is a continuation of the Pennsylvanian and is a period in which the far-reaching seas of the latter were withdrawn. Where the two systems occur in the same section in North America they are almost always conformable. It is, however, more important as the transition period between the Paleozoic and the Mesozoic. In the eastern United States, the comparatively small areas of Permian rocks are separated from the underlying Pennsylvanian on the basis of their plant remains, which are more closely related to European Permian plants than to those of the underlying Pennsylvanian. They consist of about 1000 feet of sandstone, shale and limestone, and a few beds of coal.

In Nova Scotia, New Brunswick, and Prince Edward Island, strata composed of red shale and sandstone are believed to have been deposited in an inclosed basin during the Permian.

During the early portion of the period, a shallow sea extended from the Gulf of Mexico through Texas into Kansas and Nebraska, as is shown by the presence of marine fossils in the rocks. Later in the period the sea withdrew, leaving a great region dotted here and there with salt lakes which left beds of gypsum and salt, upon drying. The aridity of the climate of this area is shown not only by the presence of the salt and gypsum, but by the sun-cracked and ripple-marked red sandstones and soft red shales or "red beds." It was a region not unlike the Great Basin of Utah of today. These desert conditions continued into the Triassic, and it is, consequently, difficult and in many cases impossible to determine the dividing line between the two systems. Portions of the Pacific border were covered with seas in which marine life abounded.

Before the close of the period, the epicontinental seas had, with one or two exceptions, withdrawn from the continent.

Permian Glaciation—One of the surprising features of the Permian is the evidence of widespread glaciation during the period. The location of the glaciated areas is also remarkable. They occur on both sides of the equator and within 18 to 21 degrees of it; that is, they extend slightly within the torrid zone. The limits of these ancient glaciers are not definitely known, since the evidence has been largely obliterated, but the proof at hand implies an area greater than that during the "Great Ice Age."

The proof of this ancient glaciation is conclusive and consists of boulder clay, often containing smoothed and striated boulders which in places rest upon a striated and polished pavement of older rocks.

The glaciated areas were not in the polar regions nor, in many places, at a high altitude, as is shown by the relation of the glacial deposits to strata containing marine fossils. In Australia, for example, the glacial formations are interbedded with marine sediments. Moreover, coal beds occur in the formation.

In particular, the Permian glacial deposits occur in the following countries. In India ancient boulder clay rests, in places, directly upon a striated, roche-moutonnée surface, and some of the glaciated areas extend nearly to sea level. This is remarkable when taken in connection with the fact that the glaciated area is within the tropics. In South Africa, the boulders of the glacial deposit are often striated and rest on a striated rock pavement. In Australia, several boulder beds point to repeated advances of the ice or to several stages of glaciation. Thin glacial moraines of Lower Permian age resting upon a glaciated surface of Upper Carboniferous rocks have been discovered in Germany, and Permian boulder beds in England have been interpreted as of glacial origin. Boulder clay of glacial origin occurs in Permian strata in Brazil and Argentina, and conglomerates near Boston, Massachusetts, believed to be in part of glacial origin, are thought to be either of Pennsylvanian or Permian age.

The exact age of the glacial deposits of India, South Africa, and Australia is somewhat in doubt. They are usually called Permo-Carboniferous and occurred either at the close of the Pennsylvanian or in the Lower Permian. The development of annual rings in Upper Permian trees of certain regions has given rise to the belief in warm summers and cold winters for at least a few thousand years near the close of the period.

Permian Deserts—Over large areas of the Earth's surface deserts existed in the Lower Permian, as the ripple-marked and sun-cracked red sandstones and shale and the interbedded salt and gypsum testify. Central and western Europe, England, and western North America are known to have been so affected.

Igneous Activity—Numerous volcanoes broke out in England during the period, and the rocks were broken by earthquake shocks, as is shown by earthquake fissures filled with what appears to be Permian sandstone.

Appalachian Deformation—In the discussion of the geography of the various periods of the Paleozoic, attention has been called repeatedly (1) to the continent of Appalachia, a broad upland, sometimes high and sometimes low, and (2) to the Appalachian trough west of it in which much of its waste was poured. Two points have been emphasized: (1) that the trough or geosyncline sank as it was weighted with sediments, and (2) that, as Appalachia was worn down by the streams, a compensating rise took place. With the exception of comparatively short periods of emergence, sediments were accumulating in the Appalachian trough from the beginning of the Cambrian until the Permian, during which time more than 25,000 feet of sediments were laid down. One of the most important upward movements of the trough occurred near the close of the Ordovician, apparently at about the time the Taconic deformation was taking place. Others occurred in the Silurian and between the Mississippian and Pennsylvanian periods.

With these and other minor exceptions the great Appalachian trough was the site of deposition during the long periods of the Paleozoic, and a thickness of more than five miles of sediment accumulated.

Towards the close of the Carboniferous, the most striking event in the geological history of eastern North America was consummated. At this time the sediments of the Appalachian trough yielded to the strain that had long been accumulating and folded into a great mountain system, the axes of the folds extending in a northeast-southwest direction, one range reaching from Nova Scotia to Rhode Island, another from New York to Alabama, and a third in Arkansas forming the Ouachita Mountains.

The probable cause of the yielding of this particular portion of the crust to lateral pressure was the fact that the geosyncline was a zone of weakness just as the bend in a crooked stick determines the point at which it will break when pressure is applied at the ends. The rocks in all portions of the trough were not equally deformed: those in Pennsylvania and West Virginia have been, for the most part, compressed into gentle folds, while those in the southern Appalachians in Tennessee and elsewhere were broken by so many thrust faults that the reconstruction of the region is often difficult. The intensity of the folding diminished from east to west. In eastern Pennsylvania, for example, the folds are more compressed and faults are more common than in the central part of the state, while in the western portion, the rocks were little disturbed and are almost horizontal. The greater deformation on the eastern side of the trough is also seen in the character of the coal in eastern and western Pennsylvania. In the former, it is metamorphosed to anthracite, while in the latter it is bituminous. The effect of lateral pressure on competent strata, such as quartzites and limestones, and on incompetent, such as shales, is well shown. Where the former were thick, the strata were either thrown into great folds or, when broken, were thrust over the adjacent rocks. The effect on shales is in marked contrast, for they were crumpled into minute folds and crushed.

Not only were the sediments of the Appalachian trough folded, but the rocks of the continent of Appalachia were also deformed as they had indeed been a number of times before. As a result, those portions of this old land which are at present exposed at the surface are extremely complex.

Although the Permian was the period during which the principal folding occurred, some deformation had previously taken place. In the Ordovician folding occurred, and in the Middle Devonian mountains were formed in Maine, New Brunswick, and Nova Scotia.

Age of the Deformation—The time at which the Appalachian deformation took place is known from the usual evidence. The youngest rocks which are infolded are Pennsylvanian. Upon the upturned edges of these the Triassic rocks rest unconformably in certain places. Since the Permian strata are absent in eastern Pennsylvania, it is probable that the deformation took place during the latter and that it continued into the Triassic, since the oldest rocks of that period seem to be everywhere lacking.

Other Continents—The western half of Europe was part of a large continent that extended from Russia far into the Atlantic. In the southern part of this continent, lakes and swampy depressions existed in the Lower Permian, in which rank vegetation grew and large and small amphibians and primitive reptiles dwelt. These swampy areas were drained by an elevation in the Upper Permian which converted them into broad plains separated by hills and mountains. The climatic effect of these changes was marked, some portions of the region becoming very humid and others, from which the moist winds were shut off by the mountains, becoming arid. Volcanic activity was prevalent during a portion of the period. The epoch of elevation was followed by one of subsidence and later by disturbances which cut off a great lake, like the Caspian Sea, and salt and gypsum were deposited as it dried up. During the Upper Permian, the thickest known salt deposits were accumulated, one of which, near Berlin, has been penetrated 4000 feet.

Permian strata cover large areas in southern Asia, in Australia, in southern Africa, and in South America, the unique features of which are the extensive glacial deposits.

Invertebrates of the Carboniferous

Protozoans—During the Carboniferous, for the first time in the Paleozoic, Foraminifera became abundant and varied. Certain genera built up limestone deposits, occasionally of considerable thickness. One of the most characteristic forms of the Mississippian (Fusulina) was like a grain of wheat in form and size (pg. 288, top, *H, I*).

Coelenterates and Echinoderms—*Cup* (below, *A-C*) and *honeycomb corals* continued to be important in the Carboniferous, and contributed largely to the formation of thick limestone strata. *Blastoids* (pg. 286, *F, G*) were so abundant in the Mississippian that some beds are largely made up of them, but their extinction was reached before the end of the Pennsylvanian. Where favorable conditions existed, *crinoids*

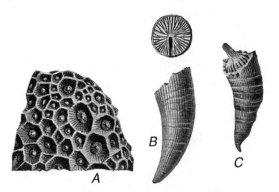

Carboniferous corals: *A, Lithostrotion canadense* (Mississippian); *B, Hapsiphyllum calcareforme* (Mississippian); *C, Lophophyllum profundum* (Pennsylvanian).

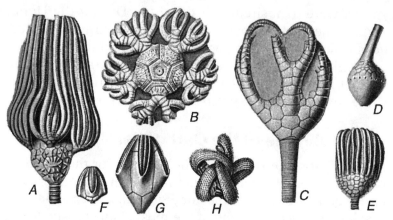

Mississippian echinoderms. Crinoids: *A, Actinocrinus multiradiatus; B, Platycrinus discoideus; C, Onychocrinus exsculptus; D, Batocrinus (Dizygocrinus) rotundus* (with arms removed); *E,* the same as *D* with arms. Blastoids: *F, Pentremites robustus; G, Pentremites pyriformis.* Brittle Stars: *H, Onychaster flexilis.*

(above, *A-E*) were unusually abundant; especially was this true in the Mississippian. Sea *urchins* (echinoids) were more abundant and larger than ever before, but were subordinate to the crinoids in numbers.

Molluscoids—Bryozoans lived in considerable numbers. Among many less striking forms was one genus with a peculiar habit of growth about an axis which gave it a screwlike shape, hence the name Archimedes (right). The Carboniferous was rich in *brachiopods* (pg. 287, *A-M*), but before its close the leading Paleozoic genera had

Carboniferous bryozoan: Archimedes wortheni (Mississippian).

disappeared. One characteristic genus (Productus) (pg. 287, *A, G*) had one large, convex, spinose valve and one concave one.

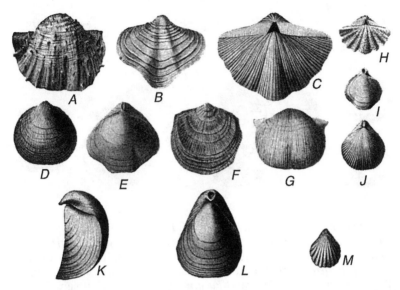

Carboniferous brachiopods: *A, Pruductus costatus* (Pennsylvanian); *B, Athyris lamellosa* (Mississippian); *C, Spirifer cameratus* (Pennsylvanian); *D, Rhipidomella burlingtonensis* (Mississippian); *E, Seminula argentea* (Pennsylvanian); *F, Derbya crassa* (Pennsylvanian); *G, Productus burlingtonensis* (Mississippian); *H, Spiriferina spinosa* (Pennsylvanian); *I, Seminula subquadrata* (Mississippian); *J, Eumetria marcyi* (Mississippian); *K, L, Dielasma bovidens* (Pennsylvanian); *M, Hustedia mormoni* (Pennsylvanian).

Mollusks—*Gastropods* (pg. 288, top, *A-G*) and *pelecypods* (pg. 288, bottom, *A-F*) continued much as in the Devonian. The *cephalopods* (pg. 289, *A-E*), on the other hand, showed considerable advance in the complexity of the suture lines. The angle-sutured goniatites were common, and, before the close of the Permian, ammonites with their complex sutures appeared. Some of the straight, simple orthoceratites continued throughout the Paleozoic into the Triassic.

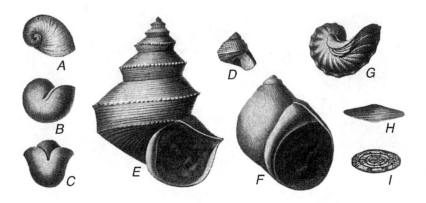

Carboniferous gastropods: *A, Plastostoma broadheadi* (Mississippian). *B, C,* two views of *Bellerophon subloeris* (Mississippian); *D, Pleurotomaria nodulostriata* (Mississippian); *E, Worthenia tabulata* (Pennsylvanian); *F, Naticopsis altonensis* (Pennsylvanian); *G, Bellerophon percarinatus* (Pennsylvanian). Foraminifera: *H, Fusulina secalica* (Mississippian); *I,* section of *Fusulina,* showing structure.

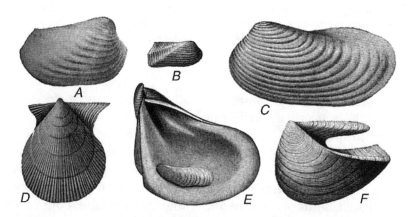

Carboniferous pelecypods: *A, Grammysia hannibalensis* (Mississippian); *B, Pleurophorus tropidophorus* (Pennsylvanian); *C, Allorisma terminale* (Pennsylvanian); *D, Aviculopecten occidentalis* (Pennsylvanian); *E, Myalina recurvirostris* (Pennsylvanian); *F, Monopteria longispina* (Pennsylvanian).

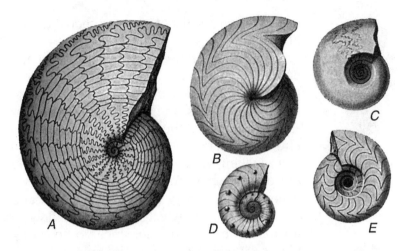

Carboniferous cephalopods: *A, Medlicottia copei; B, Aganides rotatorius; C, Waagenoceras cumminsi; D, Temnocheilus forbesianus; E, Muensteroceras oweni.*

Arthropods—*Trilobites* (pg. 290, top) and *eurypterids* continued into the Carboniferous, but became extinct at its close.

Insects—The earliest insects of which any fossils have yet been found lived in the Carboniferous (pg. 290, bottom, *A, B*), and from that period about 1000 species have been described. They appear to have been more generalized than those of subsequent periods; that is, they were simple in structure and united characteristic features of two or more distinct groups, and were therefore probably the ancestors of those insects whose characters they combine. One extinct order (*Paleodictyoptera*, old, netted wing) is especially interesting because it is believed to be the stock from which all insects were descended. Carboniferous insect groups seem widely different when the external form only is considered, but a more careful study shows that the differentiation had little depth. The wings were all membranous, none having yet been developed for protective covering, as in the beetle. The number of wings in

Carboniferous
trilobite:
Phillipsia major.

every case was four, none having been dropped at that time.

Two groups were especially prominent, the cockroaches (below, *B*) of which there were large numbers, some being as large as a finger, and many species; and the dragonflies (below, *A*), which reached the great size of almost two and a half feet across the wings. The absence of all sucking insects, such as bees, wasps, and butterflies, whose food consists of the nectar of flowers procured by a sucking apparatus, and in fact of all insects which depend upon flowers for food is not surprising, since no flowering plants were in existence at this time.

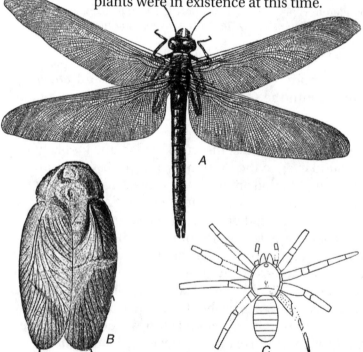

Carboniferous insects: dragonfly, *A, Meganeura monyi* (some of these were two and a half feet across the wings); *B, Adeloblatta columbiana; C,* spider, *Arthrolycosa antiqua.*

Vertebrates of the Carboniferous

Fishes—During the Mississippian, sharks existed in an abundance which has not been equaled before or since, as is shown by the large number of species, of which nearly 300 have been described. Before the close of the Carboniferous, however, the number had become very small, only about 20 species being known in the Permian. As in the Devonian, the sharks are known from their fin spines and teeth (right), the latter being of the crushing type. The sharks were, for the most part, small, seldom being more than five feet long, while none attained the size of their modern relatives.

A back spine, tooth, and scale of a Mississipian shark.

Only sharks and typical ganoids (Actinopteri) were important, the armored lungfish and fringe-finned ganoids having reached their climax in the Devonian.

Amphibians—Next to the appearance of the fishes the most important event in the history of vertebrates was the rise of amphibians, of which the salamanders, newts, and frogs are modern representatives, since the ancestors of reptiles, of mammals, and even of man himself are to be found among them. Amphibians resemble reptiles, but differ from them in the fact that in the earlier period of life, they breathe in water by means of gills, like fishes; and it is only in the later period of life that they breathe air by means of lungs, like reptiles. The most important difference between fishes and amphibians is in the organs of locomotion, fishes having fins and amphibians legs, in the adult stage.

It is hard to point out briefly all the differences between amphibians and reptiles, and it is indeed sometimes extremely difficult, if not impossible, to tell from the skeleton alone of extinct forms to which class a specimen belongs. One important difference, however, is to be found in the articulation of the skull to the backbone, which in amphibians is by means of two knuckle-like projections of bone (condyles) and in reptiles and birds by one. In fishes there is no movable articulation.

All Carboniferous amphibians belonged to the Stegocephali (below) (Greek, *stege*, roof, and *cephale*, head), in which the skull was covered with bony plates, and the teeth were conical with walls that were sometimes highly infolded (pg. 293, top). The limbs of most were weak and adapted more for crawling than for carrying the body well above the ground.

Carboniferous amphibians varied greatly in size and shape, some attaining a length of almost eight feet. One characteristic genus, Eryops (pg. 293, bottom, and below), had a rather large,

A Permo-Carboniferous landscape. The characteristic vegetation; two figures of the amphibian *Eryops* upon the land, each about seven feet long, are shown; a reptile (*Limnoscelis*) in the water; and a gigantic dragonfly in the air.

broad, flat head, and unevenly spaced, conical (labyrinthine) teeth, no neck, and a thickset body with broad, five-toed feet that were probably webbed. The tail was flattened vertically. Even with its roofed skull, Eryops would not look unlike an overgrown modern Japanese giant salamander, since the bones of the skull were covered with skin. It was able to crawl clumsily and slowly over the land, but must have been far more at home in the water. That this clumsy, small-brained beast should be

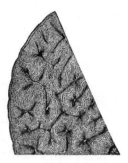

Cross section of the tooth of a *Stegocephalian*. Note the complicated labyrinthine structure.

one of the highest types of living beings of its time may help us to realize how remote the period was, and to what an extent vertebrate life has been evolved since then.

Some of the Carboniferous Stegocephali were armored and some had no protection; some had skulls nearly two feet long, while the skulls of others were not larger than one's thumbnail; some had stout limbs, while the limbs of others were atrophied and the body elongated and snakelike.

The most common amphibian of the Carboniferous, Branchiosaurus (pg. 294), did not differ greatly in appearance and habits from those today. The teeth were small and conical,

Top view of the skeleton of *Eryops*, a Permian *Stegocephalian*. The head is covered with the thick, bony plates characteristic of the order.

the eyes were protected by a movable ring of bony plates, and the lower surface of the body was covered with thin scales. The presence of gills in immature specimens shows that they lived in the water at least a portion of their life.

Certain Permian amphibians (Lysorphus) resembled a modern salamander (Amphiuma) in size, shape, and habits so strongly that it seems actually to have been related to it. So abundant are the skeletons of these amphibians in certain localities in the Permian strata of Texas that hundreds have been found embedded in nodules.

Amphibians are known from their footprints to have lived in the Mississippian.

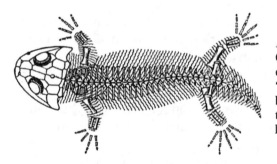

Branchiosaurus, a small Carboniferous amphibian occurring abundantly. The "roofed" head, the rings of bone about the eyes, and the scaly covering of the lower side are shown.

Origin of Amphibians—Several lines of evidence show that amphibians may have been descended frorn the fringe-finned ganoids (crossopterygians): (1) the teeth of both are often labyrinthine (pg. 293, bottom); (2) the bones of their skulls are similar in position and arrangement; (3) some primitive amphibians have rings of bony plates (sclerotic plates) about the eyes; (4) the structure of the fin of the fringe-finned ganoids was such that a leg might have been formed from it by modification.

Rise of Amphibians—The rise of amphibians was a momentous step in the evolution of life, but it was not a surprising one. Fishes were the predominant race of the Devonian, and it was to be expected that the structure of some of them would in time become modified to take advantage of the realm of the lands where food was either more easily obtainable or of a more nutritious quality than in the seas, or where the competition was less keen. The extensive swamps of the Devonian and Carboniferous and the shiftings of the seas have been assigned as the immediate causes of the rise of the amphibians from fishes. It seems more probable, however, that during every period of the Paleozoic, swamps and shallow water were present, in which amphibians would have been evolved had fishes been present which were so constructed that by slight modifications they could become adapted to land conditions. How this was accomplished is shown in the development of the individual amphibian, which in the tadpole stage is physiologically a fish, but which later breathes by means of a simple sack-like lung instead of gills. It is important to note that the Carboniferous amphibians, more than their modern relatives, possessed characters closely allying them to the fishes, and that their fishlike characters were more like those of Devonian than of modern fishes.

Some doubtful amphibian tracks have been found in the Devonian, but no bones earlier than the Carboniferous have been discovered. However, the well-developed limbs of the earliest species indicate a line of ancestors that lived in the Devonian. Amphibians attained their greatest importance in the Carboniferous (Pennsylvanian and Permian) and have taken a very subordinate place since the Triassic. Even before the close of the Pennsylvanian they had begun to give place to the reptiles.

Reptiles—The reptiles of the Permian were even more varied than the amphibians and have been placed in three groups. The first group (cotylosaurs) includes the most primitive reptiles known, its members differing from other reptiles, among other particulars, in having a roofed-over skull. All the members of this class that are known had very short necks; short, stout limbs; rather heavy bodies, and usually long tails (pg. 291). Some had long, sharp, curved teeth (Labidosaurus,

The head of a Permian reptile, *Labidosaurus.*

left) arranged in two or more rows, which were probably used for prodding in the mud for soft-bodied invertebrates, and for crushing; some had conical teeth in front and crushing teeth behind; and some had a mouth filled everywhere on jaws and palate with short, stumpy teeth, suitable only for crushing shellfish; while others had slender teeth indicating insectivorous habits. In some, the claws terminated in flattened nails, while in others the claws were sharp and curved. The habits of this group varied somewhat, some being more terrestrial than others, but all probably lived in swampy places and about lagoons.

A second group of Carboniferous reptiles (pelycosaurs) had lighter skulls than those of the first group; larger necks; longer, better-formed legs and feet, and usually longer tails. The best known of these (Dimetrodon, below) was about 10 feet in length and was especially characterized by a finlike crest on the back formed by spines of its vertebrae. The skull had strong, sharp, carnivorous teeth, indicating that the creature was a fierce, predaceous animal. The use of the crest on the back is unknown, and aside from its presence the animal was very primitive in structure. Some members of this group (Varanosaurus, pg. 297) were swift-running reptiles, living in the forests, hiding under logs, and feeding on the numerous cockroaches and other insects.

A third group, although not very conspicuous in the Permian, were of great importance both because of their great development in the

Dimetrodon, a spiny but primitive reptile from the Permian of Texas.

Triassic and also because they were the probable ancestors of the mammals. They were small reptiles with short but strong legs and large heads. The mammalian characters are to be seen in the skull, the teeth, and other parts of the skeleton. This important group is described more fully in the following chapter.

Varanops, a Permian reptile forty-four inches long, on a *Sigillaria* log.

Rise of Reptiles—The rise of reptiles from amphibians was a logical sequence. On account of their aquatic larval life, amphibians were restricted to the vicinity of the water, while reptiles, because of the development of a firm eggshell and the omission of the aquatic stage of the young, were able to populate the dry lands and thus to take advantage of many kinds of food. Amphibians may indeed be considered as transitional forms by which reptiles were evolved from fishes.

The rapid development of this class after it was once well-started was probably due to its higher organization, to the lack of competition in the new surroundings, and to the abundance of food. It has been suggested that the purification of the air as a result of the withdrawal of carbon dioxide in the formation of coal produced an atmosphere which was more favorable for the development of air-breathing animals, but the amount of carbon dioxide withdrawn does not seem to have been sufficient to have made a great difference.

Carboniferous Plants

The forests of the Carboniferous were very different in appearance from those of today. None of the trees common at present in the forests and swamps were in existence, flowering plants were conspicuous by their absence, and even grasses and mosses were lacking. The living relatives of the Carboniferous trees are for the most part lowly, inconspicuous plants.

The land plants of the Carboniferous belonged to six great groups: (1) ancestral ferns, (2) seed ferns (pteridosperms), (3) club mosses (lycopods), (4) an ancient extinct group, the sphenophylls, (5) the horsetails (Calamites), and (6) the Cordaites and possibly some conifers.

(1) **Ancestral Ferns and** (2) **Seed Ferns**—The shale overlying coal beds is often full of the fronds of fernlike plants (left), and

A Paleozoic fernlike plant, *Pecopteris.*

so perfect is the preservation of some of them that the finest venation is shown. The beauty of such fossils is especially striking when the shale is light in color, since under such conditions the delicate outline of the black fossil frond is brought out with great distinctness. It was formerly thought that these fernlike leaves were the remains of ferns, but a more careful study and further discoveries have shown that few are true ferns; on the contrary, the great majority belong to an extinct family, the seed ferns or pteridosperms.

The important difference between these families lies in the

reproduction, the true fern producing spores and the pteridosperms, seeds.

Ancestral Ferns—The most important family of Paleozoic ferns (Marattiaceae) has descendants living today, but the Paleozoic members of the family were tree ferns, reaching in some species a height of upwards of 60 feet. This great family has dwindled to a few genera which are now confined to the tropics; one of which, however, the elephant fern, sends up huge fronds to a height of 10 or 12 feet. In addition to the tree ferns there were doubtless low, herbaceous ferns, living under the same conditions as the ferns of today.

Seed Ferns (Pteridosperms)—The members of this extinct group, if living today, would probably be called ferns by the casual observer but a careful examination would show that instead of having small sporangia on the back of the fronds, they bore seeds, sometimes as large as hazelnuts, which were surrounded by a thick, fleshy outer coat. The pteridosperm group (left) is interesting as a connecting type, since it is a link between the ferns, on the one hand, and the cycads, which were the dominant plants in the Mesozoic, on the other. Whether it stands as a connecting link between the ferns and the great groups of higher plants, or whether it leads to the cycads and stops there, cannot as yet be affirmed.

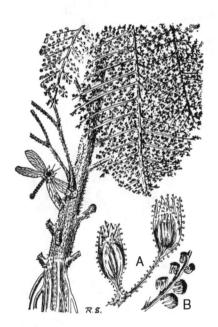

Lyginodendron. Restoration showing the stem, roots, and foliage; *A*, seeds; *B*, disks and pollen sacks.

(3) **Club Mosses (Lycopods)**—The conspicuous trees of the Carboniferous were gigantic club mosses, some of which grew to a height of more than 100 feet. One of these, the Lepidodendron (Greek *lepis*, scale, and *dendron*, tree) was freely branched (right) and had, consequently, somewhat the general outline of our forest trees, but with that the resemblance ended. The leaves were numerous and slender and were commonly arranged in oblique rows about the trunk and branches. When shed, their bases left diamond-shaped scars which gave a characteristic appearance to the bark (below). The largest Lepidodendron trunk yet described was found to be 114 feet long up to the point where it began to branch; the diameter of the base was about three feet, while at a height of 114 feet it was one foot, showing that it was a tree of slender proportions. Other species, however, were of a somewhat sturdier build.

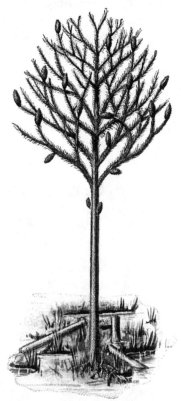

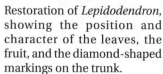

Restoration of *Lepidodendron*, showing the position and character of the leaves, the fruit, and the diamond-shaped markings on the trunk.

Impression of the bark of the *Lepidodendron*, showing the leaf bases and characteristic diamond-shaped scars.

Bark of *Sigillaria*, showing the vertical arrangement of the leaves and the fluted surface.

The Sigillaria (Latin, *sigillum*, seal) differed externally from the Lepidodendron, especially in two particulars; it branched sparingly (right) and the leaves were arranged in vertical rows (above), the leaf scars on the bark giving rise to the name "seal tree." In some species the leaves were a yard long. Sigillarian trunks almost as slender and high as those of the Lepidodendron have been found, but, as a rule, they were shorter and stouter, one specimen six feet in diameter and 18 feet high having been described.

Restoration of *Sigillaria*, showing the position and character of the leaves and the fruit, and the peculiar bark. Some of them grew to a height of a hundred feet, with a diameter of six feet.

The fruit of both the Lepidodendron and Sigillaria was in the form of well-defined cones that were usually borne at the ends of the smaller branches. In the clay underlying coal seams the "roots" or underground stems of the lycopods often occur and are called Stigmaria.

The trunks of the lycopods consisted of a hard, woody rind and a soft, cellular interior which quickly decayed. As a result of this structure the fossil trunks seldom show their original cylindrical form, but are usually flattened into thin sheets.

The great lycopods of the Carboniferous are now represented by the insignificant ground pine and Selaginella.

The coal beds of the Carboniferous are largely composed of Lepidodendron and Sigillaria remains. Some coal of this period (cannel), however, is made up chiefly of spores of Carboniferous plants.

(4) **Sphenophylls**—This extinct group is interesting because it suggests a common ancestor for the lycopods and horsetails (Equisetales). The plant had a slender, ribbed stem, seldom more than a quarter of an inch in diameter, which bore delicate, wedge-shaped leaves (right) attached in whorls to the stem by their ends. Sometimes the leaves were deeply cut, making them almost hairlike in appearance. These plants probably had a trailing habit, or perhaps supported themselves on stronger plants. Sphenophylls bore cones somewhat like those of the Calamites.

Stem and leaves of *Sphenophyllum*, a slender plant, the stem seldom exceeding two fifths of an inch in diameter. The sphenophyllums probably supported themselves by limbing.

(5) **Horsetails (Calamites)**—Except for their greater size, the horsetails of the Carboniferous had much the appearance of the lowly horsetail or scouring rush of today and were, indeed, related to it. The tree horsetails of the Paleozoic are called Calamites (pg. 303, left), from the most important genus of that time. They were trees which reached a height of 60 or more feet and were almost as conspicuous as their Lepidodendron and Sigillaria neighbors. Their habit of growth appears to have been similar on a glorified scale to that of some of their living relatives. Some were simple shafts, while others were probably gracefully

A   *B*

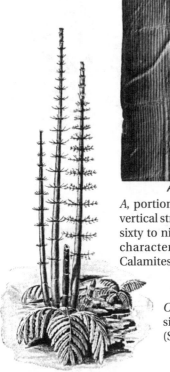

A, portion of trunk of a *Calamites*. The nodes and vertical striations are characteristic. Some grew to be sixty to ninety feet high. *B, Annularia*, showing the characteristic arrangement of the leaves of the Calamites.

Calamites. They seem to have had habits similar to those of the horsetails of today. (See above, *A* and *B*.)

branched trees with many boughs.

The bark (above, *A*) of the Calamites had characters which readily distinguished it from other trees of its time. The stems were ribbed, the ribs ending at each "node" and a new set continuing beyond the node to the next, when an alternate set again appeared. The leaves (above, *B*) were simple and were attached to the nodes in whorls. Conelike fruits were borne at the ends of the twigs and contained spores of one kind.

Calamites began in the Devonian and went out with the Paleozoic, but the horsetails of the Mesozoic were transitional, both in size and in structure, to the modern horsetails.

(6) **Cordaites and Other Gymnosperms**—The trees of this group were the most highly developed of the Paleozoic forests.

They were large trees (right) which sometimes grew to be 100 feet in height and were easily distinguished from the other trees of the time by the large, sword-shaped leaves which were borne in a crown on the top of the main trunk. Some specimens of Cordaites leaves exceed three feet in length and closely resemble the leaves of such plants as the lily and Indian corn, although not related to them. The fructifications, which were borne in a poorly developed cone, were of two kinds, male and female, the latter having a fleshy cover somewhat like a plum. Cordaites were on a level with the seed ferns as regards seeds, but in the structure of the wood and in other respects they were more highly organized. Cordaites became extinct before the close of the Paleozoic, unless the ginkgo or maidenhair tree is a descendant.

True conifers possibly appeared before the close of the Paleozoic, as the presence of the genus Walchia (right) in the Permian appears to show.

Conditions under which the Coal Plants Grew—The most abundant plants of the Carboniferous swamps, the Calamites, Lepidodendron, and

Cordaites. Large trees with long, narrow leaves sometimes a yard in length. They are allied to the conifers as well as to other orders of plants.

Stem and leaves of *Walchia*, a characteristic Permian conifer.

Sigillaria, had narrow leaves with a small surface exposed to the sun. At the present time plants with leaves of this character grow in the bright sunlight, while the leaves of shade plants are large, the greater size being necessary in order that they may be acted upon by a larger amount of light. Some of the coal plants, such as the true ferns and seed ferns, had fairly large leaves, but they were not of unusual size, indicating that they were only partially shaded by the small-leaved Calamites and Sigillaria. From this evidence it has been held that the Carboniferous plants did not live in a misty atmosphere through which the sun's rays penetrated with difficulty, but in one which was not unlike that of the present.

The character of the foliage of the coal-making plants may, however, have been an adaptation to conditions which more than counterbalanced the effect of bright sunlight. Their roots were those of water plants, and their leaves were not only narrow but were supplied with various devices for preventing the loss of water by rapid transpiration.

The evidence at hand points to the existence of extensive swamp areas which slowly sank as the half-decayed vegetation accumulated on them, and which were so near sea level that a slight sinking killed the vegetation growing there and buried them under sand, clay, or lime ooze. It is probable, therefore, that the coal plants (Calamites, Lepidodendron, Sigillaria) of the Carboniferous lived not only in fresh but even grew out in the brackish water of the shallow interior seas.

Coal

Coal occurs in very thin beds in the Devonian and in thicker beds in the Mississippian, but it is in the Pennsylvanian or Coal Measures that it occurs for the first time in beds or *seams* thick enough to be of commercial value. The thickness and purity of the coal beds of this period are such as to make it the most important of all coal-bearing systems.

Mode of Occurrence—The total thickness of the Pennsylvanian or Coal Measures is 4000 to 5000 feet in the Appalachian Mountains and l8,000 feet in Arkansas, but of this great accumulation of sediment seldom more than two percent is coal, the remainder being sandstone, shale, limestone, and iron ore. In the section shown on pg. 276, it is apparent that there is no regular order of succession of the beds, except that often a bed of fire clay immediately underlies a coal seam. It is also usual to find shale immediately overlying the coal, although this does not invariably happen. In different portions of the same field the same order is usually found, but in separate basins the order may vary greatly and is probably never the same in all particulars.

Origin of Coal—Coal is of vegetable origin, as is proved (1) by a microscopic examination which, even in dense anthracite, shows the cellular structure of plant tissue, and (2) by stumps of trees with their roots penetrating the underclay which sometimes underlies the coal seam. In South Wales, for example, there are 100 coal seams in which such stumps are embedded. In Nova Scotia, of 76 coal seams 20 have upright stumps with spreading roots penetrating the clay; in the United States few such occurrences are known. (3) Leaves are often beautifully preserved in the shale immediately overlying the coal. (4) Fire clay often, although not invariably, underlies coal beds. The character which a fire clay possesses of withstanding intense heat is due to the absence of alkalies, such as potash and soda, whose withdrawal was brought about by the plants whose roots removed the soluble salts which they required for food or which were removed by the leaching action of the water in the lakes or lagoons.

It is evident, therefore, that coal is compressed bituminized or mineralized vegetable matter.

Necessary Conditions for Coal Formation—(1) *How Vegetable Tissue is Accumulated*—It is generally believed that coal originated, for the most part, from vegetation that grew in swampy or marshy places, although evidence has been advanced recently which shows that much coal was formed from organic matter, spores, wood, and leaves, carried into swamps and lakes. It is a matter of common observation that wood decays much less rapidly below water than above it. This is shown by piles and posts which may be entirely rotted away where exposed to the air, while they are well preserved where continually soaked with water. The reason is to be found in the fact that vegetation in the open air is readily attacked by destroying fungi, the carbon is oxidized to carbon dioxide and the hydrogen to water, and as these are volatile the entire substance of the plant may disappear; while in water the oxidation proceeds much less rapidly and completely, and wood-destroying organisms cannot flourish in water. Of the vegetation of luxuriant forests only thin layers of humus remain, and the abundant vegetation of dry, fertile plains fails to accumulate, although the slow-growing bog moss (sphagnum) of cold regions may accumulate to form thick beds of peat.

(2) *How it was Kept from Decay*—The chemical changes which take place in vegetable tissue (which has a composition approximately of $C_6H_{10}O_5$) when deposited in water, result in the formation of marsh gas (CH_4), carbon dioxide (CO_2), and other gases. The effect of these changes consists in (1) a reduction in volume, (2) a reduction in the volatile constituents, (3) a reduction in the amount of water, and (4) a relative increase in the percentage of carbon, since although the greater part of the hydrogen and oxygen are removed, the carbon is only moderately reduced.

The proof in support of the assumption that the great coal deposits were developed in swamps, the vegetation accumulating where it grew, is to be found (1) in the basin-shaped seams which are often thickest in the center and thin out to black shale at the edges; (2) in the remains of aquatic animals in the midst of the coal; (3) in the roots of trees embedded in the underclay in the position in which they grew;

and (4) in the purity of the coal. If the coal was formed from vegetation that had drifted together, it would contain sand or mud in appreciable amounts. It is not unusual, however, to find coal with no more impurities (ash) than the wood from which it was derived would have contained. The nearly uniform thickness of the coal beds over hundreds of square miles is also offered as an objection to the theory that the vegetable matter was drifted together.

(3) *How it was Changed to Coal and what Varieties Resulted*—The principal varieties of coal are peat, the partially decayed vegetation of swamps; lignite or brown coal; bituminous or soft coal; and anthracite or hard coal. All have been derived from peat, lignite being the second stage, bituminous the third, and anthracite the fourth. The last stage is graphite, in which all the volatile constituents have disappeared and pure carbon only remains. Anthracite coal occurs in regions where the strata have been much folded and faulted. It is therefore generally believed that the heat and pressure of dynamic action are essential processes in coalifaction or bituminization. It has also been suggested that in regions of great folding the fractures which have been produced facilitate the escape of gases from coal and thus hasten the process. In Rhode Island, dynamic metamorphism has been so intense that the coal has gone beyond the anthracite stage and contains so much graphite as to be of little value. In Colorado and elsewhere, bituminous coal has been converted to anthracite where cut by dikes, and in Mexico, coal has been baked to graphite by heat. Such graphite is of great value in the manufacture of lead pencils. Some varieties of coal result from the kind of vegetation of which it is composed. Cannel coal, for example, is made almost wholly of the spores of Carboniferous plants.

Conditions Favoring Coal Formation in the Pennsylvanian—To understand the great accumulation of coal during the Pennsylvanian one must picture to oneself the conditions at that time. The land appears to have been low, and sluggish streams meandered through extensive fresh-water marshes. The great inland seas, shut off on the east by the continent of Appalachia, were bordered by wide fresh and salt water marshes in which vegetation flourished. Similar conditions are to be seen today in the Dismal Swamp of Virginia, on the coast of New Jersey, the Carolinas, and Florida.

Since less than five percent of the Coal Measures (Pennsylvanian) consist of coal, the greater part of the system being composed of sandstones, shales, clays, and in some localities limestones, it is evident that subsidence accompanied deposition. The submergence was not continuous, however, but was interrupted by many halts, with occasional slight elevations. When the sea bottom was built up sufficiently, plants grew on it, and salt water marshes, which eventually became fresh, appeared. In the course of years the trees fell, and upon their fallen trunks others grew up. In the process of time their remains made thick beds of peat. A too rapid subsidence inundated the swamps, killing the vegetation, and the peat was then covered with sediment. If the water was far from shore, beyond the reach of mud and sand, limestones were deposited; if close to shore, mud and sand were laid down. When the downward movement ceased, the bottom of the sea was built up until it again became shallow enough to permit plants to grow on it. The order of deposition shown (on pg. 276) is thus explained. As has been stated, an elevation sometimes occurred, as is shown by unconformities. Some of the unconformities, however, were produced merely by the shifting of the stream channels in the swamps.

The number of coal beds in any vertical section varies greatly: in Pennsylvania and Nova Scotia as many as 30 are known, while in Illinois there are often less than 10. Some of these beds are workable, but many are not.

Extent and Structure of Coal Beds—Individual coal swamps of the Pennsylvanian were very extensive. The Pittsburgh coal bed, one of the greatest in the world, extends over an area of at least 12,000 square miles in Pennsylvania, Ohio, and West Virginia. The extent of some modern peat bogs compares favorably with those of the Pennsylvanian, but their thickness is much less. One extends across Holland and Belgium into France, and the Alaskan tundra has a much greater continuous area than the largest of those known in the past. All coal beds are not of great extent, some corresponding to the small peat bogs of today: one basin 200 yards in diameter was found to have two coal beds, one two and the other 16 feet in thickness; and another one 115 yards in diameter was found to have a coal seam eight feet in maximum thickness. A given thickness of coal represents only about five percent of the original thickness of the peat bed, consequently a coal bed 16 feet thick represents a peat bed about 320 feet in thickness. None of the peat bogs of the present have the great thickness of peat necessary to make great beds of coal.

The commercial importance of Great Britain and much of the remarkable development of the United States are due to the presence of abundant and accessible supplies of coal.

Climate during the Deposition of Coal—Since coal occurs not only in the temperate zones and in the tropics but even in the polar regions, it has been assumed that the climate of the Pennsylvanian was uniform throughout the world. This is further borne out by a study of the structure of the wood, which shows no rings of growth such as are developed in plants living in a climate in which there are dry and wet or cold and warm seasons. The question as to the temperature and the amount of moisture has given rise to some discussion. It has been generally assumed that the large size of many of the trees and the accumulation of their remains in swamps are proofs of a warm, humid climate. The thickness of the coal seams has also been considered confirmatory evidence. Several objections have been offered to this belief, however. (1) At present the great accumulations of peat are in cold, temperate climates. (2) Peat is rarely formed of rapid-growing plants but chiefly of the remains of such plants as sphagnum moss. (3) It has been pointed out that, as a whole, the leaves of Carboniferous plants bear a resemblance to those of living plants that are adapted to dry (xerophytic) conditions, being narrow and possessing devices to prevent the rapid evaporation of water. In the tropics, peat accumulates more from tree trunks and leaves which have been floated into lakes and marshes, and little from moss or trees that grew *in situ.*

When all the evidence is considered, there seems little doubt that the climate of the regions in which coal accumulated was moist and warm, although not tropical.

Problems of the Permian

No other period of the Earth's history offers so many unsolved problems as the Permian. These problems have to do with the climate and the life of the period.

(1) Why was the Permian so fatal to marine life? During this time the invertebrate life, for the most part, either became extinct or was much modified in important structural features. The impoverishment of the life is shown in the estimate that the number of Permian species was only two percent of that of the combined Mississippian and Pennsylvanian. One factor in the extinction of such a large percentage of species was doubtless the emergence of the continent and the consequent withdrawal of most of the epicontinental seas. This drove the life of the warm, shallow seas into the coastal waters of the ocean, where it was not only obliged to compete with other species but was compelled to live under conditions to which it was not accustomed. Such a radical change in environment was, doubtless, fatal to many species, and it is consequently not surprising that a large number disappeared. The faunas in the restricted epicontinental seas were crowded, and competition was severe. Where such seas persisted, as in India and California,

the change in the fauna was gradual and no satisfactory dividing line can be drawn.

A further result of the elevation of the continents (or withdrawal of the seas) was probably the changing of the position of the ocean currents, which were forced to take new courses. The extinction of some of the plant food upon which the animals of the epicontinental seas ultimately depended would have had a marked effect on the life. These and other causes may have combined to bring about the sweeping changes in the invertebrate life at the close of the Paleozoic.

The land vertebrates did not suffer as did the invertebrates. This may have been due (a) to the greater land area over which they could spread and the greater variety of conditions open to them; (b) to the greater variety, or more suitable varieties, of plants and insects which they could use for food; or (c) to their better organization. During this time, so fatal to the marine fauna, the amphibians continued in abundance and the reptiles became supreme.

The plant life also suffered a great change: the important Carboniferous groups became extinct, or nearly so, and their places were taken by plants of a more modern type. Whether this was due to the draining of the swamps, with the resulting death of the swamp vegetation, or to the spreading of upland trees which existed in the Pennsylvanian but of which nothing is now known, cannot be stated. Once well-established, however, the more highly organized upland plants probably became in time suited to swamp conditions and occupied the places formerly held by the lycopods and horsetails.

(2) Why was the Permian a period of glaciation, and in particular, why were the areas affected not only near the equator but near sea level?

Various explanations have been offered, but none has a general acceptance. One is based on the assumption that the carbon dioxide contents of the air were decreased. The depletion of carbon dioxide is believed, by the adherents of this theory, to have been the result of a combination of causes, indirectly because of the elevation of the continents and an increase in the land area. As a result of a greater land surface being exposed to the agents of the weather, the rocks upon weathering extracted much carbon dioxide from the air. The depletion of this gas was further hastened by the oceans, which are believed by the supporters of this theory to have absorbed great quantities of carbon dioxide. The air was also freed of its carbon dioxide during the accumulation of coal. One doubtful element in the theory is the efficacy of carbon dioxide in retaining heat.

Another solution of the problem is found in the amount of water vapor in the atmosphere, since water vapor is known to act as a blanket in retaining heat. The enlarging of the land area decreased the amount of water vapor in the atmosphere and thinned the thermal blanket.

(3) Why was the climate generally arid in the northern hemisphere during the Lower Permian? The great extent of land and the narrowing of the oceans undoubtedly had a marked effect in producing an arid climate, since less moisture would have been evaporated and it would have been precipitated over a wider area.

(4) Why was the great Appalachian system raised at this time? It is usually stated that strains had been accumulating in the Earth's crust throughout the Paleozoic and that these strains were relieved by the folding of the Appalachian trough at its close, but this does not fully answer the question.

Summary of the Paleozoic Era

The Building of the Continents—The continent of North America was probably covered by seas in every part during some portion of the Paleozeic, but two areas seem to have been especially free from epicontinental seas during all but perhaps a small part of the era. These areas lie in the Laurentian region of eastern Canada and in the southeastern United States, where the continent of Appalachia formerly stood and of which the

Piedmont Plateau is a part. During this era, the seas varied greatly in extent and in position. In only a few areas, notably in the Appalachian trough and in the Great Basin region, did they persist through the greater part of the Paleozoic.

Evolution and Extinction of Life—A study of the accompanying table (pg. 315) brings out some important points concerning the life of the era. It is seen that certain classes began, or at least are first known, in the earlier periods, culminated in the later periods, and then after several periods of struggle became extinct. If one were to make a careful study of each of these classes it would be found that the genera of each class had a shorter life than the class as a whole, and that the species had a still briefer one. It would also be found that striking evolutionary changes took place during their life histories. It is also to be seen that some classes gradually increased in importance throughout the era, while others were inconspicuous in the Paleozoic, but if a Mesozoic table were examined it would be found that these inconspicuous forms became prominent in the latter.

Climate—Our knowledge of the climate of the Paleozoic is not extensive. As a whole, the evidence points to uniform conditions and no well-marked climatic zones. There were, however, glaciers in certain regions in the Cambrian and Permian, and perhaps in other periods. Certain areas were arid during portions of the Paleozoic, as their red sediments, gypsum, and salt show, the aridity having probably been caused in most regions by elevated land areas which shut off the moist winds of the oceans.

		PALEOZOIC				
		CAMBRIAN	ORDOVICIAN	SILURIAN	DEVONIAN	CARBONIFEROUS
PROTOZOA	FORAMINIFERA					
COELENTERATA	GRAPTOLITES					
	CHAIN CORALS					
	CUP CORALS					
	HONEYCOMB CORALS					
ECHINODERMS	CYSTOIDS					
	CRINOIDS					
	BLASTOIDS					
	STAR-FISH					
	SEA-URCHINS					
MOLLUSCOIDEA	BRYOZOA					
	BRACHIOPODA					
MOLLUSCA	PELECYPODS					
	GASTROPODS					
	STRAIGHT CEPHALOPODS ORTHOCERATITES					
	GONIATITE CEPHALOPODS					
ARTHROPODA	TRILOBITES					
	EURYPTERIDS					
	INSECTS					
VERTEBRATES	FISHES					
	AMPHIBIANS					
	REPTILES					

Table showing the distribution and relative abundance of the life of the Paleozoic.

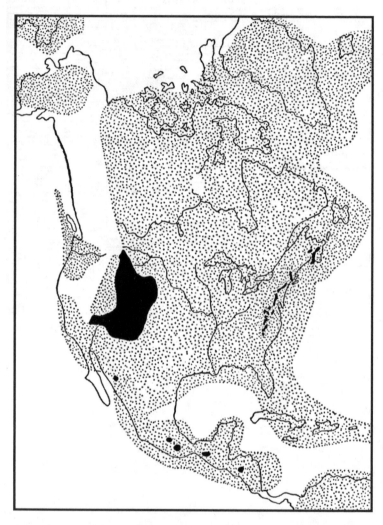

Map showing the probable outline of North America during a portion of the Upper Triassic. The continental deposits are shown in solid black.

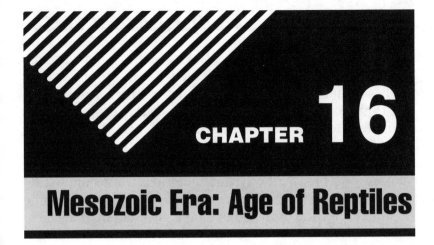

CHAPTER 16

Mesozoic Era: Age of Reptiles

The Mesozoic is divided into four periods, as given below:

Cretaceous Periods
{ Upper Cretaceous

Lower Cretaceous
(Comanchean)

The word comes from the Latin *creta*, meaning chalk, because of the great thickness of the chalk of this period in England and France.

Jurassic Period

So named because of the fine development of the strata of this period in the Jura Mountains.

Triassic Period

So named because of the three-fold development in Germany where the strata were first carefully studied.

Triassic

Atlantic and Gulf Coasts—A number of points seem to be well established concerning the distribution of land and water in North America during the Triassic (pg. 316). (1) The complete absence, so far as known, of marine sediments from the eastern half of the continent indicates that the coast line was farther east than now and that during the entire period the lands were being reduced by erosion. Indeed, it is possible that not only Newfoundland but even Greenland and Iceland were united to the continent. (2) The presence of Triassic rocks in long, narrow bands roughly parallel to the Atlantic coast, that stretch from

Nova Scotia to North Carolina, the longest of which extends from the Hudson River across New Jersey, southeastern Pennsylvania, through Maryland and Virginia, formerly gave rise to the opinion that these deposits were formed in tidal estuaries whose waters for the most part were brackish or nearly fresh. It seems more probable, however, that the deposits were not formed in continuous water bodies but in river basins analogous to the Great Valley of California. These continental deposits were formed by the confluence of alluvial fans made by streams flowing from higher land at the margin of the area; deposits formed by rivers meandering over the lowland; lake deposits in places where the drainage was obstructed, as in Tulare Lake, California; and it is possible that parts of the area were covered by tidal waters and that in such places estuarine deposits were laid down. Since the region was in an arid or semiarid condition, deposits of wind-blown sand were doubtless laid down on land, some of which probably constitute a part of the Triassic sandstone.

These basins were separated from the Appalachian Mountains on the west by ridges of crystalline rocks. The presence of high land between the basins and the Appalachians is shown by the composition of the sediments laid down in the depressions, which were not derived from sedimentary rocks, as would have been the case if the Appalachians had drained eastward through them, but are granitic and were derived from metamorphic and igneous rocks. The present thickness of these deposits is very great, it being estimated that some of those of Pennsylvania and Connecticut are several thousand feet thick. The liability to error in estimating the thickness of these deposits, because of the concealed faults, is so great that no

Section through the Connecticut valley and adjacent region in Massachusetts. *Js* is Triassic sandstone.

figures can be considered more than provisional. The sediments north of Virginia are usually red sandstones and shales, with occasional thin beds of black shale and limestone.

In Virginia and North Carolina coal conditions prevailed, but, with the exception of these and the abundant footprints of the Connecticut valley, fossils are rare. Fish and plant remains in thin beds are, however, occasionally found. Since no marine fossils have been discovered in any of the Triassic deposits of the east, the exact age of these deposits is somewhat in doubt, but the evidence points to the Upper Triassic as the time at which they were laid down. This formation is known as the Newark because of its development near the city of that name in New Jersey.

During the deposition of these sediments lava flows of considerable extent and thickness were poured out; dikes and sills were intruded: and in a few places volcanoes were in eruption. The evidence of this gaseous activity is especially well shown in the Connecticut valley (pg. 320, *A*), where in certain localities there are three distinct lava flows. The lava forming the Palisades of the Hudson, which varies in thickness from 300 to 850 feet and stretches for 70 miles from north to south, is an intrusion. The faulting and subsequent erosion (pg. 320, *B*) of the Triassic sediments and lavas of Massachusetts, Connecticut, and New Jersey have resulted in hills and mountains of (for these regions) unusual shape.

Western Interior—In the western interior of North America (map, pg. 316), the deposits are also for the most part red and, as a rule, devoid of fossils. Some of the sediments were deposited in freshwater lakes, others in saltwater lakes, and some are probably of eolian origin.

Pacific Coast—In the early part of the period, the Pacific coastline, with the exception of a comparatively narrow bay that stretched west from California to Wyoming, was probably farther west than now. Later in the period the sea spread over

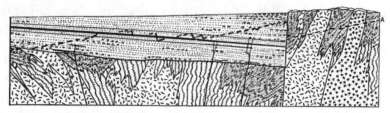

A, Section across the Connecticut valley, showing the thick strata of sandstone (dotted) and the lava beds (solid black). The dotted line shows the present outline of the surface. The complex structure of the underlying rock and the rock of the highlands is well shown.

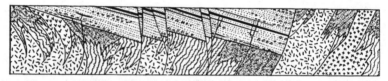

B, Section across the Connecticut valley, showing the same region as above after faulting had occurred, and after erosion had worn the region to a peneplain.

the land until it covered a large area in Alaska, Canada, British Columbia, Washington, Oregon, Nevada, and California, and smaller areas in Mexico. The fossils from the Triassic deposits of the west are in many cases very abundant.

Triassic in Other Continents—Triassic rocks have a wide distribution in Europe, where both marine and continental deposits occur; the marine being, in general, in the south of Europe and the continental in the northern and central portions. The Triassic in Germany has a threefold division (hence the name Triassic). The lowest consists of deposits which were laid down in fresh and salt lakes, not unlike those of the Triassic of the western interior of North America, and to some extent are of eolian origin, as the dune structure of some of the sandstone shows. Sun cracks, raindrop impressions, and tracks of animals also occur. The relation between the fertility or barrenness of a soil and the rock from which it was derived is well illustrated by this formation in Germany. Since the rocks are composed of quartz sand containing little plant food, the region underlain by them is not cultivated, but has been allowed to remain forested, and hence this formation has been called the "forest formation."

During the Middle Triassic (Muschelkalk) an inland sea connected with the ocean spread over a large part of the area of the earlier formation. This sea was later shut off from the ocean and soon became a salt lake, as the deposits of gypsum and salt show. In the Upper Triassic (Keuper), marine beds, thin coal seams, gypsum and salt deposits, and in the last stage (Rhaetic) marine deposits again show a few of the fluctuations of the period.

Jurassic

Atlantic and Gulf Coasts—In eastern North America the emergence of the continent seems to have continued from the Triassic, no trace of marine sediments being known except in Mexico where the Gulf of Mexico extended west of its present position. Probably near the close of the Triassic the continent was warped in such a way that the Triassic sandstones and shales of the Connecticut valley were tilted to the east, and those of New Jersey and farther south to the west. The Jurassic, like the Triassic, appears to have been a period of continued erosion in the eastern half of the continent.

Western Interior—No early Jurassic rocks are known with certainty to occur in the western interior, but later in the period an arm of the sea of great width extended south (pg. 322) from Alaska to Wyoming, Utah, and the Black Hills of South Dakota. The prevalence of sandstones, with only occasional limestone beds, shows that the sea was a shallow one. Since the fossils of some of the beds are of marine species, closely

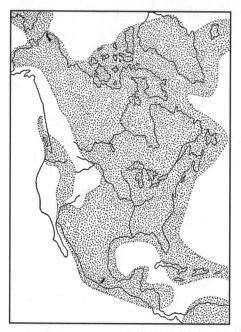

Map showing the probable outline of North America during a portion of the Upper Jurassic.

resembling those of Siberia rather than those of California of the same age, we must suppose that a mediterranean sea was connected at the north with the ocean, and that a long land barrier separated it from the Pacific. After a comparatively short existence, this great bay was drained by elevation of the land; and its site was covered in the southern portion by a widespread, continental formation (Morrison) which contains skeletons of dinosaurs and other reptiles and a few mammalian remains. Because of the absence of marine fossils, it is not yet certain whether these beds are late Jurassic or early Cretaceous, or whether the lower portions belong to the earlier period and the upper to the later.

Mountain Forming in the West—In the west, where the Sierra Nevada Mountains now stand, Jurassic sediments derived from extensive lands on the east had been accumulating in a great subsiding trough (geosyncline), until they had attained a maximum thickness of five or six thousand feet. The sediments deposited in this trough, including Triassic, Jurassic, and Paleozoic, attained the enormous thickness of nearly 25,000 feet. Near the close of the period, this huge accumulation of sediments began to yield to great lateral compression and was folded and upheaved into the first Sierra Nevada Mountains, which perhaps rivaled in height any in existence today; they may, however, have been eroded almost as rapidly as they rose and therefore never have reached a great elevation. During the folding great quantities of igneous rocks, especially granites, were forced into the folded sediments, forming upon cooling batholiths and stocks. The muds and sands in the neighborhood of the intrusions were also changed to schists and other metamorphic rocks; and even at a distance shales were metamorphosed to slates. At about the same time, the Coast Ranges, the Cascades, and farther north the Klamath Mountains began their growth. It should not be inferred from the above that the present height of the Sierra Nevadas was the result of these movements: the Sierra Nevadas of today, as will be explained later, are the result of a great fault on the east, which occurred at a much more recent date.

Jurassic of Other Continents—The greater portion of Europe and Asia was above the sea during the earlier part of the period, but a progressive submergence soon began, culminating in the Upper Jurassic, at which time the two continents were traversed by straits and seas which cut them into a number of large and small islands. The submergence of central and northern Russia is of special interest, since in this basin was developed a peculiar fauna which spread into the great mediterranean sea of the western interior of North America. An arm of the sea covering the site of the Himalayas separated India from northern Asia. This Upper Jurassic submergence in Europe and Asia was one of the greatest in all the recorded geological history of these continents.

Lower Cretaceous (Comanchean)

Atlantic and Gulf Coasts—No marine deposits of the Lower Cretaceous (below) have been found on the Atlantic coast, but a belt of continental sediments, stretching from Martha's Vineyard island to Georgia (the Potomac group), occurs, which only rarely attains a thickness of more than 600 feet. The lesson which these deposits teach of the physical condition at, and immediately preceding, the time of their formation is interesting. They show that at the close of the Jurassic, the eastern portion of the continent had been eroded down to a comparatively level plain over wide areas, the surface of which was covered deeply with weathered rock, which resulted from the decay of the formations lying underneath, which the slow-moving streams of

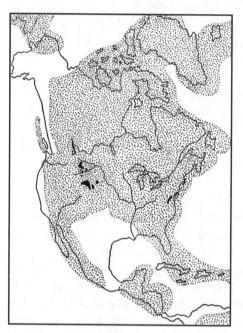

Map showing the probable outline of North America during a portion of the Lower Cretaceous. Continental deposits are shown in solid black.

that time were unable to transport. At the beginning of the Lower Cretaceous, however, the Piedmont and Appalachian regions were raised, perhaps along the axis of the Appalachian tract, while the land nearer the coast was but little disturbed, either remaining comparatively level or being depressed into long troughs, somewhat similar to those of the Triassic. Under these new conditions, the streams in the higher regions began again to erode. On account of the abundance of loose, weathered material, the streams in their courses to the sea soon had all the sediment they could carry and as soon as a lower gradient was reached dropped their loads. This resulted in the formation of deltas, flood plains, marshes, and shallow lakes. The deposits formed in this way were gravels, composed of the quartz of quartz veins and quartzites, clay from the decayed feldspar, shales, and slates, and arkose in the immediate vicinity of feldspar-bearing rocks.

The most striking feature of the Lower Cretaceous geography is the expansion of the Gulf of Mexico towards the west and northwest and the deep subsidence of its floor, upon which were deposited a great thickness of limestones. Large areas in Mexico, Texas, and New Mexico were covered at this time. In this sea, the Ouachita Mountains stood out as a promontory, as is shown by the ancient shore line which has been traced around their foot. The sediments have a thickness of 5000 feet on the Rio Grande and are even thicker in Mexico. At the base of this marine formation is one which is in part a littoral deposit but is mostly marine (Trinity).

Western Interior—In the western interior non-marine formations, sometimes including coal beds, occur (Morrison which may be Jurassic, Kootenai, Cloverly, Lakota, Fuson).

Pacific Coast—On the Pacific coast the conditions were very favorable for erosion, because of the newly raised Sierra Nevadas which were being rapidly cut away, the material derived from them forming a thick deposit in the Sacramento valley. In addition to this area, other narrow strips were submerged east of the present coast of British Columbia and Alaska, during portions of the period.

Lower Cretaceous of Other Continents—In Europe, as in North America, the Lower Cretaceous formations are largely of continental origin and are not as widespread as those of the

Upper Cretaceous. In general, it can be said that important geographical changes occurred in various parts of the earth at the close of the Lower Cretaceous, as are recorded in the unconformities between the Lower and Upper Cretaceous strata and in the difference in their distributions.

Upper Cretaceous (Cretaceous)

The Upper Cretaceous was a period of great subsidence (below), no other in the earth's history since the Paleozoic being comparable to it. Not only were portions of the Atlantic and Pacific coasts submerged, but a vast inland sea covered for a time the central portion of North America, from the Gulf of Mexico to the Arctic Ocean, separating it into two land masses.

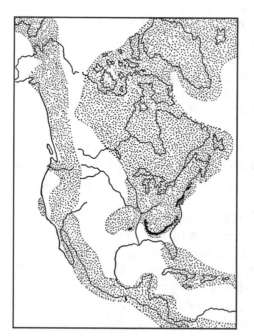

Map showing the probable outline of North America during a portion of the Upper Cretaceous. The inland or epicontinental seas were widespread. Continental deposits are shown in solid black.

Atlantic and Gulf Coasts—The seawater spread over the coastal plains of the Atlantic and Gulf states, and strata composed of sands, clays, chalk, and the "green sands" (glauconite) became accumulated. The map (left) displays the supposed distribution better than a written description. At this time the eastern half of the continent was most likely a comparatively flat plain (Kittatinny peneplain) to which it had been reduced during the long ages of the earlier Mesozoic, notwithstanding occasional warpings. Across this plain the Potomac, Susquehanna, and Delaware rivers meandered, probably in very much the same courses as today.

Pacific Coast—On the Pacific coast, Upper Cretaceous sediments occur in California and northward at points as far distant as Alaska, where they are sometimes conformable and sometimes unconformable. Usually they are not thick, but in one locality (California), at least, they apparently reach the great thickness of 25,000 feet.

Western Interior—The geography of the western interior can be roughly divided into (1) an epoch when the sea was excluded, and beds, mainly of non-marine (Dakota) sediments, were laid down; (2) an epoch of pronounced extension of the sea during which a great thickness of marine sediments (Colorado and Montana) accumulated; and (3) an epoch during which the land was so low that slight oscillations produced conditions which resulted in the formation of bodies of salt, brackish, or fresh water (Laramie).

(1) The sediments of the first epoch (Dakota) probably covered an area 2000 miles long by 1000 miles wide, which stretched from Canada on the north into Texas on the south, and from Minnesota and Iowa on the east to beyond the present site of the Rocky Mountains on the west. The formation is largely sandstone, though it contains much conglomerate and clay, and some lignite. At this time marshes and lagoons existed near the shores, while inland, sluggish streams were depositing fine sediment over the bottom. The presence of brackish water fossils in beds of this age in Kansas indicates marine conditions at certain times, or at least a low shore an which fresh and salt water were mingled. The porous beds thus formed are now the great water-bearing strata of the Great Plains, the water of these porous sandstones being derived from the rains which fall upon

and the streams which flow over their upturned edges in the mountainous regions. Although of such wide extent, the Dakota sandstones have a fairly uniform thickness of only 200 to 300 feet.

(2) This epoch (Dakota) was followed by one of extensive submergence which resulted in the formation of a great sea (Colorado), stretching from the Gulf of Mexico to the Arctic Ocean and covering the Great Plains of Canada and the United States and the site of the Rocky Mountains, with the possible exception of some large and small islands. As the sea encroached upon the land, muds were first deposited, but as it deepened and widened, the waters became clearer, and chalk and limestone were laid down locally, while in the extensive bordering swamps peat accumulated to form important beds of coal which are best developed in Wyoming and Utah. This larger sea gave place later to a somewhat more constricted one (Montana), along the edges of which the conditions were favorable for coal formation. This is shown by thick coal beds of this time in Montana, Wyoming, California, Utah, and New Mexico. It is probable that neither of these seas was of great depth.

(3) The closing epoch of the western interior was the Laramie. The evidence points to a land so low that a slight oscillation either raised or submerged it. When the latter occurred, the sea overspread the land, and marine sediments were deposited; as the sea was filled in by sediments or was partially drained by elevation, swamps and marshes were formed and peat accumulated in sufficient quantities to form later some workable beds of coal. There is probably as much coal in the Cretaceous formations of the west as in the Carboniferous of the eastern United States, though usually of a poorer quality.

There is some disagreement as to where the line between the latest Cretaceous (Laramie) and the earliest Tertiary (Eocene) should be drawn. A formation (Lance) in Wyoming, containing dinosaurian remains and other typical Mesozoic animals but with plants that have a Tertiary aspect, is separated from the Laramie below by an unconformity involving, it is thought, the removal of over 20,000 feet of strata. This formation is placed by some in the Tertiary, although others believe that it should be included in the Mesozoic.

The maximum thickness of the Upper Cretaceous in the western interior is about 24,000 feet, making it one of the great periods of the earth.

Upper Cretaceous of Other Continents—Not only was the Upper Cretaceous a period of great submergence in North America, but in other continents as well. Large tracts of Europe were beneath the sea at this time.

Limestones were deposited in southern Europe, and chalk to a thickness of several hundred feet, accumulated in France and England; the latter has, however, little development elsewhere on the continent. In Asia, the land area was much smaller than now, the Himalaya region, as well as large tracts in India and elsewhere, being covered with water. Australia and South America show a similar extension of the seas. The summits of much of the eastern Andes of South America, to a height of 14,000 feet or more, are formed of Upper Cretaceous beds.

The Cretaceous Peneplain—Before the close of the era, not only North America but Europe and Asia appear to have been reduced by erosion to low, monotonous plains upon which few, if any, elevations of great height remained. This being the case, the era as a whole must have been one of great quiet, during which crustal movements were uncommon. One should remember, however, that at the close of the Triassic the faulting and elevation of the sandstones and shales of that period occurred; that at the close of the Jurassic, the Sierra Nevadas were raised, but that before the end of the Upper Cretaceous even these elevations had for the most part disappeared. The Appalachians were largely worn down to base level, and the

Laurentian region of Canada was a comparatively flat plain. Under these conditions the streams of that time flowed in meandering courses to the ocean, and the climate was probably uniform, warm, and humid.

Mountain-making Movements at the Close of the Mesozoic— During the closing stages of the Upper Cretaceous, great crustal disturbances began which resulted in the formation of mountain ranges from Alaska to the southern tip of South America. These movements, following the long period of quiet just described, were not sudden, but were anticipated by upwarping in Colorado, Wyoming, and other places, as the presence of more abundant coarse sediments indicates. The great Rocky Mountains of Canada and the United States had their birth at this time, but not their full growth until later. The structure of these mountains is in marked contrast to that of the Appalachians, whose elevation was the result of great lateral movements which folded and crowded together the strata. Although horizontal compression was important in the formation of the Rocky Mountains, the result of the vertical movements is much more conspicuous. The growth was also assisted by faulting. In Utah, the Wasatch and Uinta mountains and in British Columbia, the Gold Range were also raised.

That the deformation did not take place previous to the Laramie is proved by the fact that the strata of this stage and those of greater age are folded with equal intensity, while the overlying Tertiary rocks are less disturbed, showing that the deformation took place before the deposition of the latter. In some areas, however, the lowest Eocene (Fort Union and Wasatch) show as steep dips and were apparently as much disturbed as the underlying Cretaceous, indicating later deformation. These disturbances were accompanied by volcanic eruptions and intrusions of lava. The laccoliths forming the Henry Mountains of Utah were elevated by lava which was forced into the strata at this time. At this time, too, a large part of the continent of North America was affected by movements of greater or less strength, so that at the close of the era, dry land extended from the Sierra Nevadas on the west to the "Fall Line" on the east.

With the formation of the Rocky Mountains, the fourth great range of the North American continent came into existence; the Taconics being formed at the close of the Ordovician, the Appalachians at the close of the Paleozoic, and the Sierra Nevadas at the close of the Jurassic.

Duration of the Mesozoic—Many facts point to a great duration for this era. (1) The erosion of an immense thickness of rocks from the Appalachian Mountains and the reduction of the continent to a peneplain was accomplished during the era and must have taken an almost inconceivable length of time. (2) The first Sierra Nevada Mountains, although formed in the latter half of the era, were not only raised—possibly to a great height—but were also, later, worn down to a peneplain. (3) During the Upper Cretaceous alone, 24,000 feet of sediments— almost five miles—were deposited, being worn from the land and carried little by little to the seas by the streams. (4) The evolution in the animal and plant life of the era is very striking, from the standpoint of both form and structure. It does not seem possible that, under the conditions existing at that time as we understand them, these changes could have been brought about rapidly.

No matter upon what basis the estimate is made; whether the time necessary for the erosion of thousands of feet of strata, or that required for the deposition of great piles of sediment, the length of the era must have been enormous. An estimate of 9,000,000 years has been suggested, but should be taken merely as an approximation. It may be too large, or several millions of years too short.

Life of the Mesozoic

In the early days of the study of geology it was believed by many that life ceased to exist at the close of the Paleozoic and was recreated at the beginning of the Mesozoic. This belief was based upon the great dissimilarity of the life of the two eras and upon the apparent absence of fossils in the intermediate strata. As a more careful study of these rocks was made, and new exposures were discovered, fossils were found which, though rare, proved that the life was in many respects transitional. The change in vegetation between the two eras was not cataclysmic, as was formerly supposed, though it was rapid or almost sudden. The animal life suffered even more than the plant, very few of the Paleozoic genera surviving in the following era. The transition, in other words, apparently took place with great rapidity and affected all classes of life. If a change in the character of the rocks is made a basis for separation, it is found that although great unconformities occur, yet in many places it does not seem possible to tell where the dividing line should be drawn. For example, in America (Kansas and Wyoming) between horizons yielding Permian fossils and those yielding Mesozoic there are "at least one thousand feet of continuous, conformable, uninterrupted, and homogeneous deposits of red sandstone which may belong to one period or to both," and in Europe the Permian in many places merges into the Mesozoic (Triassic) so insensibly that it is impossible to state where one ends and the other begins.

Comparison of the Life of the Paleozoic and the Mesozoic— The dominant plants and animals of the Paleozoic disappeared, for the most part, with the Permian. The lepidodendrons, sigillarias with the exception of a few stragglers, Calamites, Cordaites, sphenophylls, and a number of important genera of ferns had vanished; and their places were taken by a flora of very different character, so that the forests of this era were very unlike those of the preceding in general appearance.

With the close of the Paleozoic, the abundant corals of that era had disappeared and were replaced by a new type, differing both in structure and appearance. No cystoids or blastoids survived. The crinoids are, with the exception of two genera, of a type quite different from those of the Paleozoic. The race had reached its zenith and its decline had begun, though now and then a species made its appearance which by its local abundance seemed, for a time, to give promise of regaining the former importance of the race. Sea urchins, which for the millions of years of the Paleozoic had remained in a subordinate position, were replaced by forms of modern structure and occupied, to some degree, the place vacated by the crinoids, cystoids, and blastoids.

The brachiopods of the Paleozoic were, as far as external appearance is concerned, of two classes; those with long hinge lines, giving them a square-shouldered look, and those with short hinge lines and sloping shoulders. Of these, the square-shouldered and most characteristic type soon disappeared, and only a comparatively few genera of the sloping-shouldered type survived. The brachiopods, as was true of so many other classes, after attaining considerable importance gave way to other classes of animals. In this case, as they decreased in abundance, the pelecypods and gastropods increased. The comparatively simple-sutured cephalopods of the Paleozoic, such as the Orthoceras and the angled goniatite, were quite suddenly replaced by an abundance of cephalopods with complicated sutures. The Orthoceras, which lived throughout the whole of the Paleozoic, had a few survivors in the Triassic, but these soon became extinct.

The fishes of the Triassic resemble those of the Permian in most particulars, but many of the Permian genera are wanting. The Age of Amphibians passed with the Permian; and although the Stegocephalia lived on into the Triassic, they disappeared before its close; and the insignificant frogs and salamanders of the present are the sole representatives of that once varied and conspicuous race.

The cause of the revolution in life at the close of the Paleozoic, as has been seen, must be found in the very different physical conditions which were present at this time, since not one or two but many orders of animals and plants either became extinct or were profoundly affected. The formation of great mountain ranges, the withdrawal of epicontinental seas in America, Europe, and elsewhere, must have produced a climate markedly different from that of the Carboniferous, since the ocean currents, with their great stores of heat, would be forced to take courses different from those which they formerly held. Moreover, the circulation of the air would be affected by the high mountain ranges. Besides these more evident causes, it is possible that a radical change in climate resulted from the withdrawal of carbon dioxide during the Carboniferous, and that this, combined with the above-mentioned and other physical changes, caused the extinction of those species which, because of their lack of variability, could not adapt themselves to the new conditions.

Plan of Study—For the sake of continuity in the study of the various groups of animals and plants described, the life of the four periods of the Mesozoic will not be studied separately, but the periods in which the genera and species occur will often be referred to.

Invertebrates

Chalk—Chalk is composed largely of the remains of Foraminifera. Although these unicellular organisms have been found in Paleozoic strata, being abundant in the Lower Carboniferous (Mississippian), it was not until the Jurassic that they attained a great development. Although conditions were very favorable for their increase in the Jurassic of Europe (but not of America) and still more so in the Cretaceous, they were even more important as rock builders in the Tertiary.

The best known chalk deposits are those of which the cliffs of Dover, England, and Dieppe, France, form a part; and because of their conspicuous character, the name Cretaceous—Age of Chalk—was given to the period in which they occur. In the United States also, Cretaceous chalk is extensive. Chalk and chalky limestone many hundred feet in thickness are found in the Lower Cretaceous series of Texas; and another deposit in the Upper Cretaceous extends from Texas northward through the Great Plains region, in Kansas, Colorado, and Nebraska. However, this name is not altogether appropriate, since by no means all the rocks of that period are composed of chalk. It seems probable that the chalk of the Cretaceous was not deposited in seas of great depth as is true of the Globigerina (chalk) ooze of today, which in portions of the ocean is being laid down at a depth of 12,000 feet or more, but that the water was only moderately deep. The occurrence in the chalk of certain mollusks which do not seem to be of deep-sea species indicates this. Therefore, there is little reason to believe that the great chalk beds of the Cretaceous were deposited at depths of thousands of feet, and that the ocean bottom was later raised to form dry land. An explanation for the purity of the chalk, if deposited in comparatively shallow water, is to be found in the conditions existing at the time. As a result of the low relief of the land with its thick covering of vegetation, there was little erosion, and the scanty sediments were laid down but a short distance from the shore. Consequently, the Foraminifera which throve in great abundance near the shores as well as in the waters of the deep seas, as they do today, were not upon their death covered with clastic sediments, but in the course of time built up thick deposits of lime, composed largely of their own remains. The genera and species of Foraminifera have generally, as might be expected from their low organization, a long range in time: some of the species which occur in the Paleozoic are still living.

Flint nodules, varying in shape and ranging in size from that of a walnut to two feet in length, are of common occurrence in certain portions of many chalk beds. They are composed largely of siliceous protozoa (Radiolaria), sponge spicules, and silica that has no organic form. These flint nodules were probably formed by concretionary action, the silica scattered rather uniformly throughout the deposits being brought together to form masses of varying size.

Sponges—Sponges appeared in the Triassic of Europe in small numbers and became so numerous in certain localities during the Jurassic as to form thick strata with their remains. They were still more abundant in the Cretaceous of Europe, though not common in America.

Corals—The Paleozoic corals (Tetraeoralla) did not immediately give place to the modern type (Hexacoralla, below, *A, B*), but a few lingered for a short time in the Triassic. Before the close of that period, the new type (Hexacoralla) became so thoroughly established as to build coral reefs where conditions were favorable. It was not, however, until the Jurassic that extensive reefs were formed by their remains. The general appearance of the Cretaceous corals is not unlike that of the corals of today.

Mesozoic corals: *A, Thecosmilia trichotoma* (Triassic and Jurassic); *B, Thamnastraea prolifera* (throughout the Mesozoic).

Crinoids—This class (below, *A*, *B*) was rare both near the beginning and with some exceptions (Uintacrinus of the Niobrara Chalk of Kansas), near the close of the Mesozoic, but became abundant though not diversified, in the Jurassic, at which time the crinoids attained their greatest size and beauty. The stem of one has been traced seventy feet without reaching either end. The "head" in some individuals (below, *A*) is as large as a feather duster and similar to it in appearance. The genus to which these large specimens belong (Pentacrinus) is still found in the West Indian seas. In America, the class appears to have been rare throughout the era.

Although the structure of the Mesozoic and Tertiary crinoids differs markedly from that of the Paleozoic, perhaps the most conspicuous external difference lies in the great development and subdivision of the arms and the relatively small body (calyx) of the later type. All Paleozoic crinoids were attached to the sea bottom by stems, and this was also true of the great majority of the Jurassic genera, but a few free-swimming forms began then and have continued to the present. In these unattached forms, the animal begins its existence fixed to the bottom by a stem, as did its ancestors, but later becomes free.

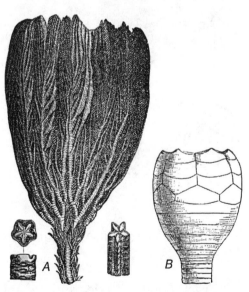

Mesozoic crinoids: *A, Pentacrinus fossilis; B, Apiocrinus parkinsoni* (without arms).

Sea Urchins (Echinoids)—A new type of sea urchin (below, *J-C*), which had a few forerunners in the later Paleozoic, soon entirely replaced the older type. One marked difference between the two groups lies in the number of rows of plates forming the "shell," which was variable in the old, but in the new was constant. Early in the era, a fivefold symmetry (below, *A*) was the rule, but later a twofold or bilateral symmetry characterized the greater number of species. Sea urchins with club-shaped spines (below, *A*) were abundant in the Jurassic and Cretaceous. Inconspicuous, rare, and little changed throughout the ages of the Paleozoic, sea urchins had a rapid development early in the Mesozoic, became abundant in the Jurassic and Cretaceous, assuming the place formerly held by the crinoids, and finally culminated in the Tertiary.

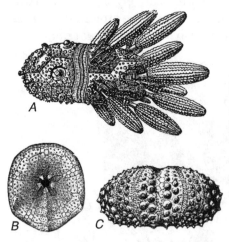

Mesozoic echinoderms: *A, Gidaris coronata* (Jurassic); *B, Cassidulus subconicus* (Upper Cretaceous); *C, Diplopodia texanum* (Lower Cretaceous).

Starfish—Starfish are not a conspicuous race in the Mesozoic.

Brachiopods—Aside from the abundance of a few genera at different times in the Triassic and Jurassic, there is little of interest in this class during the Mesozoic. Most of the species (below, *A-D*) belong to the genera that are living in the seas of today and are almost exclusively of three families. A few of the Paleozoic long-hinged type (Spiriferina, etc.) survived for a short time, but were inconspicuous. After their long period of ascendancy in the Paleozoic, brachiopods became unimportant and have remained in a subordinate position ever since.

Mesozoic brachiopods: *A, Terabratula humboldtensis* (Triassic); *B, Rhynchonella aequiplicata* (Triassic); *C, Rhynchonella gnathophora* (Jurassic); *D, Lingula brevirostra* (Jurassic).

Pelecypods—Almost in proportion as the brachiopods declined the pelecypods (pg. 340 *A-J*) increased, both in numbers and variety. This rapid development is shown in the Triassic, where only about one fourth were of Paleozoic genera. In the Jurassic, the oyster tribe is conspicuous and is represented not only by the true oyster but by others that, although belonging to the same order, have a different external appearance (Gryphaea, pg. 340 *C*, Exogyra, etc.) and are much more characteristic. In the Jurassic and still more in the Cretaceous, a number of pelecypod genera appear which depart radically from the typical forms in which the two valves are alike. In some of these (Diceras, pg. 340 *F*) each valve is horn-shaped; in others (Requienia, pg. 340 *J*) one valve is long and spirally twisted and the other flat, with a low spiral; in others (Radiolites, pg. 340 *G*) one valve has the appearance of a cup coral and the other is flat with prolongations extending into the horn-shaped valve. These irregular, unsymmetrical bivalves are usually firmly attached by one valve, their irregular development being due, to some degree, probably largely, to this fact. It is interesting to note that these extraordinary forms appear contemporaneously with the extravagantly modified cephalopods. A characteristic

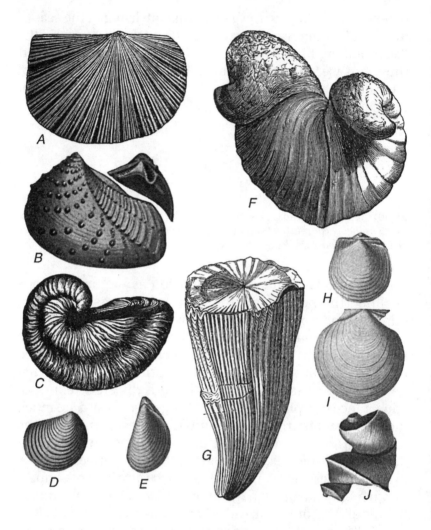

Mesozoic pelecypods: *A, Halobia (Daonella) lommeli* (Triassic); *B, Trigonia clavellata* (Jurassic); *C, Gryphaea arcuata* (Mesozoic); *D, Inoceramus vanuxemi* (Upper Cretaceous); *E, Aucella pioche* (Lower Cretaceous); *F, Diceras arietinum* (Jurassic); *G, Radiolites cornu-pastoris* (Upper Cretaceous); *H, Eumicrotis curta* (Jurassic); *I, Camptonectes bellistriatus* (Jurassic); *J, Requienia patagiata* (Lower Cretaceous). *C, F, G,* and *J* present forms very unlike the typical pelecypod.

Cretaceous genus is Inoceramus (pg. 340 *D*), also found in Jurassic deposits.

Gastropods—The Mesozoic gastropods (below, *A, B*) were, as a whole, less simple than those of the Paleozoic, although many of the older type, in which the mouth of the shell is a complete ring, lived on. In one branch, a tube was developed through which the waste waters of the body were carried and emptied some distance from the opening into which the fresh waters entered, a structure the advantage of which is obvious. Forms of this type were lacking in the Paleozoic, but became common before the close of the Mesozoic. Towards the close of the era many of the genera which reached their highest development in the Tertiary and recent times appeared.

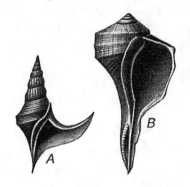

Mesozoic gastropods: *A, Anchura americana* (Cretaceous); *B, Pyropsis bairdi* (Cretaceous).

Cephalopods—The Paleozoic types of cephalopods (pg. 229 *A-D*) are represented in the Triassic strata by orthoceratites and goniatites and occur with the fringe-sutured ceratites (pg. 343 *I*) and the complex-sutured ammonites (pg. 343 *A*), but soon disappear.

Ammonites—Ammonites developed with wonderful rapidity from the first rare members (in the Upper Silurian or Devonian) into numerous families, hundreds of genera, and thousands of species, reaching their acme in the Jurassic. In the Cretaceous they gradually declined, dropping off one at a time, until all were gone before the end. In numbers, diversity of form, and ornamentation, ammonites are remarkable. Especially towards the end of the race (in the Upper Cretaceous) unusual forms appeared. At this time—and occasionally in the Triassic and Jurassic—many began to uncoil; some were coiled during the early part of their life, but as they approached old age became less coiled (Scaphites, pg. 343, *L*); others formed open coils (Crioceras, below, *A*); some were turreted (Turrilites, pg. 343, *H*); one common form (Baculites, pg. 343, *J*) became straight like the Orthoceras; others assumed forms which seem to have been entirely a matter of accident, as is shown especially

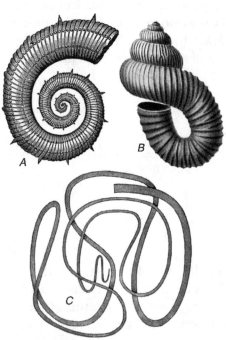

well in a specimen from Japan (*Nipponites mirabilis*, left, *C*). Many suggestions have been made to account for the "death contortions" of the ammonites, but none is satisfactory. The one most in favor is that they mark the senility of the race.

The descent of ammonites from goniatites is shown in two ways: (1) by comparing specimens from successively older formations and noting the progressive changes, and (2) by studying the oldest and youngest portions of the shell of an individual (pg. 343, *B*). In these shells every stage in the growth of the individual is preserved,

Cretaceous ammonites: *A, Crioceras; B, Nostoceras stantoni* (Cretaceous). In this specimen the death of the animal was probably caused by its own growth, as the edge of the living chamber is almost in contact with the lowest whorl of the spiral; *C*, Restoration of *Nipponites*, a remarkable genus from Japan.

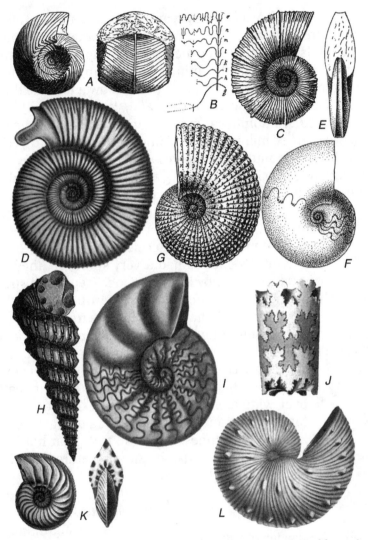

Mesozoic cephalopods: *A, Tropites subbullatus* (Triassic), side and front views; *B,* Development of sutures in *Tropites; C, Lytoceras fimbriatum* (Jurassic); *D, Perisphinctes achilles* (Jurassic); *E, F, Meekoceras gracilitatis* (Lower Triassic), front and side views; *G, Sagenites herbichi* (Upper Triassic); *H, Turrilites catenatus* (Cretaceous); *I, Ceratites nodosus* (Triassic); *J, Baculites,* showing the complexity of the sutures (one segment is shaded to emphasize this); *K, Cardioceras cordatum* (Jurassic), side and front views; *L, Scaphites nodosus* (Cretaceous).

so that if the shell of a full-grown ammonite is separated along its septa from the apex to the living chamber and its sutures studied, it is found that they increase in complexity—the first suture or two made by the animal when young being simple, like those of the Silurian nautilus; then follow a few like those of the Devonian gonitites (pg. 260, bottom, *A, B*); the complicated ammonite sutures beginning when the whorl is only two or three millimeters in diameter. In other words, each individual ammonite recapitulates the history of its race. It is consequently possible by studying a well-preserved individual to tell what its genealogical tree was. In no other animal can the evolution of the race be so well studied. It should be borne in mind, however, that the record of some of the stages of development is often omitted by "acceleration."

Since many of the species had a very short life, they are especially important in showing that widely separated strata are of the same age. It should be remembered, in this connection, that ammonites were free-swimming or crawling animals, and also that upon their death the gases of putrification caused them to float. They were consequently moved, either by their own volition or by the ocean currents after their death, over wide areas, and hence are excellent "index fossils." The ammonites contributed largely to the Jurassic limestones, but of all this great horde of shelled cephalopods, the simple-sutured nautilus alone survived the Mesozoic.

Restoration of a Belemnite. The solid, cigar-shaped "guard" (lower end) is a common fossil in the Jurassic and Cretaceous. Next above the "guard" is the phragmocone, and above this the proöstracum.

Naked Cephalopods (Belemnites)—In Jurassic deposits, straight, cigar-shaped fossils (left) are sometimes found in great abundance. These belemnites are usually 3 to 5 inches in length, although some specimens several feet long have been found. Ink bags are associated in some specimens, showing that their possessors darkened the water to escape their enemies, as do the squids of the present. These are the internal shells of naked cephalopods which resembled the squids of today in general appearance. This class is first known from the Triassic, but once started, the race rapidly increased, culminating in the Jurassic and declining rapidly in the Cretaceous. Only one surviving genus (Spirula) is living today. The solid internal shells of the belemnites constitute a considerable part of some Jurassic limestones.

Crustaceans—Crustaceans (below, *A, B*) of a very modern appearance took the place of the trilobites and eurypterids. In America few fossils of this group have been found, but in the Jurassic lithographic limestone of Bavaria many beautifully preserved specimens have been collected. It is possible that this class was as abundant in America as in Europe, but if so, the conditions for the preservation of the remains were not favorable. Ancestral long-tailed crustaceans of the lobster and shrimp type (Macrura) began in small numbers in the Triassic, and a few survivors of these ancient forms are living in the deep seas of

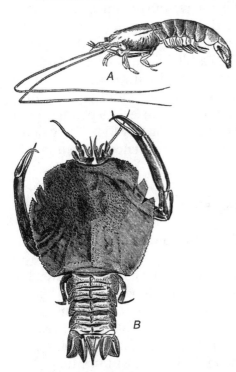

Mesozoic crustaceans: *A, Penoeus meyeri* (Jurassic); *B, Eryon propinquus* (Jurassic).

the present. Crabs are, in general, crustaceans of the lobster type in which the tail is abbreviated and turned under the body and the shell widened and otherwise modified (Brachyura). Crabs did not appear until the Jurassic and were derived from the long-tailed series, as numerous specimens intermediate between the two show.

Insects—Insects are better known from the Jurassic than from any other portion of the era, probably, as in the case of the crustaceans, because the conditions favorable for the preservation of their remains were better than at other times. True cockroaches and beetles are known from the Triassic; and practically all of the groups of today were present in the Jurassic, with the exception of those depending upon flowering plants for their food. Crickets, locusts, and cockroaches (Orthoptera), May flies, dragon flies, and caddis flies (Neuroptera) occur. Wood beetles are found associated with driftwood in the Jurassic; and flies (Diptera), plant lice, and aquatic bugs (Hemiptera) are known. The absence of insects depending upon the pollen and nectar of flowers is probably indirect evidence that flowering plants were not yet in existence in the Jurassic.

Fish and Amphibians

At the beginning of the Mesozoic, less modification in structure is noticeable in this class than in others to be considered, but the changes were by no means inconsiderable.

The shark tribe (pg. 347, *A*) has had a long and varied history. It began in the Silurian and abounds still in the warm seas of the present. These fish were abundant, both in species and individuals, in the Lower Carboniferous (Mississippian), but during the Permian declined rapidly, almost to extinction. In the early Mesozoic, however, they once more began to increase and were common before its close. The cobblestone-pavement toothed shark lived on and is represented today by one genus,

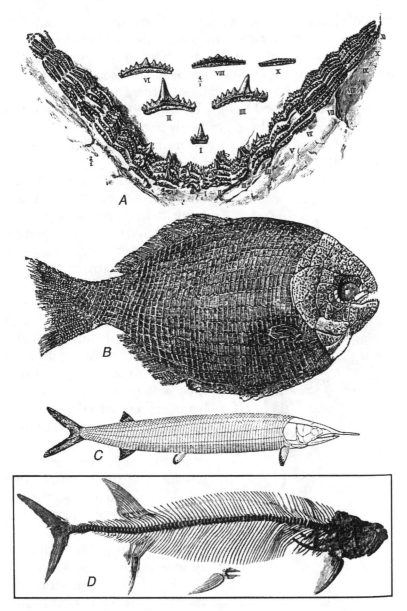

A, jaw and teeth of the shark, *Synechodus* (Cretaceous); *B*, ganoid fish, *Dapedius* (Triassic); *C*, ganoid fish, *Aspidorhynchus* (Jurassic); *D*, teleost fish, *Portheus* (Cretaceous).

the Port Jackson shark (Cestracion). A possible explanation of this curious fluctuation is as follows: in the Carboniferous, being the most powerful animals and having no enemies, they multiplied until their increase was checked by their very numbers. Then the overspecialized forms and those that failed to respond to changed conditions dropped out, leaving the best to survive. These, then, gradually increased to the Middle Tertiary when, through a change in climate, they became again comparatively rare.

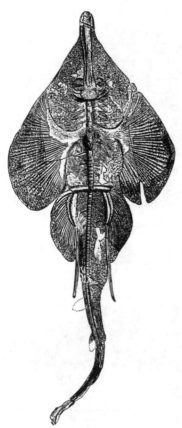

An ancestral skate, *Rhinobatus* (Jurassic).

The skates and rays (left) are sharks in which the body has been admirably adapted to bottom living and probably should be regarded as the culminating forms of the specializing, bottom-living sharks of the Mesozoic. The skates of the Mesozoic and Tertiary, without doubt, mimicked the color of the ocean bottom and glided along inconspicuously, just as their living descendants do. The teeth of all skates are simple, crushing, pavement teeth, suited for crushing the shellfish and crustaceans upon which they live. They are first known in the Jurassic and are abundant in the seas of today.

The lungfish almost disappeared from the seas of the Mesozoic, but a few (the best known of which is Ceratodus) have succeeded in living on in small numbers to the present.

Ganoids (pg. 347, *B, C*) were the common fish of the Triassic and Jurassic. Although they had no bony skeleton, they are well preserved because of their thick, enameled scales which were exceptionally well suited for fossilization. The ganoids are, as a rule, rather small fish and never attained the size of sharks. As the modern fishes with bony skeletons (teleosts) increased during the Cretaceous and Tertiary, the ganoids gradually disappeared until, at present, only a few species are in existence. Two of the commonest living ganoids are the gar pike and the sturgeon, both of which are extremely plentiful in some localities.

The bony fishes (teleosts), descendants of the ganoids, have been found in small numbers in the Lower Jurassic, but probably began in the Triassic. They held a subordinate place, however, until the Cretaceous, when they appeared in great numbers. Among the teleosts of the Cretaceous were herring, cod, salmon, mullet, perch, and catfish. One characteristic Cretaceous type, Portheus (pg. 347, *D*), should be mentioned. It was a teleost that occasionally attained a length of fifteen feet and was provided with large, flattened, irregular teeth. The suddenness of the appearance of teleosts was due to the fact that, once they were established and able to compete with the fish of that time, there was no hindrance to their migration, and, in a comparatively few years, geologically, they had spread into the seas of the whole world. Most of the ganoids became extinct either because of their inability to compete with the teleosts in the search for food, or because of climatic and other conditions.

Amphibians—The amphibians reached their greatest development in the Permian, but were present in considerable numbers in the Triassic, after which their remains are seldom found. Individuals of this class attained their greatest size in the Triassic, Mastodonsaurus (so-called because of its bulk) having a skull four feet long and probably attaining a length of 15 or 20 feet. Although large for an amphibian, the size is not great as compared with some modern crocodiles. In general appearance, Mastodonsaurus resembled the modern salamander, but it differed in several essential points of structure. Its teeth were of the complicated labyrinthine type, and the skull was roofed over with bony plates (Stegocephalia). It is possible that bony plates protected the chest, but if so, positive proof is lacking. The Stegocephalia became extinct before the close of the Triassic, and thus far, with the exception of two specimens of frogs from Wyoming, no amphibian remains have been found in Jurassic rocks. The cause of the extinction of this great amphibian order is probably to be found in the highly developed reptiles, large and small, with which they had to compete. A few specimens of salamanders of modern type, differing little in general appearance from the salamander of today, though of a different genus, have been found in the Cretaceous. Because of the lack of fossil evidence, the ancestry of the modern amphibians is not known.

Reptiles

Reptiles with Mammalian Characters—The reptiles with mammalian characters, as far as fossil evidence shows, began, and were represented by many genera in the Permian, but, since they apparently attained their greatest development in the Triassic, their discussion has been postponed to this chapter. This group of reptiles is included under the term Theromorpha (Greek, *ther*, beast, and *morphe*, form), because of the strong resemblance, both in teeth and skeleton, to mammals. They are remarkable in possessing not only mammalian characters, but amphibian as well, and occupy a position intermediate between mammals and amphibians. It seems probable that the Theromorpha include the progenitors of the mammals. One amphibian character is seen in the backbone, the bodies of the vertebrae of which are hollow at both ends (amphicaelous) and, in some cases, are only partly connected with bone. The teeth

Skeleton of a mammal-like (theromorph) reptile, *Endothiodon.*

Skull of a theromorph with beak-like jaw *(Oudenodon).* The animal was herbivorous. The skull is one and a half feet long.

of certain genera (Cynognathus) are of three kinds as in mammals: incisors, canines, and molars. In some cases, the limbs are decidedly mammal-like in structure. Another group of theromorphs have toothless jaws (above), covered with horn like a turtle's (Oudenodon), and some possess, in addition, two long canine teeth (Dicynodon). It is possible that the former (Oudenodon) is the female of the latter.

The theromorphs were all land animals, with limbs for the support of the body, but they varied greatly in appearance and habit. Some (Pareiasaurus, pg. 352, *A, B*) were as large as rather small cattle, about nine feet in length and standing about three and a half feet high, but with short legs and small, peg-like teeth, showing that they were herbivorous. They are believed to have been tortoise-like in habits, and probably protected themselves by digging in the ground. Associated with these in the same beds are carnivorous theromorphs (pg. 353), some with skulls two

A, skeleton, and *B*, restoration of the herbivorous mammal-like (theromorph) *Pareiasaurus*. The length is about eight feet. The surface ornamentation of the restoration is entirely fanciful.

feet in length, with long, tiger-like teeth. Attention has already been called to the fact that as soon as herbivorous animals appear in any age, carnivores, often closely related to the herbivores, also occur and prey upon their less agile neighbors. So among the theromorphs we find some of massive build being destroyed and devoured by their swifter, carnivorous relatives.

The theromorphs diverged with such rapidity in the Permian that, by its close, various groups appeared, differing slightly from one another, as has been seen. They survived the severe changes which brought the Paleozoic to a close and spread over three continents, but became extinct before the beginning of the Jurassic. It has been suggested that their rapid development and great variation may have been due to a more

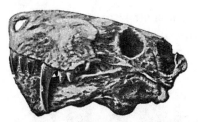

Skull of a large, carniverous, mammal-like reptile (theromorph), *Inostranservia*. The skull is nearly two feet long.

oxygenated atmosphere resulting from the withdrawal of the carbon dioxide which was abstracted from the atmosphere to form coal in the Carboniferous. It is possible that their extinction was due to competition with the better organized reptiles of the Triassic.

Dinosaurs

The preeminent land animals of the Mesozoic were the dinosaurs (Greek, *deinos*, terrible, and *saurus*, reptile), which occupied the place in nature now held by the land mammals. Some were larger than the largest animals of the present day, with the exception of a few of the whales, while others were as small as a common fowl; some walked in a more or less erect position, while others moved about on all fours; some had limbs as light as birds, while the limb bones of others were the largest and heaviest known; some were covered with a bony armor, and others were without such protection; some were very agile, others were slow-moving; some were carnivorous, others herbivorous; all were alike in having very small brains. All the continents of the world, including Australia, were occupied by them.

The dinosaurs may be separated into four groups: (1) carnivores (Theropoda), (2) unarmored quadrupeds (Sauropoda), (3) unarmored bipeds (unarmored Predentata), and (4) armored dinosaurs (armored Predentata). Of these, the first only was carnivorous, the others being herbivorous.

Carnivorous Dinosaurs—The most striking features of the carnivorous dinosaurs were the bipedal habit and the disparity in size between the fore and hind limbs, a character which increased as the race became older, until in later forms (Tyrannosaurus, right) the arms are so absurdly small that it is difficult to conjecture their use. As the fore limbs decreased in size and gradually relinquished their function, although never entirely abandoning it, the hind legs, in addition

Skull of Tyrannosaurus, a gigantic carnivorous dinosaur. The skull is four and a half feet long and the animal was sixteen feet high when standing.

to their duty of supporting the weight of the body, had to assume a grasping function as well, and the claws became, in consequence, great talons, differing thus markedly from those of an earlier type (Anchisaurus). This grasping function, however, was perhaps transferred to the teeth quite as much as to the hind limbs. The earlier forms probably walked on all fours, but as the fore limbs became smaller, they stalked about on their hind legs, or possibly leaped about in kangaroo fashion with the forward part of the body lifted from the ground and balanced by the powerfully developed tail. A second group (Compsognathus, right) differed from the above in being lighter in build, with the fore limb

A carnivorous bird-like dinosaur, about two feet long and named *Compsognathus*.

developed for grasping its prey.

The skull is very light and bird-like in some genera (Anchisaurus); and, although quite large in others (Tyrannosaurus), it is always relatively delicate. The skeleton is very light, as would be expected of animals of their habits, and the limb bones are hollow. An improvement in the teeth is noticeable from period to period; those of the earliest (Anchisaurus), although plainly for eating flesh, are not the perfect instruments possessed by those of later date (Allosaurus, below), which are long and somewhat flattened, with serrated edges. It is not known that the carnivorous dinosaurs were especially ornamented; one genus (Ceratosaurus), however, possessed a horn on the nose, and a row of small bones embedded in the skin down the middle of the back, but aside from this, ornamentation was rare. They varied greatly in size, from animals as small as a cat to the largest carnivorous land animals that ever lived. Tyrannosaurus was 40 feet long, with teeth projecting from two to six inches from the jaw. It is possible that this last was developed to prey upon the great armored dinosaurs which attained their greatest size and most perfect protection in the Upper Cretaceous, shortly before the extinction of the race.

Restoration of *Allosaurus*, a carnivorous dinosaur. The small size of the fore limbs as compared with the hind is striking.

The carnivorous dinosaurs are the earliest known, beginning in the Triassic and living throughout the whole of the Mesozoic.

Unarmored Quadrupedal Dinosaurs (Sauropoda)—These were the largest animals of the time. A study of the skeleton and restoration of Brontosaurus (below) or an allied form gives a truer conception of the animal than any written description. The long neck with its absurdly small head, the large body, stout limbs, and long tail make an animal differing from any now living. Certain characters of the skeleton are unusual. The leg bones, ribs, and tail bones are solid and heavy; the head and the vertebrae of the neck and back, on the contrary, being constructed so as to combine minimum weight with the large surface necessary for the attachment of the huge muscles. The significance of the remarkably heavy bones of the lower portion of the skeleton, combined with the unusual lightness in the

Fig. 135. Skeleton and restoration of *Brontosaurus*. These herbivorous dinosaurs grew to be sixty feet long.

upper portion, is that the animals lived in the water a large part of the time. Under such conditions, the greater the weight of the bones, the greater would be the ease of walking with the body partly submerged in water. The lightness of the head and the vertebrae of the neck would be of advantage in making rapid movement of these members possible.

The teeth are long and either cylindrical or somewhat spoon-shaped and are set rather far apart, a shape and arrangement fitting them for biting, but not for mastication. The brain is smaller than the spinal cord.

The reptiles of this family grew to be as much as 80 feet long and stood 16 or more feet high, and some are believed to have weighed 35 to 40 tons. They lived on flat plains, such as those at the mouth of the Amazon today, occupied by interlacing streams and small lakes in abandoned river channels; in a warm climate with luxuriant vegetation. That the water was fresh is shown by the fossil remains with which they are associated, such as freshwater plants, and shells, fish, crocodiles, turtles, and other dinosaurs. They went on land occasionally but not habitually, since the great weight of the solid bones would impede their movements, thus rendering them less able to escape their enemies. In the water they could swim with ease, propelled by their long tails. Their food was either floating plants, or such as were loosely attached to the bottom or banks; but they probably sometimes cropped foliage growing 20 feet above the water, which their long necks enabled them to reach. The character of the teeth precludes the possibility of hard, tough vegetation, since these are weak and not adapted to grinding. The lack of grinding teeth made it necessary for them to bolt their food, and it is interesting to note the occurrence of polished flint pebbles associated with the remains, which may have been "stomach stones" or "gastroliths," used in grinding the food after it had been swallowed.

These huge, four-footed creatures were probably descended from the carnivorous dinosaurs, either before or after the latter acquired the bipedal habit. When the carnivorous race became widespread and competition more severe, certain of them probably had a mixed diet at first, which in time became entirely herbivorous. After this change was established, the increase in size was largely a matter of abundance of food and lack of enemies. Although the body increased in bulk and changed in structure, the teeth failed to be modified to a great degree, but retained many of their ancestral characters to the end of the race. The unarmored quadrupeds first appeared either at the close of the Triassic or at the beginning of the Jurassic, and survived into the Lower Cretaceous. Their extinction may have been caused by a change in climate; by starvation as the result of the disappearance of the water plants upon which they fed; by the arrival or development of powerful enemies; or in other ways.

Unarmored Bipedal Herbivorous Dinosaurs (Unarmored Predentata)—The dinosaurs of this group were similar in general appearance to the carnivores, but differed in their less graceful build. Some of them attained a large size, being as much as 30 feet in length and standing 15 feet high (Iguanodon and Trachodon). The hind legs of some were twice as long as the fore. The heads varied considerably in different genera, being long and rather slender in most, but flat and ducklike in one specialized form, the duck-billed dinosaur. They were all alike in having the front of the jaw toothless and covered with horn. The rear portion of the jaws, however, was in one genus (Trachodon) provided with a battery of chopping and shearing teeth, composed of 45 to 60 vertical and 10 to 14 horizontal rows (right), though the rows were not all in use at the same time, the total number of teeth in some individuals being more than 2000. Others (Iguanodon and Camptosaurus) had only one row of

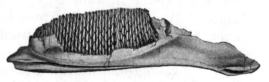

Portion of the lower jaw of *Trachodon*. The numerous teeth form a kind of pavement.

shearing teeth in use at one time. The teeth were replaced as rapidly as they were worn out. Teeth of this sort indicate that their possessors chopped or sheared their food and were able to live on tough, hard vegetation, such as the cycads and, perhaps, even the siliceous horsetails of the period.

They had three toes on the hind feet, terminating in hoofs (Trachodon), or claws (Camptosaurus). The fore limbs had three well-developed fingers, with one, or sometimes two other rudimentary ones. In a mummified specimen (Trachodon) found in Wyoming, the epidermis, which is covered with flat, bony scales, is seen to be extremely thin and the markings exceedingly fine and delicate for an animal of such dimensions. The same specimen shows that the fore feet were webbed, the skin reaching beyond the fingers and forming a sort of paddle. Since these animals had strong, powerful hind legs and were without armor, it is evident that their existence depended upon their ability to escape their carnivorous enemies by speed. Some (Camptosaurus and Iguanodon) apparently lived on the dry land, while others (Trachodon) were amphibious. The latter were able, when on land, to run rapidly, and when in the water to swim, perhaps, with the speed of the crocodile, as is indicated by the great, flattened, crocodile-like tail.

Armored Dinosaurs (Armored Predentata)—The reptiles of this group were of two very different types. A representative of one family is Stegosaurus (Greek, *stegos*, roofed, and *saurus*, reptile), an animal of greater bulk than an elephant. The restoration (pgs. 360, 361) shows two rows of broad plates on either side of the back bone, varying from a few inches to two feet in height and less than an inch in thickness, except where they were embedded in the skin, and with spines near the end of the tail six inches to over three feet in length. The stout fore limbs are much smaller than the hind, but a study of the joints shows that the creatures were quadrupeds. The front of the jaw was toothless and covered with horn as in the preceding group. The teeth in the back of the mouth were weak shearing teeth,

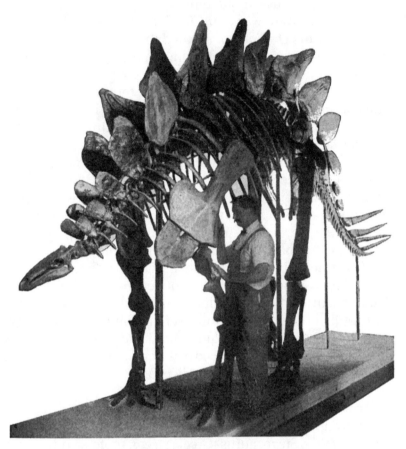

Skeleton of *Stegosaurus*, an armored herbivorous dinosaur.

not strong enough to masticate the coarser vegetation of the time; and they must, therefore, have fed, for the most part, on succulent plants. It is possible that they lived on the land bordering marshes. One of the most remarkable features of this unique reptile is to be seen in its nervous system. The brain is estimated to have weighed only about two and a half ounces (about one-fiftieth that of an elephant of smaller size), while the enlargement of the spinal cord above the hips is twenty times larger than the brain. This "hind brain" was probably the nervous center for the great muscles of the tail. It is likely that Stegosaurus did not face its enemy, but protected itself by swinging its long, powerful tail, which, however, was not very flexible. The long hind limbs suggest the possibility of considerable speed, and the fore limbs are so constructed as to make it possible for the animal to pivot the body rapidly so as to keep the tail to the enemy. These reptiles were descended from unarmored, herbivorous dinosaurs with a bipedal habit.

Two causes of the extinction of this family are suggested: the change in the vegetation to modern plants, and the senility of the race, indicated by the spinose character. It is certainly true that an animal of such bulk, so ornamented, would not be likely to vary to such an extent as to meet radically new conditions. Stegosaurus and its armored ancestors have been found only in the Jurassic.

Restoration of *Stegosaurus.*

Another family of armored dinosaurs, differing widely from Stegosaurus is represented by Triceratops (Greek, *tri*, three, and *ceras*, horn), one of the largest dinosaurs of the time (Cretaceous), with a length of about 20 feet. The noticeable feature of Triceratops is the skull with its two enormous horns, three feet long and six inches in diameter at the base, one above each eye, and a shorter one on the nose. (In a closely related genus the horn on the nose was long while those above the eyes were short.) The skull projected over the neck like a great bony frill and was fringed with short, bony points. The front of the jaw was sharp and parrot-like, and covered with horn, while the rear of the jaw was provided with shearing teeth. One genus of horned dinosaur (Torosaurus) had the largest head of any land animal, the skull being nearly nine feet long. The body, as well as the head, was protected, as is indicated by various spines and plates found associated with the skeleton, which were evidently embedded in the skin during life, doubtless for protection. The toes, five in front and three behind, were provided with hoofs.

Triceratops, unlike Stegosaurus, faced its enemies, as do cattle today. Punctures of the skull and frill over the neck, and broken horn cores are frequently found, showing that Triceratops often had combats with other animals. These creatures had the largest heads and smallest brains for their bulk of any of the reptiles, and were unquestionably extremely stupid, depending upon their size and armor for protection. During the life of the race, the animals increased in size and developed longer horns and a more complete frill over the neck. Triceratops had a relatively brief career, beginning in the Cretaceous and disappearing with its close.

Summary of Dinosaurs—Dinosaurs are first known from the Triassic, at which time they were numerous and diversified, as is shown by the great number and variety of footprints in the Triassic sandstone, although few skeletons have been found. This class became more abundant, larger, and more varied in the Jurassic, culminating either in that period or in the Cretaceous. During the Mesozoic, they became more and more specialized, the specialization culminating in Stegosaurus and Triceratops among the herbivores and in Tyrannosaurus among the carnivores. After becoming adapted to widely different conditions of life, assuming many strange forms and spreading over all the continents of the world, they disappeared with the Mesozoic and left no descendants.

Migration and Extinction of Dinosaurs—Our knowledge of the reptilian life of the Mesozoic lands is almost entirely confined to that which lived in the delta and coastal swamps of the era: the brontosaurs, the trachodons, the tyrannosaurs are all dinosaurs that lived either in swamps or on their margins. Of the upland reptiles little is known. At the close of the Jurassic, the gigantic dinosaurs were almost completely wiped out, doubtless because of the draining of the swamps in which they lived and their inability to adapt themselves to other conditions. Early in the Cretaceous, however, the swamps were again populated by other huge dinosaurs as well as many of smaller size. This fauna was a new one and was not descended from that of the previous period. Either it (1) migrated from some region as yet unknown or, more probably, (2) was developed from surviving small, active denizens of the uplands which are as yet practically unknown towards the close of the Mesozoic (in the late Cretaceous) the modern type of vegetation was probably associated with the dominance of mammals on the uplands and, although the dinosaurs held on in the swamp regions and had adapted themselves more or less to the new vegetation, they had probably become extinct in the uplands long before the close of the period. When the elevation that divides the Mesozoic from the Tertiary occurred, it caused the disappearance of this swamp fauna, but in the following period (Tertiary) we find a swamp fauna being developed again, not from upland dinosaurs but from the mammals which had taken their place.

Crocodiles—The Triassic ancestral crocodiles have so many characters in common with the primitive reptiles (of the Pelycosaur type) and dinosaurs that the order to which they belong is determined with difficulty. Among the changes that the crocodiles underwent during the Mesozoic, the following may be mentioned. (1) The vertebrae were biconcave (amphicoelous) in the Triassic, Jurassic, and most of the Cretaceous, as in fish, and not concave in front and convex behind (procoelous) as in modern genera. (2) The older crocodiles had the opening of the nasal passages into the mouth placed far forward, whereas living crocodiles have them placed in the extreme back of the mouth. This change in position is of advantage in that it makes it possible for the animal to breathe while it is drowning its prey. The early Mesozoic crocodiles were probably obliged to go to the land to devour their food. The marine crocodiles were doubtless descended from a group that lived in rivers, and these, in turn, from terrestrial or amphibious ancestors, although the earliest known crocodiles are marine.

Shortly before the close of the Jurassic, a side branch (Thalattosuchia) appeared, which were thoroughly adapted to a marine existence. They were covered with a bare skin, without scales, and the tail ended in a long fin. The fore limbs were paddle-like, while the hind limbs were less modified, probably because of the necessity of visiting the shore for egg-laying. After a brief existence, this family disappeared.

The crocodiles underwent a marked change early in the Cretaceous, at which time the more modern crocodiles and gavials were developed.

Marine Reptiles—One of the most significant features of reptilian evolution is the way in which the reptiles, after they had become adapted to land life, were enabled by their superior organization and greater activity as air-breathing animals, to re-invade the sea repeatedly and successfully. Hardly had the reptiles become well established upon the land, before some took to the water and became perfectly adapted to a marine existence. Members not only of one but of several classes of reptiles were so modified. It is not remarkable that some of the land reptiles should have changed their habits, when it is remembered that in the shallow waters bordering the land there was an abundant supply of fish for food, and also that there was probably some overcrowding on the land which would force the weaker species to take the food that was not to the liking of their stronger neighbors.

Ichthyosaurus (Greek, *ichthus*, fish, and *saurus*, reptile)—The most conspicuous features of reptiles of this order (below) are the heavy body, with its pointed head and numerous teeth, and the powerful tail, with its vertical fin, adapted for rapid propulsion. Some individuals grew to be 40 feet long, although the usual size was very much less. The jaws of some individuals were five feet long and were furnished with 200 conical teeth. The eyes were large, not only in proportion to the size of the skull, but in the largest species actually attained in some perhaps the size of the human skull, and were provided with a ring of radiating, bony plates (sclerotic plates), like those of the early amphibians, which were apparently for the purpose of focusing the eye, as well as for protection.

Ichthyosaurus, showing both the skeleton and the "shadow" made by the carbon of the fleshy parts of the body.

The limbs consisted of paddles, made up of three or more rows of polygonal bones, the whole being covered with a leathery membrane. The skin was smooth and without scales. The vertebrae were biconcave, as in fishes. That Ichthyosaurus was carnivorous is shown by the contents of the abdomen, which often contains fish scales and the remains of shelled cephalopods (Belemnites).

They were remarkably well adapted to aquatic life, as is shown by the paddle-like limbs; by the outline of the body, which was so modified as to permit movement through the water with as little resistance as possible; by the sharp teeth for the catching and retention of slippery prey. The occurrence of undigested, immature young within the ribs of a number of specimens indicates that their offspring were produced alive.

Although only the later stages of the evolution of the ichthyosaurs are known, yet it is evident that they were descended from land reptiles. This is shown by the structure of the limbs of the earlier forms, which were more like the legs of land animals than were those of the later species.

The following progressive changes, fitting the animal for marine existence, have been traced: (1) The limb became more paddle-like and less leg-like, both in the structure of the skeleton and in the external shape. (2) The head became longer and better adapted for catching fish and other slippery animals. (3) The eyes became larger and more efficient for seeing in the water. (4) The neck became shorter. (5) The body gradually became more fishlike in shape and could move through the water more rapidly and with less resistance.

Ichthyosaurs began in the Triassic, culminated in the Jurassic, and lived, for a short time, in the Upper Cretaceous. During the Jurassic, they appear to have been very abundant and to have occupied every sea.

Plesiosaurus (Greek, *plesios*, near, and *saurus*, reptile)— These marine reptiles are characterized (pg. 367) by a short, stout body, a short tail, and usually by a long neck and small head. The tail was probably of greater use in steering than as an organ of propulsion, the powerful, paddle-like limbs being for that purpose. These paddles had five digits, but each digit was made up of a large number of small bones, in some cases as many as 20. Plesiosaurs varied greatly in size, some being 30 to 40 feet long, but they usually did not attain a greater length than 6 to 15 feet. One American species (Elasmosaurus), for example, was 40 feet long, with a small head and a neck 22 feet in length. The other extreme was Pliosaurus, equally huge in bulk, but with a skull nearly 5 feet long and a neck of only a foot and a half. Most of the smaller Plesiosaurs had small heads. The skin was smooth, without scales. The sharp, flaring teeth show that the creatures lived on animal food, possibly on small fish or some of the cephalopods which were so numerous in the seas of the time. Judging from the shape of the body, they probably swam slowly, depending upon stealth rather than speed in capturing their prey.

It has been shown by a study of the neck vertebrae that the neck was too stiff for very quick movements, but would, nevertheless, be of great assistance both in capturing prey and in enabling an animal quickly to reach the surface for air. Within the body cavity of some skeletons, a large number of polished pebbles have been found— in one case a peck of them—from the size of a hen's egg to that of a baseball. These "gizzard stones"

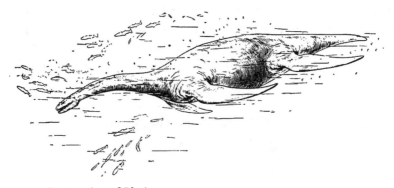

Restoration of *Plesiosaurus.*

were doubtless of use in grinding the food, which was swallowed whole. If the plesiosaurs fed, to any extent, on the shelled cephalopods, some such apparatus must have been extremely useful. Plesiosaurs ranged from the Triassic to the end of the Mesozoic and reached their greatest size in the Cretaceous, and, perhaps, their greatest abundance in the Jurassic. They were not closely related to the Ichthyosaurs and were probably descended from a different race of land reptiles.

Mosasaurus (Sea Lizards)—As the ichthyosaurs disappeared in the Upper Cretaceous, their place was taken by the mosasaurs (below), long, slender reptiles, with a scaly skin like that of modern snakes which attained a length of 35 feet or more, although usually smaller. The heads were pointed and provided with sharp, stout, pointed teeth. The jaws were so constructed as to make it possible for the animal to swallow an object of almost the diameter of itself.

This was accomplished by a hinge in each half of the lower jaw (below) which permitted it to bow outward when open. The articulation of the jaw with the skull also assisted in this process. The limbs were not as greatly modified as in the Ichthyosaurs, but were completely paddle-like and resembled those of the whale. The great speed with which it could be propelled by its tail made the catching of its fish food an easy matter. Mosasaurs were descended from land animals and may have sprung from the same stock as modern reptiles. They were not well established until the Upper Cretaceous, in which period they rapidly diverged and swarmed the Atlantic and Gulf coasts and the interior seas. They had a wide distribution, being found in North and South America, Europe, and as far south as New

Skeleton of a Cretaceous *mosasaur* about sixteen feet long.

Zealand. They disappeared with the Mesozoic, after having had a comparatively short life.

Turtles—It is an interesting fact that, although turtles are so widely different from other forms at the present, yet, even when first known—in the Upper Triassic—they are as typically turtle-like as now. Jurassic turtles were abundant, had a world-wide distribution, and were closely related to existing genera. The first strictly marine turtles (in which the feet are modified to form "flippers") have been found in the Cretaceous, one of which, Archelon, was of great size, the head measuring three feet in length, the total length of the animal being 12 to 14 feet. In this case, the shell proper had disappeared and the broadened ribs were possibly covered with a soft skin, as in some living marine turtles (Dermochelys). Land turtles did not appear until the Tertiary.

A number of suggestions as to the origin of turtles have been offered, but since the earliest known species are far from being generalized, the whole matter is, as yet, in doubt.

Flying Reptiles (Pterosaurs)— Either because of the overcrowding of the land, or for some other reason, a race of flying reptiles was developed during the Jurassic and Lower Cretaceous, and occupied the realm of the air, in which there was no competition.

The pterosaurs (right, pgs. 370, 371) are as extraordinary, in many ways, as any animal that ever lived. They had a short body, hollow bones, a rather large but

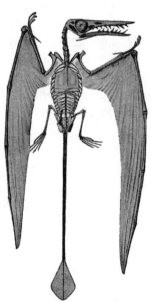

A Jurassic pterosaur (*Rhamphorynchus*). Length about twenty inches.

light head, and jaws which at the beginning of the race were provided with slender teeth, but which in some highly specialized later genera were toothless and sheathed with horn, as in modern birds. The most remarkable and characteristic features, however, were the large, membranous wings, supported by one greatly elongated finger, the fourth. The breastbone, to which the muscles of flight were attached, was large and keeled, and the shoulder girdle was strong. Some had long tails with a kind of rudder at the extremity, and others were tailless. The pterosaurs varied greatly in size; some were as small as sparrows, some were the size of partridges, while others were the largest flying creatures that ever lived, the wings measuring over 20 feet from tip to tip (pg. 371).

One of the best known and least specialized genera of the Jurassic pterosaurs (Dimorphodon; Greek, *dimorphos*, two-formed, and *odont*, tooth) (below) had, as the name implies, two kinds of teeth, those in front of the jaw being sharp and strong and fitted for tearing, while those in the back of the jaw were small and sharp, with a sawlike edge. This pterosaur could probably walk on all fours or on its hind legs alone. When standing on its hind legs, it was less than two feet high, and its wings had a spread of a little more than four feet. The wings (below) were formed by a naked membrane, without feathers

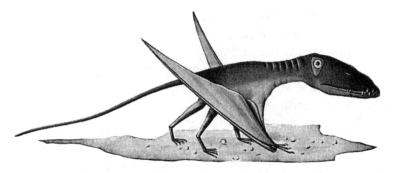

A Jurassic pterosaur (*Dimorhodon*). The extreme length from the tip of the nose to the end of the tail was a little more than three feet.

or hair, stretching from the body to the greatly elongated fourth finger. Although the least specialized of the pterosaurs, they possessed few characters connecting them with other reptiles.

Perhaps the most highly specialized animal that ever existed was a pterosaur (Pteranodon; Greek, *pteron*, wing, and *a-odont*, without a tooth) that lived in the Upper Cretaceous. In this animal (below) it would seem that everything possible was sacrificed for flight. The upper portion of the body, the wing, shoulder, and breast were all extraordinarily strong, while the lower portion of the body and hind limbs were very weak. The head was highly developed, being long and slender, with a dagger-like beak and toothless jaws. The head was about four feet long, the body only slightly longer. It is thought that, notwithstanding its large size, it was so lightly built that in life it did not weigh more than 25 pounds. In fact, the bones of the largest specimen, even as petrified, do not weigh more than 5 or 6 pounds. When not sailing in the air, pteranodons probably spent their time suspended from cliffs or trees by their slender, clawed fingers. Pteranodons lived upon fish, as is shown by the fishbones and scales found within their skeletons. Because of the small pelvis, we must suppose that if they laid eggs, the eggs were very small.

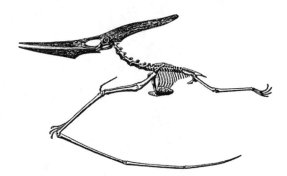

Skeleton of *Pteranodon*, the most highly specialized of the pterosaurs (Cretaceous). Everything was sacrificed for flight and feeding. The wings measured almost 20 feet from tip to tip.

Because of the high degree of specialization of the earliest pterosaurs, nothing definite can be said as to their ancestry, but it is possible that pterosaurs, carnivorous dinosaurs, and birds all sprang from a common ancestor (such as Euparkeia). Although flying animals, they were not the ancestors of birds. The first evidence of their appearance has been found in the later Triassic, but they did not reach North America until after the middle of the Jurassic, at which time they swarmed over the epicontinental seas. None lived into the Upper Cretaceous.

Toothed Birds

Archaeopteryx—If the skeletons of the earliest known bird had not had feathers associated with them, it is probable that they would have been described as belonging to the Reptilia, with some birdlike characters. This oldest bird (Archaeopteryx; Greek, *archaios*, old, and *pterux*, a wing) (below) was about the size of a small crow, with a small, stout, birdlike head and a birdlike brain, but its jaws, instead of being of horn as in modern birds, were provided with sharp, conical teeth. The wing was peculiar in having three reptile-like claws, by means of which the bird could crawl about the trees, instead of flying. The hind limb was much like that of modern birds and had four

Restoration of Archaeopteryx (Jurassic). The long vertebrated tail, clawed wings, and teeth are well shown.

digits. The vertebrae were biconcave, as in fish and some reptiles. The tail was one of the most peculiar features in that it was vertebrated, with a pair of feathers springing from each joint. In modern birds, the feathers are arranged like the sticks of a fan. Archaeopteryx was not well adapted for flying, as is shown by the poorly developed breastbone. With the exception of occasional short flights, it probably soared somewhat as flying squirrels do today. Birds probably did not have dinosaurian ancestors, but were presumably derived from a group of primitive, dinosaur-like reptiles that were capable of running on their hind legs. Archaeopteryx is not known to have lived in America; and only a few specimens have been found in Europe, all of which are from the Jurassic.

Hesperornis—This bird was adapted for life in water instead of in air. It was the largest bird of its time, attaining a length of nearly six feet. The jaws were supplied with small teeth which, instead of being set in sockets as in Ichthyornis were in grooves. As in snakes, the jaws were so constructed as to permit the bird to swallow large prey. The tail was vertebrated, but was intermediate between Archaeopteryx and modern birds. Hesperornis was perfectly adapted for aquatic life. Wings were wanting, and only a rudimentary bone was left to show that a wing existed in its remote ancestors. The feet were modified in a manner not found in any other bird, living or fossil, being so joined to the leg as to turn edgewise as the foot was brought forward. The resistance of the water was in this way lessened. This adaptation to aquatic conditions may have been so perfect that not only flying, but walking as well, was abandoned. The bird was covered with soft feathers, as fossil impressions show. Hesperornis lived only in the Upper Cretaceous.

Ichthyornis (Greek, *ichthus*, fish, and *ornis*, a bird)—This bird (pg. 374) was about as large as a pigeon and must have looked very much like a modern bird. It was, however, radically different in some particulars. Its slender jaws were toothed, the teeth being small and set in sockets, twenty on each side below and fewer above. The vertebrae were biconcave, like those of fishes and many extinct reptiles but no modern bird. The tail was about midway between the vertebrated tail of Archaeopteryx and those of the birds of the Tertiary and today. The strongly keeled breastbone for the attachment of the muscles proves that it was a powerful flyer. Although Ichthyornis shows a distinct advance over Archaeopteryx in its less vertebrated tail, its power of flight, and the loss of the claws on the fore limbs, an equal or greater change is to be seen between the Cretaceous birds and those of the Tertiary.

It is interesting to speculate upon the cause of the abandonment of teeth for a horny jaw both by birds and pterosaurs. A toothed jaw would insure the retention of every fish captured, but would prove a hindrance to its being swallowed quickly. Possibly toothed birds and pterosaurs were obliged to go to land before being able to devour their food, but those with horny beaks could bolt their food on the wing.

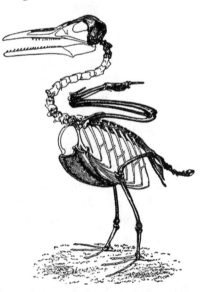

Fossil birds are comparatively rare, even from the rocks of periods when birds were abundant, because of the lightness of the skeletons, which caused the carcasses to float on the seas for a long time before sinking to the bottom, with the result that the skeletons were usually devoured by fish or beasts of prey before they had a chance to be buried in the sediments. Bird fossils are rare in the Mesozoic also since they are not now and were not then to any

Ichthyornis, a small, toothed bird with strong powers of flight (Cretaceous).

extent swamp dwellers, and what is known of Mesozoic land life is chiefly limited to the fauna of the swamps. Because of this, it is probable that only a small part of the bird life of the Cretaceous is known.

Mammals

Jaw of either a primitive mammal or a theromorph reptile (Triassic), twice the natural size.

A few very small lower jaws have been discovered in Triassic deposits of America and Europe (Dromatherium, left), which have been considered by some students as reptilian and by others as mammalian. If they are reptilian, they are theromorphs; if mammalian, either monotremes (egg layers that suckle their young, like Platypus), or marsupials (mammals that produce their young in an immature condition, like the opossum). It is suggestive, however, to note that creatures that are either reptiles with strong mammalian characters or mammals with well-marked reptilian characters existed before true mammals made their appearance. With one exception, the few mammalian remains found in the Mesozoic are of small size, indicating animals not larger than rats. Those of the Jurassic and Cretaceous appear to be insectivores and to be either monotremes (egg-laying mammals) or marsupials, but none clearly belong to the highest type of mammals (Eutheria, common mammals of today) nor are they closely related to other forms.

The probable relationships of the mammals and other vertebrates are shown in the table (pg. 376).

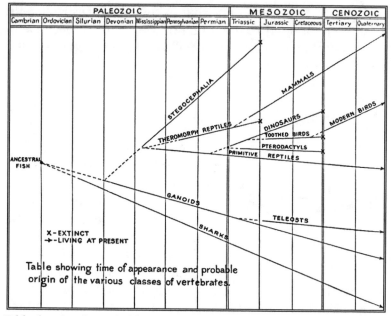

Table showing the probable relationships of vertebrates, with their geological distribution.

Plants

The vegetation of the Mesozoic is of great interest, since it was during this period of world history that the now dominant types of plants were introduced. Mesozoic plant life, as indeed does the plant life of all geological ages, affords a reliable clue to the climatic and physical conditions which prevailed during the several periods, and incidentally offers, to some degree, an explanation of the striking changes which took place in the animal life.

In discussing the vegetation of the Mesozoic, a division into Lower and Upper should perhaps be made, because of the introduction of modern plants (angiosperms) in the Lower

Cretaceous and the subordination of the typical early Mesozoic plants in the Upper Cretaceous.

Owing to considerations, physical and otherwise, concerning which there is not complete agreement, the lower part of the Triassic affords but scant remains, and it is not until we come to the upper part (Rhaetic) that the plant remains can be really dignified as a flora. In North America there are less than 150 species, and the entire Triassic flora of the world probably does not exceed 300 or 400 forms.

Horsetails—The *horsetails,* which entirely replaced the calamites of the Carboniferous, do not appear to have differed markedly from those now living, except that they were often of larger size, some having been reported that are from five to eight inches in diameter. It is presumed that they formed dense growths, like canebrakes, in or along swamps, marshes, or lakes, as do certain of their living representatives today, the largest of which—a South American species—is an inch in diameter and 20 or 30 feet in height.

Cycads—Among the most characteristic and abundant plants of the Triassic and Jurassic was the great group of *cycads* (using the term in the broad sense to include the Bennettitales and Cycadales). They were similar in general appearance to those of the present, but differed in some important characters. Fossil cycad trunks (pg. 378) are generally short and stout, never apparently reaching a greater height than three or four feet, and usually much less. As in modern cycads, a crown of long, stiff leaves sprang from the top of the trunk, which was scarred throughout by the leaf-bases of previous leaves. Some fossil cycads from South Dakota are so completely preserved that such delicate structures as immature leaves, flowers, pollen, and some seeds with their contained embryos, are retained in remarkable perfection. Because of this perfection of preservation, almost as much is known of the structure of this extinct group as of its living relations. The position of the seed-bearing cone and the large leaves bearing the pollen sacks of a fossil cycad is well shown in the diagram (pg. 379). Cycads, which appear to have grown on the dryer lowlands about the swamps, had their origin in the Permian, reached their greatest abundance in the Jurassic, and are rare after the close of the Mesozoic.

A group of cycad trunks (Bennettites).

Ferns were common throughout the era, wherever the conditions were favorable for their growth.

Gymnosperms—The conifers (evergreen trees of today) lived on the higher lands during the Mesozoic and were represented by pines, cypresses, yews, and araucarias (monkey-puzzle), the last being especially abundant in the Jurassic. The sequoia (redwoods and "big trees" of California) had a notable development in the Upper Cretaceous. Because of the resinous character of the wood of the coniferous trees, it was often preserved, sometimes in a remarkable degree of perfection. On the whole, the Mesozoic conifers were not very different in general appearance from those of today. The early Triassic forms, however, were somewhat dwarfed, while those of the later Triassic were gigantic trees, often over 100 feet in length and from four to eight feet in diameter.

The maidenhair tree, *ginkgo*, was abundant and had a world-wide distribution in the Lower Mesozoic. This once numerous family is now represented by but one species, which probably would have been long since extinct had it not been preserved by cultivation about the Buddhist temples in Japan and China. The modern ginkgo comes of a long-lived family. Evidence has been found to indicate that, if not the existing

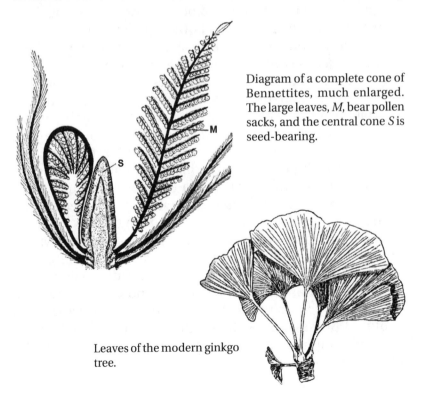

Diagram of a complete cone of Bennettites, much enlarged. The large leaves, *M*, bear pollen sacks, and the central cone *S* is seed-bearing.

Leaves of the modern ginkgo tree.

genus, at least a closely related one lived in the Paleozoic. The impressions of the leaf, seeds, and male cones of Jurassic trees are very similar to those of the trees now living (pg. 379, bottom).

Angiosperms—The *flowering plants* are, at present, the commonest of all plants, four sevenths of the existing species belonging to this class. They have, however, a much shorter known history than the conifers and various other groups, since no positive evidence is at hand of their existence prior to the Lower Cretaceous. At the beginning of this period, the horsetails, cycads, conifers, and ferns were the common and conspicuous forms; but, before its close, flowering plants of both divisions (monocotyledons, represented today by palms, lilies, and grasses, and dicotyledons, of which the elm, rose, and clover are examples) had become prominent. The sassafras, fig, willow, magnolia, tulip tree, laurel, and others have been recognized. In the Upper Cretaceous, the flowering plants became even more conspicuous, and are represented, among many others, by palms, beeches, birches, chestnuts, and poplars.

The introduction of flowering plants was, perhaps, the most important and far-reaching event in the whole history of vegetation, not only because they almost immediately became dominant, but also because of their influence upon the animal life of the succeeding periods. Hardly had flowers appeared before a great horde of insects which fed upon their honey or pollen seem to have sprung into existence. The nutritious grasses and the various nuts, seeds, and fruits afforded a better food for non-carnivores than ever before in the history of the world. It was to be expected, therefore, that some new type of animal life would be developed to take advantage of this superior food supply. As we shall see in the discussion of the Tertiary, the mammals, which kept a subordinate position throughout the Mesozoic, rapidly took on bulk and variety and acquired possession of the earth as soon as they became adapted to this new food, quickly supplanting the great reptiles of the Mesozoic.

The flowering plants (angiosperms) had their origin, as far as is known, on both sides of the North Atlantic during the Lower Cretaceous. Some of the earliest of these are somewhat generalized, but do not give a positive clue to the group from which they were descended. Just when and where they began we do not know, but once started they spread rapidly and widely, and before the close of the Lower Cretaceous had reached California, Alaska, Greenland, and Bohemia.

Climate

Triassic—The climate of the Triassic of North America, Central Europe, and North Africa seems, as a whole, to have been arid, although some areas of considerable extent had sufficient rainfall to produce a luxuriant vegetation. The proofs of aridity are to be found in the widespread occurrence of gypsum and salt, and in the prevalence of "red beds" (rock of a red color). It is well known that the deposition of salt and gypsum is the result of evaporation in excess of supply, such as can happen only in arid regions. The explanation of the red color of sedimentary rocks is not so clear. If organic matter, either animal or plant, is plentiful in sediments, the contained iron will be in the form of the gray iron carbonate instead of the red iron oxide. At the present day, for example, although the rocks of the southern Appalachians are weathered to red clay many feet deep, the sediments derived from them are gray when deposited, because of the reduction of the iron oxide by the plant debris which they enclose. A less abundant flora, due to decreased rainfall, might readily result in the deposit of red sediments without reduction. On the other hand, attention has been called to the fact that, probably in many cases, red sediments were laid down in regions where the rainfall was undoubtedly not small.

The Triassic red sandstones and shales of the Connecticut valley, with their innumerable reptilian footprints, indicate aridity in another way. It was formerly thought that these deposits were laid down in a great estuary of the sea, under conditions similar to those of the Bay of Fundy today, in which the difference between high and low tide was great. As a result, during several hours of the day, extensive mud flats were uncovered, upon which the saurians of that time walked or ran in search of food or water and left their tracks. It has been shown, however, that a number of hours at least must elapse under known conditions, before mud can dry sufficiently to form sun cracks and to retain the footprints of animals. If these conclusions are correct, we must believe that the deposits of the Connecticut Valley and New Jersey were laid down in river valleys analogous to the Great Valley of California and other structural basins. In the shallow lakes which occurred, such for example as Tulare Lake in California, the depth of the water was greatly reduced by evaporation during longer or shorter periods, and the bottom of the shallower portions of the seas was exposed for several days or weeks at a time.

There is abundant proof, however, that in certain regions the rainfall during the Triassic was plentiful, due probably, as today, to the presence of mountain ranges which caused abundant precipitation on one side and deserts on the other. In Virginia, for example, beds of coal aggregating 30 to 40 feet in thickness indicate long-continued swamp conditions. Horsetails four to five inches in diameter, ferns of large size, some of them tree ferns, prove that the climate was favorable for luxuriant growth. The petrified trees of Arizona, some of which were eight feet in diameter and more than 120 feet high, do not indicate aridity, nor, for that matter, do they prove a moister climate than that of Arizona today, in which the great pines south of Flagstaff flourish. The complete or nearly complete absence of rings in the tree trunks indicates that there were no, or but slight, seasonal changes, due to alternations of heat and cold or wet and dry periods.

Jurassic—The presence of luxuriant ferns, many of them tree ferns, horsetails of large size, and conifers, the descendants of which live in warm regions, all point to a moist, warm, subtropical climate during the greater part of the Jurassic; although arid regions unquestionably existed. The animals also indicate a warmer climate in the northern regions than at present. Saurians and ammonites lived within the Arctic circle, and corals 3000 miles farther north than now. The presence in the late Jurassic of rings in the tree trunks of northern species shows that slight seasonal changes occurred.

Cretaceous—The climate of the Lower and Upper Cretaceous seems to have been milder than at present, even that of Greenland being temperate or warm temperate. The distribution of marine fossils indicates the existence of climatic zones according to latitude, but the vegetation does not show this so clearly; for example, oaks, maples, and magnolias grew in Greenland and nearly as far north as Alaska, in the Lower Cretaceous. If a cold, but not frigid, polar sea existed from which currents extended southward, the apparent contradiction in the evidence of the plants and animals would be explained.

Coal

Triassic—Coal beds occur in the four systems of the Mesozoic. In the United States, coal of the Triassic Age was worked as early as 1700 in the Virginia-North Carolina coal fields, but these deposits are of more interest historically than economically. Coal of this age occurs also in Germany, Sweden, South Africa, and Australia, and, as in North America, is composed of horsetails, ferns, and cycads. Coal in commercial quantities occurs in Hungary, in several of the countries of Asia, in Australia, and New Zealand, in Jurassic formations.

Cretaceous—The Lower Cretaceous rocks bear coal locally in British Columbia and Alaska. The great coal-producing system of western North America is the Upper Cretaceous, the total quantity and extent of the coal formations being comparable to those of the Carboniferous. The quality is, however, usually inferior to that of the Carboniferous coal, being largely lignite, although some bituminous coal of excellent quality is produced, and in a few localities anthracite coal, made from bituminous and lignite coal by the intrusion of igneous rocks, is worked. It is interesting in this connection to note the presence of charred wood and charcoal in some of the Cretaceous beds, showing the existence of fire during the period. Although workable coal is found in all the stages of the Upper Cretaceous of western North America, that of the Montana and Colorado Stages is most important. The so-called Laramie coal has been found to belong largely to the Montana stage of the Upper Cletaceous and to the lowest stage (Fort Union) of the Tertiary.

CHAPTER 17

Cenozoic Era: Age of Mammals

Tertiary Period

Comparison of the Life at the Close of the Mesozoic and the Beginning of the Cenozoic—The Age of Reptiles apparently came abruptly to a close, and the Age of Mammals began. In the last stage of the Upper Cretaceous the dinosaurs were in the climax of their specialization and grandeur. The bulky Triceratops with his great horned head, the amphibious duck-bill dinosaur (Trachodon), as well as other armored dinosaurs, roamed about in the Rocky Mountain region. At the same time lived the swift and powerful Tyrannosaurus which doubtless preyed upon some of these herbivorous relatives. Associated with these great reptiles were small mammals of lowly organization and of small size. One of the most dramatic moments in the life history of the world is the extinction of the reptilian dynasty which occurred with apparent suddenness at the close of the Cretaceous, the very last chapter in the Age of Reptiles. This does not mean that the reptiles were wiped out of existence by some great cataclysm, but that, as measured by geologic time, the wane was rapid. What cause or causes produced this great result cannot be stated definitely.

(1) Change in vegetation has often been called in to account for the extinction of various groups of animals, but we find much the same vegetation after the extinction of the dinosaurs

as when they were abundant and at the summit of their specialization. Such trees as the fig, banana, sequoia, ginkgo, oak, and sycamore passed from one period to the other without alteration. This being the case, a change in food, unless under exceptional conditions, could not have been a cause of dinosaurian extinction. Moreover, since the vegetation remained so nearly the same at the critical time, it is not probable that the climate had been greatly modified. (2) It has also been suggested that the cause of their extinction was their inability to compete with the more agile and intelligent mammals, and the fact that their young, not having the maternal care of these higher vertebrates, were easily captured and destroyed by carnivorous mammals. Whatever the cause or causes, the great reptiles—marine, flying, and terrestrial—disappeared; and mammals soon occupied all the places in nature formerly held by them, the only reptiles surviving being those whose habits or inconspicuous form saved them from their competitors.

The Mesozoic types of birds with toothed jaws and vertebrated tails were replaced by the toothless birds with which we are familiar.

The difference between the invertebrate life at the close of the Mesozoic and at the beginning of the Tertiary is not great, although the species are different. The most noticeable feature, perhaps, is the absence of an abundant and varied cephalopod fauna which was so conspicuous in the Cretaceous seas.

Subdivisions of the Cenozoic Era—The Cenozoic (Greek, *kainos*, recent, and *zoe*, life) is the last era in the world's history. It is also called the Age of Mammals because of their predominance and importance from the beginning of the era to, and including, the present.

This era is separated into two periods, Tertiary and Quaternary, the first lasting until the appearance of the great ice sheets and the second from that time to the present. They were of very unequal duration, the former being several millions of years long, the latter probably less than one million. The life of the Tertiary became more and more modern as the end was approached, and the period is subdivided into four epochs, as is shown by the table below. In determining the age of the rocks of the Tertiary, however, the percentage of modern species is not computed, but the separation is based on certain species which had a short life and are characteristic of a single epoch.

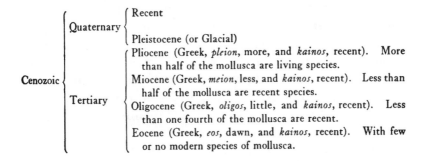

Cenozoic
- Quaternary
  - Recent
  - Pleistocene (or Glacial)
- Tertiary
  - Pliocene (Greek, *pleion*, more, and *kainos*, recent). More than half of the mollusca are living species.
  - Miocene (Greek, *meion*, less, and *kainos*, recent). Less than half of the mollusca are recent species.
  - Oligocene (Greek, *oligos*, little, and *kainos*, recent). Less than one fourth of the mollusca are recent.
  - Eocene (Greek, *eos*, dawn, and *kainos*, recent). With few or no modern species of mollusca.

Physical Geography of the Tertiary—Eocene

The deformations that raised the Rocky Mountains and drained the western interior of North America apparently affected the continent as a whole, and for a time, the Atlantic and Pacific coasts were farther out than in the period under discussion (pg. 390); thus portions of the Cretaceous sea bottom were exposed to erosion. This is shown by the old land surfaces (unconformities)—not, however, universal—between the Eocene and the underlying formations, on both the Atlantic and Pacific borders of the continent. Since, when traced eastward, the Cretaceous peneplain disappears beneath Eocene deposits, we know that the beginning of the latter epoch was marked by submergence. An important point to be kept in mind in the discussion of the physical geography of the Tertiary is that North America has been a relatively stable continent since the close of the Cretaceous.

Atlantic and Gulf Coasts—On the Atlantic coast, deposits occur on Martha's Vineyard island, but not on the mainland of New England or Canada (Newfoundland was probably a part of the continent at this time), and extend from New Jersey into Texas, by way of Alabama and Mississippi, then up to the mouth of the Ohio River and thence southwest. The Atlantic deposits of this period are usually loose and incoherent sands, clays, and green-sand marls, derived largely from the Cretaceous formations but also to some extent from older formations. In the Gulf regions the rocks are more consolidated, sandstones, limestones, and shales being common. Extensive lignite deposits occur in Texas and Louisiana, which may become valuable at some future day when bituminous coal is more costly than now. These lignite beds were formed from the peat bogs that existed on poorly drained portions of the low-lying coast, just as peat is being formed in similar regions today.

Pacific Coast—In the western portions of the continent the rocks of the period are, for the most part, sandstones and shales, with occasional conglomerates and tuffs, which rest unconformably on the older rocks in many places, but in others are conformable, the division being determined by the change in the fauna. The diatomaceous shales which occur at the top of the series (in the vicinity of Coalinga, California) should be mentioned, since they are believed to be the source of important deposits of petroleum.

During the early part of the Eocene, marine conditions prevailed over a considerable territory, but these later gave way to brackish or freshwater swamp conditions. The physical history during the latter part of the period is one of persistent but frequently interrupted submergence, in which the alternation of many coal beds (some workable) with deposits of fine shale and coarse sandstones indicates that, during this great subsidence, the depth of the water frequently changed. At times, the sinking proceeded more rapidly, and the deepened water was then filled with sediment until the tide-swept flats became marshes and, for a time, vegetation flourished vigorously in the moist lowlands, this rotation being repeated intermittently. This condition is believed to have prevailed in Alaska, western Oregon, and the Great Valley of California. Most of the coal of the west coast belongs to this epoch, making this the "Eocene Carboniferous" of the west. In the later Eocene, elevation and erosion, accompanied by volcanic outbursts and extensive lava flows, occurred in Oregon and Washington. The presence of Atlantic species in the marine deposits shows that an oceanic connection, probably in the Central American region, was in existence for a time.

Western Interior—The Eocene deposits of the western interior (pg. 390), with the exception of a few small areas in Colorado, are confined to the region between the Sierra Nevadas and the Rocky Mountains. It is thought that the region under discussion was not greatly elevated above sea level, although the summits of the mountains probably stood sufficiently high above the general level of the plains to permit the vigorous erosion which was in progress during the epoch, and which furnished the waste to form a great thickness of sediments. The mountains and hills, formed by folding, by faulting, by warping, and by volcanic debris, enclosed basins and valleys in which the streams deposited the sediments obtained from the steep slopes of the higher lands. These sediments were deposited partly in lakes and partly in alluvial fans in front of the valleys which the streams had cut in the mountain slopes. The most important deposits, however, were laid down in flood plains, in deltas, and in swamps. From time to time, the area of deposition shifted, because of the filling up of old basins or the warping of the land. Lakes were also in existence, the most famous being one in Wyoming in which the Green River formation occurs, consisting of impure limestone and thin, fissile calcareous shales, often as thinly laminated as paper. Between the leaves of these shales, remains of plants, insects, and fishes are beautifully preserved, but no remains of mammals are found, except in the form of footprints. Since these sediments were deposited in more or less isolated basins, the work of correlating them with each other, and especially with the marine deposits of the Atlantic and Pacific, has been difficult.

The Eocene was an epoch of great coal formation, especially during the earlier portion (Fort Union). The great lignite deposits that cover one half of North Dakota were formed at this time, as were also extensive areas in Wyoming and Montana.

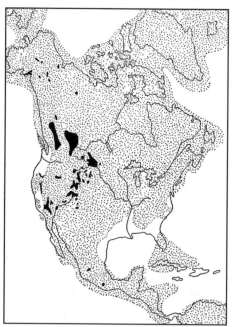

The Eocene was brought to a close by crustal movements of some importance which, on the Pacific coast, resulted in the draining of certain areas and the lowering of others to below sea level. In the same region some mountain ranges (Klamath) were again bowed up to some extent, and others (Coast Ranges) began their development. In the Great Plains region the changes were such as to bring about aggradation where degradation had

Map showing the probable outline of North America during a portion of the Eocene. The continental deposits are shown in solid black.

formerly prevailed. The interior mountain region of the west was elevated and drained, and in subsequent epochs was not a region of extensive deposition. The eastern coast remained much as before. A narrow strip of land was added to both the Atlantic and Gulf coasts.

Eocene of Other Continents—The evolution of the

continents of Europe and Asia was not so far advanced at the beginning of the Eocene as that of North America. Seas covered large areas that are now land, and there were probably extensive land masses which are now covered by the ocean. Europe was smaller than at present and at times was entirely separated from Asia by a narrow sea on the east side of the Ural Mountains. The most marked feature of the Eocene European continent was the greatly expanded Mediterranean Sea which, with its extensive arms, covered the sites of the conspicuous mountains of the present: the Pyrenees, Apennines, Alps, and Urals. Above the surface of this sea, numerous islands probably stood on the sites of some of the ranges. The greater part of Spain seems to have been separated for a time from the mainland by a sea which also covered a portion of southern France.

Not only Europe, but Asia and Africa as well, were far from having attained their present outlines. The greater part of Africa north of the equator was under water, and an extension of the Mediterranean Sea reached to the Indian Ocean. Portions of Australia, New Zealand, Patagonia, and the West Indies were also submerged.

In portions of Europe and Africa, a great thickness of limestone, made up of large Foraminifera (nummulites, p. 445), was deposited; besides which, an immensely thick mass of sandstone and shale which now outcrops on the Alps was also laid down. The nummulitic limestone was largely used in the construction of the pyramids of Egypt. The Eocene strata have since been raised to great heights, as is indicated by their presence on the Tibet Plateau at an altitude of 20,000 feet, in the Himalayas 16,000 feet above the sea, as well as high up on the Alps, Pyrenees, Caucasus, and other mountain ranges.

It will be seen from the above imperfect history that the outlines and mountainous regions of Europe and Asia were very different during the Eocene from what they are today.

Oligocene

The Oligocene, which followed the Eocene, is sometimes included in the latter, but is usually separated from it because of the distinctness of the two series in Europe, and also because they can be readily separated in North America whenever fossils occur.

Atlantic and Gulf Coasts—The Oligocene does not have a wide distribution on the Atlantic coast, but is well represented in the Gulf region, where 2000 feet of strata, rich in marine invertebrates, occur. The Oligocene in these regions rests upon the Eocene without a break, the two series being distinguished by a change in fauna. A great development of marls and limestone of this age in Central America and the West Indies shows that submergence was widespread in these regions. An island was raised in northern Florida early in the epoch which, by further arching of the sea bottom, became joined to the mainland in the Miocene.

Western Interior—On the Great Plains region continental deposits of this epoch occur at various points from British Columbia to Mexico, and outcrop from two to three hundred miles east of the Rocky Mountains. They seldom rest upon the Eocene, but on the worn surfaces of the Upper Cretaceous, showing that while deposition was taking place in the mountain basins of the Eocene, the region of the Great Plains was an open, rolling country, traversed by streams which were degrading its surface. A picture of the plains region in Oligocene times is that of broad, gentle, eastward slopes from the Rocky Mountains, plane or gently undulating and not mountainous, bearing broad streams with varying channels, sometimes spreading into shallow lakes, but never into vast freshwater sheets. Savannahs were interspersed with grass-covered pampas traversed by broad, meandering rivers. This land was dry in dry seasons, but was flooded in very high-water periods. The materials were partly erosion products of the Rocky Mountains and Black Hills, such as true sandstones and conglomerates, but they include also fine layers of volcanic dust, wind-borne from distant craters in the mountains, far out on the plains of Nebraska and Kansas.

Pacific Coast—On the Pacific coast, the Oligocene was an epoch of elevation and erosion, during which the land was not high except in a few places, as is indicated by the fine character of most of the sediments. The areas of deposition on what is now land were comparatively small.

Oligocene of Other Continents—In general, the distribution of land and water was different in the Oligocene from what it was in the preceding epoch. One important transgression of the sea covered Germany and Belgium and at the time of greatest extension joined the North Sea with the Mediterranean and Aral seas. In France and Russia, large areas were beneath the water. In the Paris basin, the presence of salt and gypsum furnishes a clue to the climate during a portion of the epoch. In various parts of Europe (Germany, Switzerland, southern France, and Bavaria) extensive swamps were present in which were accumulated the lignite deposits that are now workable to some extent.

Miocene

The outline of North America was practically the same in the Miocene (pg. 394) as in the Eocene, with the exception of the Mississippi embayment which was reduced in size, and the Florida peninsula which was formed later in the epoch.

Atlantic and Gulf Coasts—On the Atlantic and Gulf coasts the strata rest—often unconformably—on the Eocene or Oligocene, and, in general occur in a narrow, interrupted belt parallel to the older formations from Martha's Vineyard southward. The Miocene strata in some localities overlap the Eocene to landward, completely concealing it. The sediments on the Atlantic coast consist chiefly of sands, clays and marls, with occasional beds of diatomaceous earth from 30 to 40 feet thick. In Florida, Georgia, and in the Gulf region, limestones are the rule. The deposits of this epoch on the Atlantic and Gulf Coasts are comparatively thin, being only 700 feet thick in New Jersey, 400 in Maryland, and even less in North Carolina.

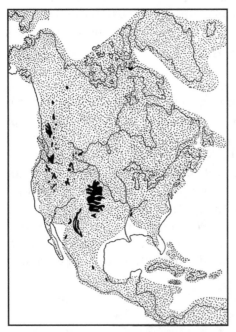

Map showing the probable outline of North America during a portion of the Miocene. The continental deposits are shown in solid black.

Economic Products of the Miocene—The economic products of the strata of this time are the phosphates of Florida, the oil of Louisiana, and the diatomaceous earth of the Atlantic coast.

Diatomaceous earth resembles chalk in color but is lighter in weight, and, since it is composed of silica, does not effervesce with acids. On account of the hardness of its constituent parts and its extreme fineness, it is used as a base in the manufacture of preparations for cleaning and polishing silver, nickel, etc. Since it is porous, it has been used as an absorbent for nitroglycerin in the manufacture of dynamite. It is also used as a non-conductor of heat.

The valuable phosphate deposits of Florida are believed by some investigators to have originated by the leaching of guano, or bone beds, and the deposition of the phosphate in the underlying limestone, either by precipitation in the pores of the rock or by replacing the limestone molecule by molecule. The phosphate may, however, have been disseminated through the beds in small quantities and later concentrated as the more soluble limestone was dissolved and carried away.

Western Interior—In the Great Plains region east of the Rocky Mountains, the conditions traced in the Oligocene continued, and were probably not unlike those now prevalent where the flood plains of the upper Paraguay, Amazon, and Orinoco rivers of South America are confluent. In this portion of South America is a region larger than that occupied by the Miocene deposits of North America, with all the conditions necessary for the deposition and present distribution of sandstones, clay, and conglomerates, together with the preservation of animal remains. North American Miocene formations are found from Montana into Texas, although largely covered to the south and east by later deposits. Sediments of this age occur also in Montana, Nevada, Colorado, Oregon, British Columbia, and Alaska.

A lake existed in Colorado at this time (Florissant) which is interesting because of the excellent preservation of many insects and plants in its deposits. It lay in a narrow valley in the vicinity of active volcanoes, whose numerous eruptions spread ashes over its surface, burying the insects and plants which had been carried into it.

Pacific Coast—The restricted seas of the Oligocene on the Pacific coast were much expanded during the Miocene, although at no time, as will be seen by consulting the map (pg. 394), was a large portion of what is now land in that region submerged. The southern portion of the Great Valley of California (San Joaquin) was beneath the sea early in the epoch (Vaqueros), and in this bay a great thickness of marine sediments, consisting of sands and clays with some conglomerates, was laid down. The variation in the lithological character of the deposits within short distances is believed to have been caused by the rather local elevation of land due to faulting and subsequent stream rejuvenation. The California earthquake rift is first known to have been a plane of movement at this time. This early (Vaqueros) sedimentation was followed by the deposition of sandstone, volcanic ash, and limestone, and a great thickness of diatomaceous material (Monterey). It was an age of diatoms. These small marine plants lived in extreme abundance in the sea and fell in showers with their siliceous tests to add to the accumulating ooze of the ocean bottom, just as they are forming ooze at the present day in some oceanic waters. It is well known that diatoms multiply with extreme rapidity. It has been calculated that, starting with a single individual, the offspring may number one million within a month. One can conceive that under very favorable life conditions such as must have existed, the diatom frustules may have accumulated rapidly at the sea bottom and aided the fine siliceous and argillaceous sediments in the quick building-up of the thick deposits of the Middle Miocene time, some of which are a mile through. These diatomaceous shales are the source of some of the richest petroleum deposits of California.

During much of the early portion of the Miocene and continuing somewhat later, faulting, folding, and volcanic outbursts of considerable magnitude occurred. Great volcanoes were active from Washington and Oregon along the Pacific ranges of California, almost as far south as Los Angeles. During the middle of the period, mountain building and great local deformations took place, the effects of which were felt from Puget Sound to southern California. Extensive faulting along the earthquake rift and other fault zones occurred, while in other regions, low, broad folds were formed. The combined result was the uplift of the Coast Ranges of California and Oregon to an altitude of several thousand feet. The Cascades of Washington were also increased in height. This stage of diastrophism was followed by subsidence (Upper Miocene), as a result of which the northern part of the Great Valley (Sacramento) was submerged and in it were deposited sands and clays and beds of diatoms.

By the close of the Miocene, the Klamath and Sierra Nevada mountains were peneplained, the material derived from them having been deposited in the Great Valley and coastal belt of northern California, forming the thick Tertiary strata now found there. These strata are composed of 8000 feet of sediments, largely belonging to the Upper Miocene, as well as an equal amount from the earlier stages. Volcanoes had practically ceased to be active over a large portion of the territory but were probably still in eruption in some localities. The Miocene deposits on the Pacific coast are much folded; and some are even overturned, being in marked contrast in this particular to those of the Atlantic and Gulf coasts, which are nearly in the position they had when first laid down.

In addition to the marine sediments just discussed, continental deposits, consisting of sands and clays with some iron and coal, were being laid down during the Lower Miocene in the northern part of the Great Valley. From the western flanks of the Sierra Nevadas, auriferous gravels were carried down by the streams and dropped in their beds during portions of the period, producing the "deep auriferous gravels" and later the "bench gravels," some of which, as now, were buried beneath streams of lava and beds of tuff.

Mountain Building—Before the close of the epoch, the upheaval of the Coast Ranges of California and Oregon and the Cascades of Washington occurred; the fault along the east of the Sierra Nevadas was made; the growth of the present Sierra Nevadas was begun and, as will be seen later, many of the great mountain ranges of the world were elevated. During this epoch, too, the plateaus of Utah and Arizona were raised so as to permit the Colorado River to begin the excavation of its great canyon. The rugged scenery so characteristic of the west is the result of elevation which, for the most part, began at this time.

Basis for Separation into Periods—In the discussion of eras and periods, attention has frequently been called to the fact that they were brought to a close by deformations, some great and some small, which produced mountain ranges, or raised or lowered large areas of the earth's surface. We have just seen, however, that one of the great times of mountain building occurred, not at the end of an era, nor the close of a period, but in the midst of an epoch. It should also be remembered that climatic and other changes thus produced had little effect on the contemporary life of the time. In other words, the separation of the history of the earth into chapters should be based, not upon the unconformities, however great, but upon the changes which the life has experienced. Fortunately, as should be expected, because of the effect of the physical conditions upon animals and plants, the sediments laid down during eras and periods are usually to be separated, not only by the rather sudden extinction of many species and the appearance of new ones, but by unconformities as well. The problem is not, however, a simple one. When, for example, a continent has been isolated for long ages, the animals and plants living on it may be largely of forms that belong to a previous epoch in other parts of the earth, just as in an age of electricity and cement some isolated tribes are still living in the Stone Age.

Igneous Activity—Perhaps no other period in the history of the earth since Pre-Cambrian times displayed such extraordinary volcanism as the Tertiary, and of the four epochs of the period, the Miocene was by far the most important in this particular.

It has already been seen that the great volcanic outbursts of the Pacific coast occurred during the Miocene—especially during the middle of the that epoch—covering that region of North America with ash which furnished the material for a great thickness of sedimentary deposits. Not only on the Pacific coast, but perhaps in every state west of the Rocky Mountains, some evidence of the igneous activity of this time can be found. It was during this period that a great quantity of lava and ash was poured into the basin of the Yellowstone National Park. Some of the forests that were buried in the ash at that time were later petrified and have been partially uncovered by erosion. Seventeen such petrified forests, one above the other, may be seen in one section (right).

The greatest area of lava in North America covers a region of between 200,000 and 300,000 square miles in Washington, Oregon, Idaho, and California, and is known, from the exposures on faulted and tilted blocks, to have a maximum thickness of at least 5000 feet. By far the largest bulk of this was outpoured during the Miocene. This enormous mass of lava was built up by successive lava flows averaging about 75 feet in thickness. On the canyon walls some of the sheets are seen to be separated by old soil beds, showing that the former lava surface had been exposed to the action of the weather so long as to be disintegrated to great depths before the overlying lava was outpoured. Lake

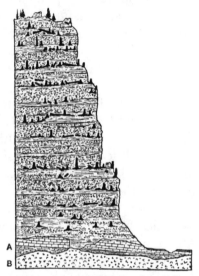

Section of the north face of Amethyst Mountain, Yellowstone National Park. Seventeen or more successive forests were covered with volcanic ash and the logs petrified. About two thousand feet of strata are shown.

beds, in one case 1000 feet thick, also rest upon one sheet and are covered by another. Although the Snake and Columbia rivers have canyons that reach a depth of several thousand feet, they have not yet succeeded in cutting their way to the base, except where they encounter the summits of the mountains buried beneath the flood of molten rock, or near the margin of the flow where it is thinnest.

Near the edge of the lava plateau, water is sometimes obtained from artesian wells, which have been sunk to the sheets of sand and gravel spread by rivers from the surrounding mountains upon the earlier lava flows whose surfaces were afterwards covered by late lavas. However, no water can be obtained in this way over large areas, because the porous lava permits the water to percolate down to great depths, where it appears as springs far down in the canyons. Because of the constant and uniform supply of water thus obtained, the volume of the rivers fluctuates less than in almost any other part of the continent.

Miocene of Other Continents—The seas that overspread Germany and Belgium in the Oligocene were withdrawn during the Miocene, but those of southern Europe not only remained extensive, but were so increased in size as to make that region an archipelago. With the exception of bays in Portugal and France and the submergence of the low lands bordering the North Sea, the shores of western Europe appear to have extended further west than now. Southern Spain was joined to Africa, probably by a wide land connection, but was, in turn, separated from northern Spain by a strait. An important and extensive sea stretched from Vienna to the region of the Black and Ara seas.

The Miocene was a period of great mountain building in the Old World as well as in the New. The Alps were upheaved and reached nearly their present altitude at this time. The elevation which produced them excluded the sea and formed basins in which rested inland seas and lakes where are preserved a record of the terrestrial life of the time. The Apennines were re-elevated late in the Miocene; and the Caucasus, on which Miocene strata occur at altitudes of 6000 feet, also date from this epoch. The Himalayas were raised either at this time or in the Eocene.

Volcanism, so stupendous in North America at this time,

seems to have been of little importance in Europe, although some of the movements appear to have been accompanied by igneous activity.

The presence of extensive Miocene beds in Australia, New Zealand, north Africa, and elsewhere tell their story of submergence.

Pliocene

Atlantic and Gulf Coasts—With a few exceptions, the eastern coast of North America had practically the same position in the Pliocene (right) as now. The Atlantic Coast from New York northward extended farther out than at present; Florida was, for the most part, under water; and a very narrow stretch along the Gulf coast from Florida to Texas and another in Mexico were also submerged. This being the case, the most conspicuous deposits are naturally those laid down upon the land, the marine sediments being now chiefly hidden from view beneath the sea. The comparatively wide

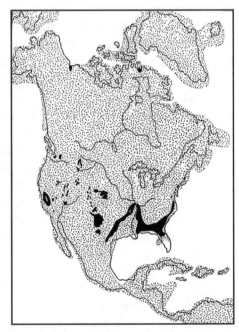

Map showing the probable outline of North America during a portion of the Pliocene. The continental deposits are shown in solid black.

distribution of these continental sediments is due in the first place to their recent age, and in the second place to the fact that some of them occupy sites of continued deposition, as, for example, in the Great Basin region. These deposits have an origin similar to those of previous epochs already discussed. Streams debouching from mountainous land dropped their sediments upon reaching a low gradient, making alluvial fans and plains. On account of the reduction of their volumes through evaporation and seepage, the rivers developed great flood plains. Shallow lakes which existed at that time, formed either by warping or by the choking up of river channels by deposits of sand and gravel, were later filled with sediments.

Mention should be made of a series of deposits (Lafayette) of Tertiary age, the exact status of which is yet in doubt (formerly supposed to be Pliocene, but some of which are Oligocene) which have an extensive distribution, occurring in many places on the Atlantic and Gulf coastal plains in the southern portion of the Mississippi Valley up to southern Illinois, and in the valleys west of the Appalachians. This formation (Lafayette or Orange Sand) commonly has a thickness of 20 to 30 feet, and is composed of gravel and sand in the lower Mississippi Valley and of clay and silt over large areas of the uplands east of the Mississippi River. It was derived from the insoluble residue of older formations and consists of chert, quartz pebbles, and other insoluble materials. The color varies, but is often red, orange, or yellow. This deposit was, probably, formed as follows. The peneplanation and subsequent weathering of the land surfaces during the early stages of the Tertiary produced a layer of loose, insoluble material. In the Oligocene an upwarping along the axis of the Appalachians began and increased during the epoch. As a result, the rejuvenated streams carried much detritus and dropped a part of it upon reaching the lower lands. With the continued rise of the mountain belt and adjacent regions, the streams removed the sediments first laid down and redeposited them farther downstream. The sands and gravel deposited not only filled up the lower portions of the valleys, but also, to some extent, covered the former divides. At present, much of the formation has disappeared in regions of strong erosion and, seaward, is more or less concealed by younger beds. In some places it caps divides but is absent from the valleys.

The marine deposits have a very limited distribution on the east coast and are of little thickness, being most important in Florida.

Western Interior—The Pliocene deposits of the western interior are widely scattered and of limited extent. Beds of this epoch have been recognized in Kansas, Nebraska, Oregon, and the Staked Plain of Texas. As already stated, it is probable that much of the Great Basin and other regions is underlain by Pliocene deposits.

Pacific Coast—Deposits of this age are less widespread on the Pacific coast than those of the Miocene. A change from marine to freshwater conditions in a portion of the area may have been due to a raising of the land near the coast, or to an elevation along faults which excluded the sea. Volcanic activity took place during the period in certain portions of northern and central California, and in the Sierra Nevadas and Cascades.

Pliocene Elevation—The deformation of the peneplain of the Appalachian region raised the Coastal Plain and shifted the coast line to the east, except in Florida, where there was a slight depression. It is possible that during this period of elevation the now submerged valleys of the St. Lawrence, Hudson, Delaware, and Mississippi were eroded.

The plateau region of the west was uplifted at various times prior to the Pliocene, during that epoch, and later, and has since been entrenched to form the great canyons for which it is famous.

Near the close of the Pliocene the Rocky Mountains and the Sierra Nevadas began a period of growth which has given them their present altitude. Instead of folding, as at the close of the Mesozoic, the elevation was chiefly due to warping and faulting. A study of a cross section of the Sierra Nevadas brings out the fact that the slope on the west is long and gradual and deeply entrenched by such great valleys as the Yosemite and Hetch Hetchy, while on the east it is very abrupt and short. This marked difference in the eastern and western slopes is due to a profound fault on the east, which was first formed in the Miocene, along which an enormous block was raised and tilted to the west, leaving its eastern edge to form the crest of the range. The movement along the fault plane has apparently not yet ceased, as is shown by a slip of 25 feet which occurred in 1872. The raising of the Sierra Nevadas enclosed the Great Basin region, shutting off the moist winds of the Pacific and making it a desert.

During the Pliocene, the Cascade Mountains seem to have been peneplained, the mountain mass being raised shortly before or after its close. The rugged scenery of these mountains is the result of comparatively recent erosion. Volcanic activity continued in the epoch and became marked at the end, many of the great volcanoes of the west dating from the close of this epoch, or later.

The close of the Pliocene was a time of widespread elevation, the outline of the continent being extended, with few exceptions, farther out than now. So marked was this elevation that for many years it was generally believed to have been the cause of the accumulation of ice which resulted in the Glacial Period.

High Plains and Bad Lands—The great sheets of clay, sand, and gravel which during the Tertiary were burying the eroded surface of the Upper Cretaceous and other rocks in the region east of the Rocky Mountains, gradually built up a great plain, in some places 500 feet thick, stretching from the foothills of the Rocky Mountains for hundreds of miles. This is known as the High Plains region. The deposition that formed the Great Plains was not continuous in any one place throughout the period, but shifted from time to time, being local and contemporary with more or less erosion. Eolian deposits (loess) were building up the level, grassed surfaces (as, indeed, they are today) and constitute a not inconsiderable part of the formation. In recent times, however, erosion has been in excess of aggradation, and the plain is being cut away. Uneroded remnants of this plain, remarkable for their level surfaces, remain in western Kansas, Nebraska, and westward.

Where the plain has been dissected by canyons and ravines, it is seen to be composed of unconsolidated gravels, sands, and clays. Since the region has a scanty rainfall, although with occasional heavy downpours (cloud-bursts), vegetation, except on the level surfaces of the plain, is sparse. The scantiness of the vegetation on the sides of the ravines, combined with the looseness of the sediments of which the country is built, affords conditions most favorable for rapid erosion when the torrents of water from the occasional heavy rains rush down the ravines. As a result, in certain places along the edges of the High Plains we find a maze of hills and ravines with almost no vegetation except on the tops of the mesas (the remnants of the former surface). These are the "Bad Lands," "Mauvaises Terres" of the early French explorers, and constitute a scenery as weird as any on earth.

Pliocene of Other Continents—The emergent condition of Europe during the Pliocene was in contrast to the widespread seas of the previous epoch. In the north of Europe, with the exception of Belgium and a little of northern France, the seas had withdrawn. Great Britain, as throughout the early epochs of the Tertiary, had a greater land area than now, since only a small portion of the southern part was beneath the sea at this time, while England, Ireland, and Scotland were probably connected; and the northern coast extended farther out than now.

The Himalayas were being eroded during the Pliocene; and thousands of feet of sandstones and conglomerates were deposited at their foot before its close, some of which, however, were laid down in the later Miocene. In South America, the coasts of Argentina and Patagonia were submerged, and the last upheaval of the southern Andes was accomplished at this time.

Life of the Tertiary

Rise of Mammals—In the present imperfect state of our knowledge of the life at the beginning of the Cenozoic, it is, perhaps, even more difficult to account for the presence of highly developed mammals, as soon as the reptiles became extinct, than to account for the disappearance of the latter.

It is improbable that the mammals on the Earth today were descended from any of the mammals whose remains have been found in the later Mesozoic rocks. Indeed, it is even doubted that the true (Eutheria) mammals were descended from the marsupials (Metatheria) mammals. Two theories are offered to explain their sudden appearance. (1) The first postulates their existence in some isolated country, in the Arctics whose climate was not cold at that time, or elsewhere, for a long period of time during which they had been developing along different lines, but from which they were prevented from spreading because of some barrier to their movement, either water or mountains. When this barrier was removed the mammals deployed over the world, and finding the new conditions favorable for their existence, rapidly took the place in nature formerly occupied by the reptiles. (2) The second theory is based upon the supposed existence of mammals on the uplands of the Mesozoic. Since practically all of our knowledge of the life of that period is obtained from coastal swamp and delta deposits in which almost no forms of life are found except those which frequented marshes, little is known of the life of the higher and more extensive areas of the Earth's surface. It is to be noted, too, that the very earliest Tertiary upland deposits contain a rich mammalian fauna. It does not seem improbable, therefore, that on the higher land of Asia and North America mammals of considerable variety were in existence during the later days of the Mesozoic; all proof of which has either been lost by the wiping out of the upland deposits by erosion, or else has not yet been discovered.

Archaic Mammals of Ancient Ancestry—In the earliest known Eocene beds (Puerco) the remains of small, archaic mammals (marsupials) associated with true mammals (Eutheria) occur, which clearly belong to animals whose ancestors lived in the Mesozoic, some of which (Plagiaulacidae) date back even to the Upper Triassic. These animals are characterized by large grinding teeth with many elevations (multituberculate), and elongated front teeth (incisors); the latter being, in some cases, chisel-shaped, as in the rabbit, and in others pointed. The "back teeth" of the lower jaw are, moreover, usually very different from those of the upper jaw. These animals were all small or of moderate size, the largest known having about the bulk of a beaver. Judging from the teeth, it seems probable that some were gnawing animals like rabbits or mice (but not true rodents), and that others were either fruit eaters (frugivores) or even insect eaters (insectivores).

This entire class became extinct before the close of the Eocene and should be considered as survivors from the Mesozoic, which lingered for a time in the Tertiary. The question naturally arises as to the cause of their extinction, since they were able to survive the many changes, not only of the Mesozoic, but of those at the close of the era as well. If the variety of their fossils in the Mesozoic formations indicates the relative abundance of these archaic mammals as compared with other forms during the era, conditions of climate or of food, or competition with the dinosaurs must have prevented their increase. If competition with the dinosaurs prevented their increase in the Mesozoic, it would have been surprising had they been able to compete with the true mammals which appeared in the Eocene. Although mammals, they were lowly in organization; and, even in the Upper Cretaceous where the vegetation was of the modern type, they were not abundant.

Amblypoda (Greek, *amblus*, blunt, and *pous*, foot)—Along with other true mammals associated with the above archaic ones of Mesozoic type, there appeared a group of heavy creatures (Amblypoda), with stout limbs ending in stumpy, five-toed feet. These amblypods (below, *A, B*) were conspicuous in North America during the Eocene but became extinct before its close.

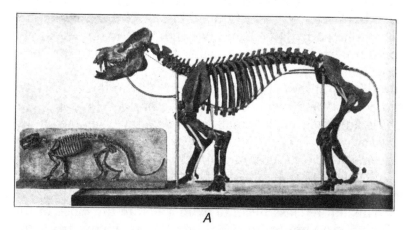

A

B

Skeleton and restoration showing the evolution of the amblypods in the early Eocene.

The early representatives (pg. 408) had few distinguishing characters except their heavy build, but before the extinction of the race, some of them (Eobasileus) not only took on greater bulk, attaining elephantine proportions, but also developed a peculiar head (below), the most conspicuous features of which were the three pairs of knobs, or horns, and the long, saberlike teeth (canines) which projected several inches below the upper jaw. One pair of the knobs was situated on the nose, a larger pair over the eyes, and the third pair above the ears at the back of the skull. It is not known whether the protuberances were covered with horn or with callous skin, but it

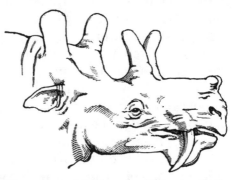

The most highly specialized of the amblypods, *Eobasileus* (Upper Eocene).

was probably the latter. The use to which the long, saberlike (canine) teeth, possessed by both males and females, were put is not definitely known, but it seems probable that they were used to pull down branches from the trees, and that the leaves were then stripped off into the mouth by a rapid side motion of the head. The brain was smaller in proportion to the bulk of the animal than in any other mammal, living or extinct, an animal weighing two tons having a brain no larger than that of a dog. Moreover, the brain was smooth, and a large proportion of it was formed of the lobes of smell (olfactory). These animals seem to have reached the climax of brute mass as compared with brain power on the mammalian stem, and are to be compared with the massive, small-brained dinosaurs of the reptilian stem. At certain times they were very abundant, as is shown by the fact that two hundred more or less complete skeletons have been collected by one museum alone.

The Lower Eocene amblypods were simpler in some

particulars than the later ones, being smaller and hornless, with shorter canine teeth, although the grinding teeth differed very slightly from those of their massive descendants. In other words, aside from increase in size and the ornamentation of the skull, the evolution of the race was slight.

It is interesting to speculate on the causes of the extinction of this race which may have had an existence of more than a million years. Its fate may have been due to two causes: (1) to the small size of the brain, and (2) to the poorness of the grinding teeth which were no more efficient in the huge forms towards the close of the period than in the earlier and smaller species. The low brain power was of disadvantage to them in their competition with other forms and also gave them little ability to protect their young from the more crafty carnivores. Bulk is a disadvantage under changing conditions and may alone have been responsible for the disappearance of the race. This great order was one of the many which, for a time, took a prominent place among the animals of the world, but which after a long span of life disappeared, leaving no descendants.

Ancestors of the Carnivores—The earliest Eocene carnivores (Creodonta) are so generalized (i.e., combine characters now possessed by widely different groups of animals) that it is difficult to tell even to what order they belong. Their teeth are rather better adapted for cutting food than for grinding it (none, however, have sectile teeth, perfectly adapted for flesh eating); and their toes are provided with curved nails that are rather clawlike but are not the sharp, retractile claws such as are possessed by the cat today. These creatures were descended from others whose feet were even more generalized, which also gave rise to the hoofed mammals, such as the horse, elephant, and ox. They were, in other words, mammals with such indefinite characters that, by the modification of their organs, their descendants could be developed into animals widely different in form and habits, such as the lion and the dog, the seal and the whale. Some of the members of the generalized carnivores (Creodonta) were larger, others smaller, than a fox. The largest form of the Eocene (Pachyaena) was the size of a small bear and had unusually blunt teeth, which are thought to indicate that it lived on decaying flesh.

The primitive carnivores (Creodonta) (below) lived through the Eocene into the Oligocene, when they became extinct. Those that passed into the latter epoch attained not only their greatest bodily size, but their greatest brain capacity as well. This bears out the general rule that the brains of surviving races are, upon the whole, larger than those of declining races. However, we shall find in our later study that certain tribes with well-developed brains, as for example certain rhinoceroses (Teleoceras) and elephants (Mastodon), failed to survive. The reason for such extinction is usually, though not always, to be found in the failure of other organs to develop to meet new conditions.

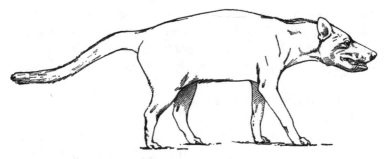

A primitive carnivorous mammal, *creodont* (Middle Eocene).

Marine Mammals—Perhaps nothing shows the rapid evolution of mammals in the Tertiary better than the appearance early in the Eocene of whales perfectly adapted to marine existence, which were not descended from the marine reptiles of the Mesozoic, but from land mammals. Whether mammals gradually acquired an aquatic habit because of the abundance of fish which they voluntarily and habitually sought, or whether they were forced to find new food on account of the competition on the land, it is not possible to state, but probably in the one way or the other, whales, porpoises, sea lions, and other animals arose. These marine mammals were not descended from a common ancestor, but some (manatee) are thought to have been derived from the same stock as the elephant, some (whales) from carnivores, and some (seals), possibly, from the same stock as the bear.

Zeuglodon—For many years, enormous vertebrae have been found in the Eocene deposits of the Gulf coast, the largest of which measure 15 to 18 inches in length and weigh 50 to 60 pounds in the fossil condition. They belong to marine mammals to which the name Zeuglodon (Greek, *zeugle*, yoke, and *odont*, tooth) has been given because of the double-rooted back teeth which present the appearance of a yoke. The head of Zeuglodon was, in some cases, 4 feet long, the length of the body 10 feet, while the tail was 40 feet long. The animal was comparatively slender, an individual 50 to 60 feet long having a thickness of only 6 to 8 feet. The teeth are very unlike those of the primitive mammals, having been modified for grasping and cutting. Behind the head were two short paddles not unlike those of a fur seal, but the hind limbs were so reduced that they were retained within the skin.

The zeuglodonts were divers and probably lived upon squids, as do the sperm whales today. The advantage of such a long tail in proportion to the rest of the body has led to two suggestions: (1) with it the animal could move at great speed through the water, perhaps 20 to 30 miles an hour; and (2), as far as definite evidence shows, the tail may have been used quite as much for the storage of fat as for propulsion.

The ancestry of the zeugolodonts has been traced back to a small whale (Protocetus) with a skull about two feet long, in which the teeth show a surprising resemblance to those of primitive carnivorous land mammals (Creodonta). This whale has the typical number of teeth (44) with one, two, or three roots, the dogteeth (canines) projecting beyond the others. Following these whales came others (Eocetus) differing from those last described, in the fine, saw-edged teeth. Probably descended from these (Eocetus) are others (Prozeuglodon) in which the teeth depart widely from those of land mammals and closely approach those of its most specialized descendant, Zeuglodon. It is thus seen that the Eocene whales were not descended from the Mesozoic marine reptiles, but from the land mammals, just as the ichthyosaurs and mosasaurs were descended from land reptiles. The specialized zeuglodonts constitute a side branch and are not true whales. They became extinct before the close of the Eocene.

Ancestors of Existing Whales— The earliest known ancestors of modern sperm whales are believed to have been small, Eocene marine mammals (Microzeuglodon), whose modified descendants in the Miocene have been called "shark-toothed" whales because they had teeth somewhat similar in appearance to those of a shark. The Miocene whale differs from its Eocelle ancestor (Microzeuglodon) in the number and simplicity of the teeth and in the skull, which resembles that of existing toothed whales. With these Miocene shark-toothed whales (Squalodonta) begins an almost unbroken series which leads to the sperm whale. By one investigator it is stated that the evolution from the shark-toothed whale to the sperm whale is sudden and almost "explosive," the entire evolution being completed in a very small section of the geological time of the Upper Miocene. Dolphins and whalebone whales are known only from the Miocene, but probably date from an earlier epoch. Sea cows (Eosiren) are mingled with the remains of zeuglodonts in the Eocene deposits of Africa.

Ancestors of the Hoofed Mammals (Ungulates)—Interest in the Eocene centers not so much upon such groups of animals as the Amblypoda, which, though the largest and most conspicuous of their time, left no descendants, as upon those animals that are either actually the ancestors of recent mammals or so closely related to them that they help us to understand the evolution and past history of the mammals living today.

Some of the earliest Eocene herbivorous mammals (Condylarthra) are so generalized that many groups seem to converge in them, even the carnivores and herbivores not being easily distinguishable. These ancestral herbivores were small or of moderate size and walked flat on the foot (plantigrade) and not on the toes (digitigrade), as do the horse and cow. The ends of the toes were not quite in the form either of hoofs or of claws. One of the best known forms (Phenacodus) (below), although not the direct ancestor of any of the modern mammals, is of great interest since it probably differed but slightly from those in the direct line of descent. It resembles

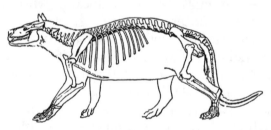

the carnivores in having an arched back, strong legs, and five toes on its feet. It walked somewhat on its toes and the toes ended in a flat "nail" which may be considered as the beginnings of a

Phenacodus, an Eocene mammal which in many particulars is like the ancestor of the hoofed mammals or ungulates.

hoof. The teeth were short-crowned (that portion of the tooth above the jaw being short) and comparatively simple, showing that their possessor was omnivorous in habit. The head is remarkably small and the nearly smooth brain is small, even for a head of this size. It apparently had no means of defense and sought safety in flight. Some species of the genus attained the size of a sheep.

It was from some such animal, so simple in structure that it might almost equally well be ancestral to the carnivores (the dog and lion) and to the hoofed mammals (ungulates,—horse, ox, camel), that the modern hoofed mammals, such as the horse, ox, rhinoceros, and elephant are descended.

Divergence of the Even and Odd-toed Hoofed Mammals (Ungulates)—The common herbivorous mammals of the present are separated into two great divisions, those with a cloven hoof (below), the even-toed ungulates (Artiodactyla), such as the pig, deer, and camel, and those with a large central toe, the odd-toed ungulates (Perissidactyla) (pg. 416), such as the horse with one toe, the rhinoceros with three, and the tapir with four toes on the fore foot and three on the hind foot. The five-toed ancestors of the earliest Eocene had already developed feet that gave promise of odd-toed and even-toed descendants; even Phenacodus, the most generalized of the early mammals, has a foot in which the central toe is rather larger than the others, and should be placed in the division of odd-toed ungulates (Perissidactyla).

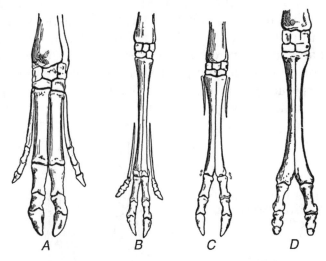

Evolution of the foot of even-toed mammals (artiodactyls); *A*, hog; *B*, roebuck; *C*, sheep; *D*, camel.

It will readily be seen that, if the weight of the body rested principally upon the middle or third toes, and if the animal raised the heel from the ground, the thumb or first finger would not ordinarily touch the ground; and if this habit of walking on the toes became better developed in successive generations, not only the first toe but the fifth as well might become of no use to the animal and might finally atrophy. A continuation of the process, accompanied by a lengthening of the foot, would result in the dropping of the second and fourth toes and in the formation of the highly specialized, one-toed foot of the horse.

If the weight, instead of being directly on the middle toe, was between the third and fourth toes, a more digitigrade habit (walking on the toes) would result in the reduction and later dropping of the thumb or first finger, leaving a four-toed foot. By the further reduction in size of the side toes, a foot like that of a pig, with two strong toes and two small ones, would result. When these side toes disappeared, the animal had but two toes on each foot, like the camel and sheep. Judging from the abundance of the even-toed ungulates (artiodactyls), it seems that, as a whole, this type of foot has proved to be the best. These modifications in foot structure apparently were the result of a change from the forest conditions of the Eocene, where soft ground and succulent vegetation were the rule, to the plains vegetation of the later times, with their siliceous grasses where a short, spreading foot would not give the animal the speed necessary to move long distances in a short time, for food and water.

Evolution of the foot of odd-toed mammals (perissidactyls): *A*, tapir; *B*, rhinoceros; *C*, horse.

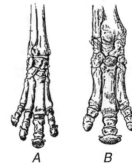

A   *B*   *C*

Factors in the Evolution of Mammals

In the course of the history of the mammals to be studied it will be found that, beginning with some such ancestor as Phenacodus, which is full of mechanical imperfections, the skeletons were modified chiefly in four particulars, each of which was of more or less vital importance to the various races affected.

(1) A race most likely to survive was one whose members had teeth enabling their possessors to grind up nutritious food, no matter how tough and hard. Particularly was this true if the teeth of successive generations developed better grinding surfaces, permitting their possessors to take advantage of new food or food that, because of inability to grind it, was unsuited to their ancestors. The efficient grinding teeth of the horse, cow, and elephant, as will be seen, are the result of such an evolution.

(2) Since the swiftest animals are more likely to escape their enemies, those that possessed limbs constructed for rapid motion were most likely to survive. The leg best suited for this purpose is one in which the foot is lengthened, the joints perfected, and the number of toes reduced. The one-toed horse may be considered the climax of such evolution.

(3) Since more sagacious animals are better able to find food, escape their enemies, and care for their young, it naturally follows that those with large brains (pg. 418, top) were more likely to survive. As the various races of mammals are discussed, attention will be called to the increase in the size of the brains, and any exceptions will be noted.

(4) Increased bulk and the strength which usually accompanies size is often a protection against enemies and, in the case of males, results in the destruction of the smaller and weaker members of the same species. As a consequence, it will be seen that the surviving species of a given order often become larger in the course of their history.

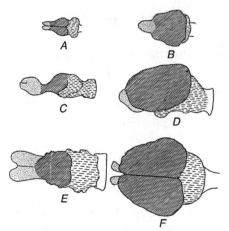

Brains of ancient (on left) compared with modern (on right) mammals: *A*, creodont; *B*, dog; *C*, early amblypod; *D*, rhinoceros; *E*, highly developed amblypod (Uintatherium); *F*, hippopotamus. Olfactory lobes (dots), cerebral hemispheres (oblique lines), cerebellum and medulla (dashes).

Mammalian Teeth—The typical or ancestral number of teeth is 44, a number which is seldom found in living forms, since some have been developed at the expense of others and some have been dropped. It is seldom that a larger number occurs. (The porpoise has 246.) The primitive type of grinding teeth (molars), from which the highly perfected teeth of the present carnivorous and herbivorous mammals were derived, had a grinding surface merely roughened by three sharp pointed cones arranged in the form of a triangle (tritubercular). From a tooth of this simple type has been developed the complicated and efficient grinders of the higher herbivores (right). Another notable change, in such races as the horse and the elephant, has been the lengthening of the tooth, adapting it to the nutritious grasses of the dry, sandy plains.

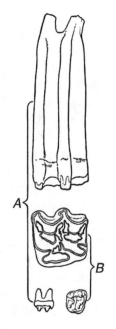

Corresponding grinding teeth of Eohippus (below) and the modern horse (above). The relative size and efficiency are shown.

Feet—The primitive foot is five-toed with the sole resting flat on the ground. From such a foot, the extremely effective one of the higher hoofed, herbivorous mammals was developed. This was accomplished (1) by the raising of the ankle and wrist joints which lifted the first and fifth toes from the ground so that these toes became useless, and degeneration set in which eventually, as in the case of the horse, caused all except the middle toe to disappear. (2) In the primitive foot the joints of the wrist and ankle were loose, but they became more efficient by the development of the "tongue and groove" structure which very effectively prevented lateral movement. The change, in general, has been from a loose-jointed limb with "ball-and-socket" joints, to one with keeled joints; from walking with the sole of the foot flat on the ground (plantigrade), to walking on the toes with the heel well elevated above the ground (digitigrade); from a five-toed foot to one with a smaller number of functional toes. It should not be forgotten, however, that along with races that were changing in structure, there lived others that have been little modified.

The feet of the carnivores seldom show a reduction in the number of toes. This is due to the fact that, since the foot must be adapted for both rending and tearing, as well as for locomotion over both rough and smooth ground, it would not be of advantage to the animal to have the number of toes greatly reduced. The principal changes from the primitive carnivores (creodonts) (pg. 410) to the modern forms, as far as foot structure is concerned, have been in the perfection of the joints, in the more digitigrade habit of walking (walking on the toes), and in the formation of sharp, retractile claws.

Limits to Evolution—There is a limit to evolution after fundamental modifications in the structure have occurred. For example, thus far no mammal is known to have been transformed from an aquatic to a land type, although numerous examples of the reverse are known. No swift-moving types have retrogressed into slow-moving forms. Animals adapted to tree life, however, are believed to have taken on terrestrial habits and to have become modified to fit them.

Lost parts are never reacquired; as, for example, if the number of toes of an animal is reduced, its descendants never have more than the minimum number possessed by its ancestors. Each part that is lost, such as a tooth or a digit, narrows down the possibility of future changes in structure to meet new conditions. A specialized organ can never again become generalized. It will readily be deduced from the above that animals highly specialized to meet certain conditions will be more likely to fail to meet changed conditions than those that have a more generalized structure.

Odd-Toed Mammals (Perissidactyls)

This division of the mammals was in the past more important than at present, being now represented by the elephant, tapir, and horse.

Titanotheres (Greek, *titan*, a giant, and *therion*, a beast)— Of the many families that, for a time, gave promise of permanence and later became extinct, none is more interesting than the titanothere, a tribe distantly related to the rhinoceros, which is first known in the early Eocene (Wind River) and became extinct with apparent suddenness in the early Oligocene (White River), just as it had, perhaps, reached its greatest abundance and variety.

Two groups of titanotheres are represented in the Lower Eocene near the beginning of the history of the race; one abundant genus (Lambdotherium) had slender limbs and was capable of swift movement, indicating that it was adapted to the open basins of the mountain regions; and the other (represented by Eotitanops) was composed of larger and stockier animals, ancestors of those in the Oligocene, that grew to be about two thirds the size of a tapir. The largest, as well as latest, forms (Brontotherium) were ambulatory creatures with an elephantine body but with legs less massive than those of an elephant. The head (below) was saddle-shaped, with a pair of large horns, placed side by side and branching off from the end of the nose, and which were probably covered with callous skin. The brain was not larger than the fist of an average man. They belong to the odd-toed division of mammals (perissidactyls), with four toes on the fore and three on the hind foot, the larger middle toe of the fore foot showing that, like the living tapir, the titanotheres belonged to the odd-toed division of mammals.

No sooner had the titanotheres reached the climax of their evolution than, with apparent suddenness, they became extinct. This is well shown in the Oligocene deposits of the Bad Lands of South Dakota, where they are magnificently represented and undergo their entire final evolution and extinction during the

Evolution of the *titanotheres*.

time taken in the deposition of the 200 feet of sediments in which their remains are embedded. The bulk and specialization of the animals rendered them more liable to extinction, since they were at a disadvantage with the smaller and more active true rhinoceroses of similar food habits which were, perhaps, able to make longer journeys between water and feeding grounds. To this should be added a growing scarcity of food, emphasized by drought at certain seasons. It is not improbable that the competition with the camels and other swift-moving forms with teeth better adapted to the conditions may have been an important factor in causing the scarcity of food which was fatal to the huge titanotheres, although not so to the less bulky rhinoceroses.

From the Oligocene on, the swifter, grazing forms tended to replace the slow-moving, browsing (feeding on the leaves of shrubs and trees) forms, although some, such as the rhinoceros, have survived to the present.

Rhinoceroses—The history of this great, odd-toed, hoofed (perissidactyl) family illustrates two points to which attention will be directed in the discussion of other families: (1) the presence in abundance in North America of members of a family that has long since been extinct in the western hemisphere but is still living elsewhere, and (2) the evolution of a number of side branches, differing widely in structure and habits. The rhinoceros family is now confined to Africa, southern Asia, and a few of the large islands of the Indian Ocean, but in the Oligocene and Miocene not only did rhinoceroses that are ancestral to existing genera live in North America, but a number of side branches were also developed on that continent which differed widely from those of today, some of which lived, at least locally, in great abundance. In a remarkable deposit of the Lower Miocene (Harrison Beds, near Agate, Nebraska) a slab of rock 10 feet by 40 feet by 18 inches was uncovered by an American Museum party in 1912, in which are 75 skulls of a species of rhinoceros (Diceratherium), together with the bones of these and other mammals. This deposit is without doubt exceptional, but nevertheless shows that, in certain localities, these rhinoceroses were extremely abundant. In the Oligocene of North America, three branches of the family are known, but in the Miocene they had evolved into a number of branches, which, however, may be united into three groups.

(1) One of these may, for convenience, be called the "swimming rhinoceros" because of the spreading, four-toed foot which was doubtless an efficient organ for swimming. This branch (Metamynodon) was apparently semi-aquatic and was fitted for life in the lakes and rivers of the Oligocene. It was stout and rhinoceros-like in shape, the eyes were placed high on the head, and the nostrils opened upward so that it could breathe when the head was partly submerged. Its canine teeth were elongated into tusks and were doubtless used for uprooting the plants from the bottom and banks of the lakes and rivers which it frequented.

(2) A second Oligocene rhinoceros branch whose career, like that of the swimming rhinoceros, terminated before the close of that epoch, was the "running rhinoceros" (Hyracodon). This animal (below) did not have the appearance which is usually associated with the rhinoceros, since it was light-limbed and agile, with horselike shoulders and limbs. It had three toes on each foot, very similar to those of the horse of its time, and was apparently adapted to the hard, dry plains of the Oligocene. It is possible that, had this animal succeeded in adapting itself to the changing conditions of the Tertiary and in competing with the other grazing mammals of the time, it would eventually have

A running rhinoceros (*Hyracodon*) showing the modification of structure for plains conditions.

dropped its side toes and have walked on one toe like the modern horse.

(3) The true rhinoceroses constitute the third group, but with two exceptions (Diceratherium and Teleoceros)—none of the North American forms had horns. None of this family are known to have lived in North America after the Pliocene, but members of the group were able to adapt themselves to the vicissitudes of the closing days of the Tertiary and roamed over Europe and Asia, some (woolly rhinoceros) being adapted even to the cold climate of northern Asia, as a carcass found frozen in the ice of northern Siberia shows.

The principal changes which the true rhinoceros group underwent in its history are (1) an increase in bulk, (2) a reduction in the toes from four on the fore foot to three on all the feet, (3) the development of horns, and (4) the development of somewhat better teeth.

Tapirs—It is interesting to find an animal living in the present which still retains the characters of animals that are more typical of Eocene and Miocene times before differentiation became marked.

The teeth of tapirs are short-crowned and differ but slightly from those of their Miocene ancestors of Nebraska and South Dakota. They are odd-toed ungulates (perissidactyls), with four toes (pg. 416, *A*) on the fore foot (the weight being on the third toe), and three on the hind foot, the fourth toe of the fore foot being small. The Eocene ancestors of the tapirs graded almost insensibly into those of the horse and rhinoceros.

Tapirs live in marshes or dense forests in proximity to water, occupying a place in nature in which there is little mammalian competition. They had a wide distribution in the past and are an illustration of a once abundant race nearly exterminated but still struggling for existence where competition happens to be least severe in their particular case. Their present occurrence only in South America and southern Asia seems remarkable unless one remembers that during the Tertiary tapirs ranged throughout the northern hemisphere, making their way to South America late in the Pliocene.

Horses—Few animals have a family history which goes so far back into the past and is at the same time so well-known as that of the horse. From an animal less than a foot in height, with a skeleton more like that of a carnivore than a horse, the changes in structure and size have been traced step by step to the present. It should be borne in mind, however, that few of the so-called ancestors are truly in the direct line, but they show us rather what the actual forebears were like.

Theoretically, the history of the horse begins with a generalized, five-toed animal which walked with the sole of the foot on the ground (plantigrade), or with the heel but slightly raised; with the normal number of teeth (44); with an arched back somewhat like a carnivore's, and with toes covered with nails which were neither hoofs nor claws; in other words, an animal similar to Phenacodus.

The earliest American horse (Eohippus; Greek, *eos*, dawn, and *hippos*, horse) of which we have a record lived in the early Eocene and was a small and unhorselike animal about the size of a fox. It still retained the normal number of teeth (44), as did practically all of the animals of its time, the teeth being simple with very short crowns, somewhat resembling those of the pig and monkey, and very unlike the long, complicated grinders of the horse of today. So generalized are the teeth of this early horse that it is often a matter of great difficulty to distinguish them from those of the ancestors of what are now widely removed orders of animals. There were four well-developed toes and a rudimentary first toe on the fore foot, while the hind foot had three toes and rudimentary first and fifth toes. The foot was a spreading one, enabling the animal to walk on fairly soft ground. From this earliest known simple form, the course of evolution consisted largely in such modifications of the skeleton as rendered the animal better fitted to secure food and masticate it and to escape its enemies. This, as will be seen, resulted in the production of a very perfect grinding apparatus. The necessity for speed in seeking safety and in going long distances for food and water, resulted in the remarkably perfect locomotive apparatus.

THE EVOLUTION OF THE HORSE.					
		Formations in the Western U.S. and Characteristic Type of Horse in Each	Fore Foot	Hind Foot	Teeth
Quaternary or Age of Man	Recent Pleistocene	SHERIDAN — Equus	One Toe Splints of 2nd & 4th digits	One Toe Splints of 2nd & 4th digits	Long-Crowned, Cement-covered
	Pliocene	BLANCO			
Tertiary or Age of Mammals	Miocene	LOUP FORK — Protohippus	Three Toes Side toes not touching the ground	Three Toes Side toes not touching the ground	
	Oligocene	JOHN DAY WHITE RIVER — Mesohippus	Three Toes Side toes touching the ground; splint of 5th digit	Three Toes Side toes touching the ground	Short-Crowned, without Cement
	Eocene	UINTA BRIDGER WIND RIVER — Protorohippus	Four Toes		
		WASATCH — Hyracotherium (Eohippus)	Four Toes Splint of 1st digit	Four Toes Splint of 5th digit	
		PUERCO AND TORREJON			
Age of Reptiles	Cretaceous Jurassic Triassic		Hypothetical Ancestors with Five Toes on Each Foot amd Teeth like those of Monkeys etc.		

Table showing the evolution of the horse.

The next horse in the line of descent (Protorohippus, above) appeared in the Upper Eocene and was four or five inches higher than the early Eocene horse (Eohippus), with longer limbs, which indicate ability for increased speed. The fore foot had four toes, but lacked the rudimentary toe, or splint, of the Eohippus, while a shortening of the outermost toe (the fifth) gave promise of a three-toed foot in its descendants. The hind foot had three toes but no splint.

The Oligocene horse (Mesohippus, above) differs from the Eocene one (Proterohippus) in having but three toes on the fore foot and a splint which represents the outermost (or fifth) toe of the earlier horses. Besides the reduction in the number of toes, the leg had lengthened as the body thickened, and the animal stood about 18 inches high. The teeth were still short-crowned and lacked the complicated structure of the later forms.

A Miocene horse (Protohippus, or Hipparion, pg. 426) shows the next stage in the evolution of the race. In these animals there was one large toe on each foot, with two smaller slender toes, one on each side of it, which were of no use to the animal, as they did not reach the ground when it walked. The teeth are very like those of the modern horse, in which the plates of enamel form curved, complex, irregular patterns, but are shorter and probably wore out at an earlier age. The average height of the animal was about three feet. Associated with this more highly specialized horse (Protohippus) were others with short-crowned teeth and with all three toes functional (Parahippus and Hypohippus).

The stage between the Miocene horse (Protohippus) and the true horse (Equus, pg. 426) is not definitely known, but was doubtless represented by an animal with a large central toe and with either very diminutive side toes or large splints, and longer and more perfect grinding teeth.

Summary of the Evolution of the Horse—The changes, therefore, which took place in the horse family during its geological history are: (1) a reduction in the number of teeth from 44 to 36, accompanied by a lengthening and perfecting of the grinding teeth; (2) a reduction in the number of the toes from five to one; (3) an improvement of the joints of the legs by means of which motion was permitted in but two directions, forward and backward; (4) a lengthening of the limbs, especially in the lower portions. This has left the center of gravity high, and the limb, though long, moves quickly like a short pendulum, combining rapidity of movement with a lengthened stride; (5) an increase in the size of the animal; (6) a proportionally greater increase in the size of the brain than the body; (7) besides the above, other changes, such as the lengthening of the neck and head to permit the animal of increased height to crop grass from the ground; (8) the gradual perfection of the body; and others of which space will not permit mention.

Probable Cause of the Evolution of the Horse—These radical structural changes seem to be the indirect result of a modification of the climate of the Great Plains region of North America and the accompanying change in the character of the vegetation. Eohippus was apparently an immigrant from Europe by way of Asia, but it was in America that the race developed, although from time to time modified representatives migrated back to Europe. During the Eocene, the climate was moist, forests covered the lands, and lakes, marshes, and streams were abundant. Under conditions such as these the early horses lived. During the Oligocene the conditions had not greatly changed, but increasing aridity caused a drying up of the streams and lakes and the development of considerable areas of prairie lands. The woodlands, meadows, and dry prairies of the time favored the evolution of several branches adapted to the different environments, and the horse remains of the epoch show that branches, fitted for the varied conditions, were developed. Some of these soon became extinct, while others gave rise to the horses of the Miocene.

The great expansion of the prairies and diminution of the forested areas in the Miocene favored the evolution of horses fitted for rapid motion on the dry, hard plains. Two explanations for the increasing length and complexity of the teeth have been offered: (1) that, as the race changed from a habitat of forest and marsh to one of prairie, the teeth became fitted to grind up the hard, nutritious grasses that covered the plains (Osborn); and (2) that on dry, sandy plains where the grass was short, the teeth wore out rapidly because of the sand grains which were necessarily caught up with the grass when it was cropped, and which wore away the teeth even more rapidly than hard vegetation would. Those who hold the latter view maintain also that the plains grasses were and are actually less hard than the vegetation of the marshes and forests, and that consequently a change to plains vegetation would have been unimportant had it not been for the presence of sand grains in the food. As a result of the above causes, we find the Miocene horses fitted for plains conditions increasing, and those with the spreading foot and short-crowned teeth fitted for forest conditions becoming extinct. After the Miocene, the race became more and more like the modern horse. By the beginning of the Glacial Period, they had become extraordinarily abundant, but at its close they had entirely disappeared from the western hemisphere, though the descendants of migrants to the Old World lived on.

Cause of the Extinction of the Horse of North America—It is difficult to assign a reason for the extinction of the horses in America. The cold of the Glacial Period has been suggested, but is hardly adequate since the climate of the continent south of the ice sheets was not unfavorable, and at the present time horses on the western plains survive a temperature of many degrees below zero, without shelter or any food other than that which they can obtain for themselves, being able, in fact, to withstand conditions fatal to cattle and sheep. The suggestion that the extinction was due to some epidemic receives some support from the discovery of two species of tsetse fly in the Miocene deposits of Colorado, similar to the African types which, in that country, render thousands of square miles uninhabitable by horses. Epidemics such as that carried by the tsetse fly, the tick, and other insects are most prevalent in wet seasons. The moist conditions which are believed to have prevailed in North America during glacial times would favor the spread of such a disease over, perhaps, the whole of the New World and might readily wipe out of existence the entire race of horses.

Elephants—The massive body and legs, the long tusks, and flexible trunk of the elephant combine to make it one of the strangest of animals and, from external appearance alone, one which might seem least likely to be descended from the generalized mammals of the early Eocene.

The earliest known fossil elephants (Moeritherium) (pg. 431) have been found in Upper Eocene deposits of Egypt. They were about three and one half feet high and, even at this time, were of stocky build, although they would hardly be recognized as belonging to the elephant family were it not for later forms which became more and more elephantlike. The structure of the skull shows that a flexible upper lip, the beginning of a trunk, was present in life. The teeth had already been reduced to 36, and one pair of front teeth (incisors) in each jaw was longer than the others, giving promise of the great tusks of the elephant and mastodon. Those of the upper jaw were sharp-pointed and curved downward, while those of the lower jaw were directed upward. The grinders (molars) were somewhat ridged. The neck was of sufficient length to permit the animal to reach the ground in feeding.

The next elephant in the line of descent (Paleomastodon) (pg. 431) appeared in the Upper Eocene and was larger and stockier, with legs similar to those of its modern relatives. The upper and lower tusks were much longer than in the earlier form, those of the upper jaw being large, with a slight downward curve. This enlargement of the tusks was accompanied by a decrease in the total number of teeth, and the three-ridged grinding (molar) teeth were better instruments for mastication than those of the earlier elephant. The structure of the skull shows that the upper lip, in this form, had probably been developed into a short trunk which, however, may not have extended much beyond the lower tusks.

An elephant from the Miocene of France (Tetrabelodon), smaller than the modern Indian elephant, shows a great advance over those of the Eocene. This is to be expected, since no Oligocene elephants have yet been found. In this form, the upper tusks are long and almost straight, while the lower are short, but since they are set in a greatly elongated lower jaw, they project almost as far as the upper. The trunk was longer than in the earlier members of the family, resting upon the lower jaw, and could only be raised and moved from side to side. As the trunk lengthened the neck shortened since the animal could feed without the mouth's reaching the ground.

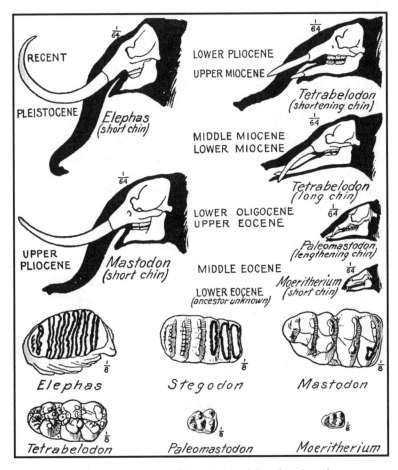

RECENT

PLEISTOCENE — *Elephas* (short chin)

LOWER PLIOCENE
UPPER MIOCENE

Tetrabelodon (shortening chin)

MIDDLE MIOCENE
LOWER MIOCENE

Tetrabelodon (long chin)

UPPER PLIOCENE — *Mastodon* (short chin)

LOWER OLIGOCENE
UPPER EOCENE

Paleomastodon (lengthening chin)

MIDDLE EOCENE

Moeritherium (short chin)

LOWER EOCENE (ancestor unknown)

Elephas

Stegodon

Mastodon

Tetrabelodon

Paleomastodon

Moeritherium

Chart showing the evolution of the elephant's head and teeth.

Moreover, as the tusks and trunk became heavier, a long neck would have been a mechanical disadvantage. The teeth in this form are quite large and have numerous elevations and ridges. This genus spread over the Old World and North America, and by the dropping of the lower tusks which permitted the trunk to hang straight down, gave rise to the mastodon (Dibelodon). The mastodon differs from the true elephant in two principal particulars: (1) in the teeth (above), which are composed of ridges covered with enamel, while the grinding surface of the elephant's tooth is made up of vertical plates of enamel and dentine, alternating with cement, the mastodon tooth being adapted for crushing succulent vegetation such as leaves and tender twigs, but not for grinding hard grasses, as is the elephant's; and (2) in the greater length of the lower jaw which in true elephants is remarkably short. The true elephants were apparently derived from a branch (Stegodon) that lived in India during the Pliocene. In this form the teeth show the first signs of developing cement between the ridges, which, by further development, formed the remarkable grinding apparatus of the mammoth and modern elephant.

Both the mastodons and the true elephants (mammoths) spread over North America, living here even after the disappearance of the last ice sheet, as is proved by the presence of skeletons in the bogs that have accumulated in the depressions of the latest glacial deposits. Paintings and carvings on ivory of mammoths, made by prehistoric man in Europe, prove the existence of elephants after the appearance of man on that continent. There is, however, no evidence pointing to their presence in North America since the advent of man.

Summary of the Evolution of the Elephant—(1) The few (three in each half jaw) long, large, many-ridged teeth of the elephant, of which not more than one and one half are in use in each half jaw at one time, were developed from small, short-crowned, simple teeth with two poorly developed ridges. (2) The tusks are greatly elongated front teeth (incisors). (3) The trunk began as a short, flexible lip which, as the lower tusks became longer, gradually lengthened to enable the animal to reach the ground for

Restoration of a Miocene elephant (*Dinotherium*) with tusks on the lower jaw.

food (pg. 432). The trunk at first rested upon the lower tusks, but when these disappeared in the course of the evolution of the race, it hung straight down and because of this new position soon developed its present characteristics. (4) Accompanying the development of the trunk and tusks was a shortening of the neck. (5) The bulk and height of the animal increased and the leg straightened to support the greater weight.

Even-Toed Hoofed Mammals (Artiodactyls)

This division of the hoofed mammals (pg. 415) is the most important at present, being represented by such animals as the camel, deer, sheep, goat, and antelope.

Camels—Evidence that the climate over large areas of the western interior of North America was dry for long periods of time is shown, as has been indicated, in the evolution of certain animals—an evolution especially fitting them for life on the plains and deserts. The camel, as is well known, is admirably adapted for arid conditions. Its two-toed foot encased in a single pad is efficient in traveling over desert sand, its long, well-nigh structurally perfect legs enabling it to move rapidly and with the minimum effort and its capacity for carrying water, were the results of a long evolution under such conditions.

Although not as well known as the ancestry of the horse, the history of the camel is fairly complete (pg. 435) from the Eocene to the present. Following a very generalized ancestor (Trigonolestes) of the earlier Eocene, which may equally well be considered ancestral to other families, there appeared in the later Eocene a very generalized camel (Protylopus), a little larger than a jack rabbit, which is possibly ancestral to the modern camels. The points in which it differs most noticeably from living members of the family are: (1) the size; (2) the small, simple teeth of the normal number (44); (3) the presence of two side toes on each foot in addition to the two useful toes; and (4) the separate bones of the forearm (radius and ulna), which did not grow together until late in the life of the individual.

The Oligocene camel (Poebrotherium) was of slender proportions, somewhat resembling a llama, though with a shorter neck. In this camel, evolution had progressed in two principal particulars: the side toes were absent and were represented by splints, and the bones of the forearm were joined when the animal was still very young.

The next in line (Procamelus) lived in the Miocene and shows a further approach to the modern camel, having a longer neck than the preceding and more camel-like contour. In this stage, the bones of the forearm were united before birth, and certain bones of the foot (metapodials, or "cannon bones") were united early in life. The splints were entirely absent. The animal was intermediate in size between the llama and camel. Some of the Pliocene and Pleistocene camels attained a larger size than any existing species. Besides these in the direct line of descent, a number of side branches arose to take advantage of various conditions of climate and food.

The camel is, in the even-toed line (artiodactyl), what the horse is in the odd-toed line (perissidactyl), each family having apparently progressed almost as far as possible in the perfection of its foot structure.

The camel family lived in the New World for perhaps 3,000,000 years and then completely disappeared from the North American continent, where its evolution had taken place. The camels (llamas and alpacas) of South America succeeded in living on in that continent, to which their ancestors had migrated in the Pliocene or Pleistocene. The entire family was confined to North America until the Pliocene, when it invaded the Old World, and in the Pliocene or Pleistocene it invaded South America. The dromedary and camel still survive in Africa and Asia, and the llama and vicuna in South America. Here again is a great family which, having developed into almost its present form in North America, became extinct on this continent, although still living on in others. The cause of the extinction of

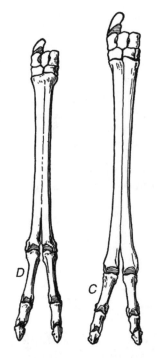

this family is as difficult to find as that of the horse; for, as in the case of the horse, the conditions in portions of America today (Arizona, New Mexico, and Mexico) have been shown by experiment to be as favorable for camels as those in their African and Asiatic homes, and they would doubtless be used in arid and semiarid regions of western North America if economic necessity demanded. Their extinction in North America may have been due to the same cause, or causes, that produced the extinction of so many species at the close of the Pliocene.

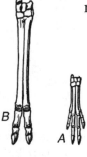

Evolution of the camel's foot from the Eocene to the present.

Deer—There are few, if any, cases in which the changes that the individual undergoes in his growth from birth to old age, so conspicuously parallel those which his ancestors underwent in the course of their geological history as in the deer. In the development of the existing deer, the males and females are born hornless; at the end of the first year the male acquires a simple, one-pronged antler; this is shed, and at the end of the second year a two-pronged antler is grown; in the next year the antlers have two or three tines, and so on until the maximum number for the species has been reached. The geological history of the deer agrees

Hornless ancestral deer (Lower Oligocene).

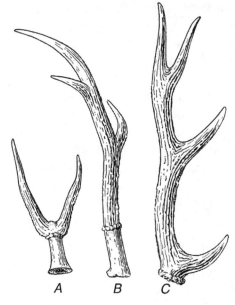

Horns of deer, showing the evolution of horns. *A*, two-pronged deer of the Middle Miocene (*Dicroceras*); *B*, the three-pronged horn of the Lower Pliocene deer; *C*, the four-pronged horn of the Upper Pliocene.

A   B   C

in many particulars with the facts of individual development from year to year. The oldest known members of the tribe (Leptomeryx) (pg. 435, bottom) in the Oligocene have no horns, as is true of their surviving relatives in Asia. The earliest deer with horns (Dicroceras) (above, *A*) of which there is any record lived in the Miocene where the antlers were two-pronged. In the Upper Miocene, deer with three-pronged antlers (above, *B*) begin, and in the Pliocene the four-pronged (above, *C*); then the five-pronged; and finally, near the close of the Pliocene, a deer appears in which the antlers are extremely branched. The deer first migrated into America after the two-pronged stage had been developed.

The teeth of the Oligocene deer are very short-crowned, but in the course of the Tertiary they become longer and more thoroughly adapted to the mastication of plains vegetation. Various side branches appeared and became extinct during the period. Among them one unusual form (Protoceras) which was about the height of a sheep had two pairs of short, bony horns and canine tusks (right). Its ancestry is unknown, and it probably reached North America by immigration.

A four-horned deer (*Syndyoceras* —Upper Oligocene).

Cattle, Sheep, and Goats— True cattle first appear, as far as known, in the early Pliocene deposits of Asia. Concerning the geological history of cattle, sheep, and goats little is known, since the difference in the ox, antelope, sheep, and goats is largely a matter of the curve of the horn and the build. This entire family (Bovidae) have their maximum development at the present time.

Swine and Related Animals—True pigs were confined to the Old World until brought to the New by Europeans, although their relatives, the peccaries, were abundant in portions of both North and South America. The entire family is very simple in structure, being the least altered of the descendants of the early, even-toed mammals (artiodactyls), and dates back to the early Eocene, although the oldest species of the true pig is not known before the Miocene. Several extinct families, distantly related to the pig, played an important part in the life of the Tertiary; and of these one (Entelodon) is especially worthy of mention. One species was as large as a rhinoceros, with a head four feet long. One peculiarity of the skull consisted in a prolongation of the cheek bones on either side of the lower jaw. They had two-toed feet, being rather highly specialized in this and other particulars.

Another generalized animal (Oreodon), in some respects combining the characters of the deer and hog but not ancestral to them, was extremely abundant at certain times during the Oligocene and Miocene and may have been partially responsible for the extinction of the titanotheres. It was not larger than a sheep, with a long tail and four-toed feet.

A Climbing Ungulate—An interesting example of a modification in structure, fitting an animal for conditions very different from those to which its relatives were accustomed, is seen in a hoofed mammal, a member of the oreodont family (Agriochaerus). In this Oligocene creature, the feet and limbs are modified in such a way as apparently to enable it to climb trees as easily as a jaguar or other large cat. The hoofs are so narrow as to be actually converted into a sort of claw, and the wrist and ankle joints are modified in such a way as to make the wrists and ankles as flexible as those of a cat.

Other Animals

Insectivores—This primitive group occupied an important place in the Eocene, since which time it has dwindled in numbers and is now represented by a few survivors inhabiting, for the most part, uncongenial regions, or else protected by spiny armor, or of subterranean habits. It is represented by the mole, hedgehog (not the porcupine), shrew, and other small animals that feed largely on insects and worms. Insectivores have remained simple in their organization since their introduction and are the least altered of the great branches. Among their generalized characters are the smooth brain, five-clawed toes, the habit of walking with the whole or greater part of the soles to the ground (plantigrade), and other less conspicuous features. The group is so generalized, in fact, that it is difficult to characterize it without a too technical description. Perhaps the most striking feature of some living genera is the elongated, or proboscis-like, nose. It is possible that this group more nearly represents the characters and habits of the primitive true mammals (Eutheria) than any other now living. There seems to be little doubt that bats are descendants of primitive members of this group.

Rodents (Gnawing Animals)—Rodents are first known from the Eocene, before the close of which epoch they had acquired practically all their present characteristics. With the exception of their powerful gnawing teeth (incisors), rodents, in the past and present, have been animals of simple structure. Their brains are smooth, and the race has apparently changed in no essential feature since Eocene times, with the exception of their teeth which have been slightly reduced in number, and the grinders (molars and premolars), in some species have become perhaps as highly developed as those of any other class. Before the close of the Oligocene, squirrels, marmots, beavers, rabbits, pocket gophers, and others were present.

A burrowing rodent with horns which appears in the Miocene and early Pliocene, is interesting as showing the possibility of variation. It seems to have been much better adapted for digging than existing gophers, but of what use the horns could have been to a burrowing animal it is difficult to imagine. They may have served as accessories to the strong claws in digging, or they may prove to be sexual characters.

In the Eocene there also appeared a race (Tillotherium) (below, *A, B*) similar in habits to the rodents, although not of this tribe, some members of which grew to be of considerable size, one species being half as large as the tapir. In South America during the Miocene, rodents were abundant, but all belonged to the great porcupine group such as live on that continent today; the forms common in North America at that time, the rats, mice, squirrels, beavers, marmots, hares, and rabbits, being absent.

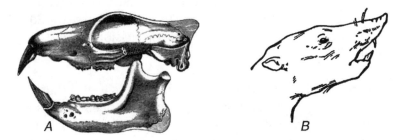

A, skull; and *B*, restoration of the head of *Tillotherium*, a peculiar rodent-like creature (Eocene).

It is an interesting fact that, notwithstanding their failure to develop a highly complex brain, rodents are, at present, the most abundant of mammals. This has been possible because of their fecundity and adaptability to varying conditions.

Edentates (Latin, *edentatus,* toothless)—The earliest Eocene edentates (Ganodonta) were so similar to the ancestral herbivores (Condylarthra) and carnivores (Creodonta) that it seems probable that the three orders were derived from a common ancestor only a short time before. The most familiar living edentates are the armadillos, the anteaters, and the sloths. The name, edentate, is misleading, since most of the order have teeth which are much alike and are without enamel. Teeth, however, are lacking in the front part of the mouth, and it was from this character that the name was given. One modification—among a number—should be mentioned. In the earliest forms the teeth were covered with enamel, as is commonly true of mammalian teeth, and had definite roots. In the later forms, the teeth become rootless, and the enamel disappears except in narrow bands. It is rather surprising to find armadillos (Metacheiromys) as far back as the Eocene, similar in many respects to the smaller armadillos of today, but apparently possessing a leathery instead of a bony shield and with different teeth.

The order did not attain great importance until the Pliocene and Pleistocene at which time it assumed a leading role in South America. Some of the South American edentates were the largest creatures on that continent. The description of the South American sloths will be taken up in the discussion of Pleistocene mammals.

True Carnivores—When traced back, it is found that such distinct families as the dog, hyena, and cat become less and less easily distinguished, until they converge in the primitive carnivores (Creodonta) of the Eocene; and these, in turn, have affinities with both insectivores and ancestral hoofed mammals

(Condylarthra). Before the close of the Oligocene, many families of the true carnivores appeared and lived in competition with the ancestral carnivores, which they entirely replaced before the close of the Oligocene.

In the Eocene and Oligocene primitive representatives of the families to which belong the dog, weasel, cat, and hyena appeared; but the families were more clearly differentiated in the latter, at which time ancestral dogs, raccoons, and weasels were common although not yet of a distinctly modern type. In the epoch following (Miocene) carnivores were abundant, and some of them so closely resemble those of today that they have been included in the same genera as living animals. Wolves, foxes, panther-like animals, saber-toothed tigers, ancestral raccoons, as well as weasels and other like forms were present. In Europe the bear and hyena were represented as well as the above. In the Pliocene, carnivores flourished and perhaps gained on the herbivores, forcing them to develop greater speed, sagacity, and powers of defense.

In South America during the Miocene, there were no true carnivores; but their place in nature was taken by carnivorous marsupials, such as live in Australia today.

Primates (Monkeys, Apes, Lemurs)—This order of mammals has special interest because it also includes man. The first known members (lemurs) date back to the earliest Eocene deposits, where their remains so closely resemble those of the generalized insectivores of that early time that it is difficult to distinguish one from the other. Monkeys and lemurs lived in North America during the Eocene, but disappeared from this continent at the beginning of the Miocene. Primate remains from the Miocene of France (Dryopithecus) are of great interest because of the similarity in some respects to the skeleton of man, and also because of the possibility that these animals were able to make rough flint implements (eoliths). How far it may be regarded as a stem from which on the one side the line led to the human race and from the other to the living anthropoids, namely the chimpanzee, orang, gibbon, and gorilla, cannot be certainly determined. The discussion of the so-called man-ape (Pithecanthropus) and others will be taken up in connection with the evolution of man in the next period.

Birds—The presence of birds in the Lower Eocene, typically modern in structure and in no sense intermediate between the Mesozoic toothed birds with vertebrated tails and the birds of today, makes the question as to the origin of the typical Tertiary life (birds and mammals) exceptionally difficult in the present state of our knowledge.

Although few of the birds of the Eocene can be referred to living genera, they are modern in all essential features. Even at this early date there were living relatives of the vultures, storks, secretary birds, sandpipers, Old World quail, sand grouse, cuckoos, swifts, herons, and pelicans. The appearance of a great, flightless bird (Gastornis) as large as an ostrich but apparently unrelated, presenting affinities to wading and aquatic birds, is interesting as showing the advanced stage of evolution at this early time.

The bird life of the Eocene of Europe (Quercy, France) gives a clue to the climate and environment of portions, at least, of Europe at that time, since it was fitted to inhabit great, warm plains, scattered over with groves. The assemblage is a tropical one and approaches that now found in tropical Africa and South America, although, as in the case of the vegetation, tropical forms are associated with others that are now typical of temperate regions.

Most of the bird fossils are from the Miocene and later formations, and belong to existing families and often to existing genera. A remarkable bird (Phororhachos) from the Miocene deposits of Patagonia shows the extreme to which bird evolution may be carried. It stood about seven feet high and had a head as long as that of a horse, armed with a pick-like projection. Its habits have been variously conjectured. The loss of wings indicates a semiarid condition.

In general, it can be said that the earliest Tertiary birds are typical modern birds, modified for various conditions of life, some being aquatic, some waders, and some land birds; and that the changes which took place during the period resulted in the production of modern genera and species. Although 2,000 species of birds are living today, less than 500 are known from the Tertiary. The reason for the rareness of bird remains, as has already been discussed, is probably due to their lightness, which causes them to float and thus exposes their carcasses, often for many days, to fish and other carnivorous animals.

Reptiles and Amphibians—The usual practice at present is to include in the North American Tertiary no formations containing dinosaur remains. This is the custom even though a great unconformity exists in the Laramie of the western interior of North America, which, were it not for the presence of dinosaur fossils in the formation (Lance) overlying the unconformity, would doubtless be considered the dividing line between the Mesozoic and Tertiary. The most conspicuous reptilian survivors of the Mesozoic were the turtles, crocodiles, and large river lizards (Champsosaurus), the last, however, disappearing early in the period. Snakes began in the Cretaceous, having doubtless been derived in the Mesozoic from lizards, by the degeneration of their limbs. A number of species have been found in the Tertiary, few of which, however, are to be distinguished from those now living, and the majority belong to the non-poisonous varieties. Some of the early sea snakes attained a length of about 20 feet.

Among amphibians, salamanders, newts, frogs, and toads occur in Oligocene deposits; and numerous impressions of tadpoles have been preserved.

Deductions as to the climate of the Tertiary can be made from the reptilian life as well as from the vegetation. Crocodiles, large and small, and of several genera, lived in abundance in the Middle Eocene (Bridger) of the western interior and suggest a climate not unlike that of Florida today, and a country similar to the bayou region of the Mississippi delta. The rivers of this time swarmed with turtles, but the presence of land tortoises, some of which were three feet long, indicates that these swampy areas were bordered by extensive stretches of dry land. Numerous land tortoises in the Oligocene deposits of the Great Plains show that dryland conditions were widespread. Since spiny lizards are largely confined today to arid regions, the presence of numerous lizards (Glyptosaurus) with skulls covered with spiny, bony plates is indicative of dry conditions in Montana during a portion, at least, of the Oligocene.

Fishes—The fish of the Tertiary were abundant and very similar to those of the present seas. Ganoids were represented by a few species, and teleosts were very much as at present, both in numbers and in appearance. The most noted deposit of fossil fish in America is that of the Green River (Eocene) shales of Wyoming, where thousands of beautifully preserved specimens have been quarried, examples of which can be seen in almost any museum. The great number of sharks' teeth (below, *A-D*) in the Tertiary deposits of the Atlantic Coastal Plain

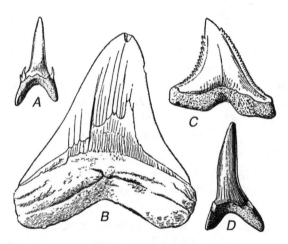

Tertiary shark teeth: *A, Odontaspis cuspidata; B, Carcharodon megalodon; C, Hemipristis serra; D, Odontaspis elegans.*

of North America and elsewhere show that sharks were very abundant in this period. Some of them must have been of great size, judging from the teeth some of which are six and one half inches long and six inches broad. A close living relative, the great white shark, has teeth one and one-fourth inches long. If the proportion of size of teeth to length of body holds true in the two species, the giant Tertiary shark (Carcharodon megalodon) attained a length of 70 to 80 feet and possessed jaws five to six feet across. No actual measurements of these sharks have been made, since their skeletons, being cartilaginous, have not been preserved.

Invertebrates

During the Tertiary, limestone strata several thousands of feet thick were built up by the accumulation of the remains of invertebrates. This is in marked contrast to the deposits formed of vertebrate remains, which are never of great thickness, and seldom form even thin beds of great extent. Limestone, locally of enormous thickness and

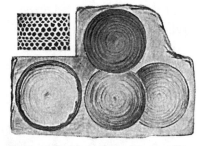

Tertiary Foraminifera: *Nummulites.* An enlargement of a portion of the shell is seen in the upper left-hand corner.

extent, covering areas in the Pyrenees and the Alps mountains, in Greece, northern Africa, Persia, China, and Japan, is often made up chiefly of the shells of Foraminifera named Nummulites (Latin, *nummus*, a coin) (above) from the shape and size. Perhaps at no time in the entire history of the world did an organism of similar size live in greater abundance. Other limestones in Europe and on the Gulf Coast of North America were formed in large part of other forms of Foraminifera.

Brachiopods and crinoids were rare throughout the period and may be considered as races about to become extinct. Sea urchins (pg. 446, top) continued to be abundant.

Coral reefs are rare in the Eocene and had a distribution different from that of today, well-developed reefs occurring on the north and south flanks of the Alps and Pyrenees. In the Miocene and Pliocene they had almost their present distribution.

Gastropods and pelecypods (below, pg. 447) were abundant throughout the Tertiary, being, as now, the most numerous of the larger invertebrates. During the period they became more and more like the living forms, until in the Pliocene they were nearly identical with those of today.

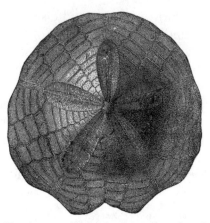

Tertiary echinoid: *Scutella aberti* (Miocene).

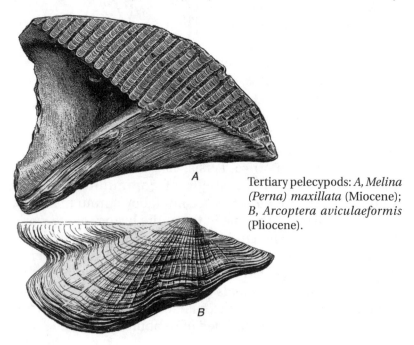

Tertiary pelecypods: *A, Melina (Perna) maxillata* (Miocene); *B, Arcoptera aviculaeformis* (Pliocene).

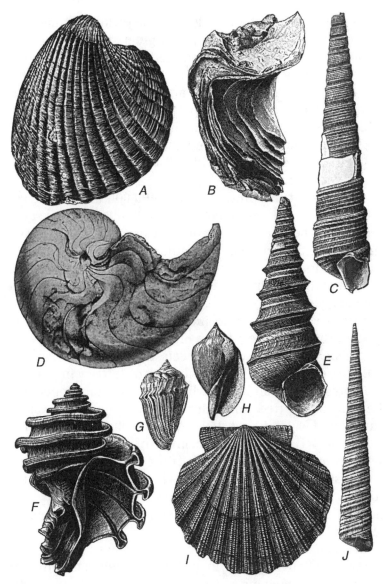

Tertiary invertebrates: *A, Venericardia planticosta* (Eocene); *Ostrea sellaeformis* (Eocene); *C, Turritella humerosa* (Eocene); *D, Hercoglossa tuomeyi* (Eocene); *E, Turritella mortoni* (Eocene); *F, Ecphora quadricostata* (Miocene); *G, Volutilithes petrosus* (Eocene); *H, Orthaulax gabbi* (Miocene); *I, Pecten madisonius* (Miocene); *J, Turritella variabilis* (Miocene).

Certain species are characteristic of the various formations of the epoch and are consequently of great stratigraphic value. Some Miocene pelecypods attained a large size; oysters 13 inches long, 8 inches wide, and 6 inches thick, as well as pectens (pg. 447, *I*) 9 inches in diameter are known. Large size was not, however, characteristic of the pelecypods of the period. Larger pelecypods are living today than in any previous period.

The cephalopods were represented by the nautilus and squid, the former having a wider distribution than now.

Insects—At the present, about 400,000 living species of insects have been described, but less than 8000 fossil species are known. The small number of fossil insects as compared with those of the present does not indicate that this class is more numerous today than at certain times in the Tertiary, but rather that few of the Tertiary species have been preserved. It has even been suggested that owing to the warmer climate and more luxuriant vegetation, and judging from the proportion of species, the total insect fauna of the Miocene of Europe may have been greater, in some respects, than it is now in any part of that continent. The greater number of Tertiary insects have been preserved in amber, in which they were entrapped when the gum of the trees on which they were crawling was first exuded and was soft and sticky. About 2000 species have been thus preserved, some of the specimens of which are in an almost perfect state of preservation, all of the external characters being as well shown as in life; others are preserved in peat; and still others in mud. The finely laminated shales of Oeningen, on the Lake of Constance, Florissant, in Colorado, and elsewhere have yielded hundreds of specimens.

Horseflies, Tsetse Flies, and Ants—The presence of horseflies very similar to living forms in the Miocene deposits (Florissant, Colorado) is interesting, as it shows that, even at that early date, the horse was probably tormented by this insect. Although the horse has changed radically, the flies have remained practically the same. In the same deposits (Florissant, Colorado) the tsetse flies occur. The exquisitely preserved ants of the Baltic amber, belonging to the Lower Oligocene formation, are in all respects like existing ants. All of them belong to existing subfamilies, most of them even to existing genera, and a few of them are practically indistinguishable from species inhabiting Europe today. That some of them were herders of plant lice is proved by blocks of amber containing masses of ants mingled with the plant lice which they were attending when the liquid resin of the Oligocene pines flowed over and embedded them. Possibly the soldier caste is a recent innovation, but the differentiation of the males, queens, and workers was as extreme, and precisely of the same character as now.

The insects of these Colorado Miocene deposits indicate that the locality was doubtless an upland or mountain with a warm, moist, but not tropical climate. The presence of such a climate in the higher lands of this epoch does not preclude the possibility of arid conditions on the Great Plains and in Texas.

There appears to have been but little important change in the insect world since the middle of the Eocene or earlier, almost no new orders or even families having appeared, although the genera and species have changed. This is perhaps not surprising, since insects of the modern type made their appearance soon after flowering plants became widespread, and had consequently perfected their structure and organs previous to the Tertiary, during the long Cretaceous Period.

Vegetation

The vegetable kingdom reached its culmination before the animal kingdom; and as far as plant evolution is concerned, it is almost arbitrary to separate the Tertiary from the Mesozoic or from the Pleistocene. Even at the beginning of the Tertiary, the general aspect of the forests was not very different from that of today, as the presence of maples, poplars, sycamores, walnuts, hazelnuts, elms, yews, cedars, and sequoias (redwoods) shows. The association, however, is rather remarkable since with the above are found figs and palms. The presence in Greenland, Iceland, and Spitzbergen of trees such as now grow in the United States indicates a warmer climate in the polar regions.

Grasses—There was, however, one very important element of the vegetation—the grasses—which at the beginning of the period apparently occupied a very subordinate position, but which before its close became widespread. Because of the part played by this type of vegetation in the evolution of the most important branch of mammals (ungulates, hoofed mammals) it will be well to discuss the evidence upon which its presence is based. If we observe the conditions of the preservation of plant remains along existing ponds, river borders, or swamps, we see at once that they are as favorable for the preservation of deciduous leaves as they are unfavorable for the preservation of grasses. Grasses are firmly attached to their roots and are not swept away either by water or wind. Leaf deposits, therefore, abound everywhere and give us sure indications of the forest flora, while we know but little of the field and meadow flora, which is of great importance in connection with the evolution of the grazing herbivorous ungulates especially. In fact, the evidence as to grasses is very limited throughout the entire Age of Mammals. The number of kinds of grasses (Graminae) found in the whole Cenozoic of Europe is comparatively small, and it is difficult to draw conclusions from fossil plant remains alone as to their relative or absolute importance. At what period grasses began to assume anything like their present dominance it is impossible to determine. The absence of native grasses in Australia is indirect evidence of their late geological development. The indirect evidence of the history of grasses, derived from the adaptation of the teeth of the hoofed mammals, disposes us to adopt the opinion that grasses attained wide distribution in both hemispheres only toward the close of the Eocene. Their evolution on favorable forestless regions was certainly a prolonged one, beginning in Mesozoic times. Proof that grasses were widespread in the Miocene is based upon the structure of the mammals; omnivorous forms were becoming grass eaters; the method of chewing was changing from a vertical, biting movement to a horizontal, grinding one; and the teeth were becoming more durable and better suited for grinding hard food. This then was apparently the time at which the grassy plains began.

Daemonhelix—Certain fossils occurring abundantly in the Miocene deposits in a restricted area of western Nebraska have given rise to some speculation. They consist of spirals of harder rock, held together by fibrous calcareous material which sometimes shows a vegetable structure, and because of their shape and size they have received the name Daemonhelix (Devil's Corkscrew). Some of them are ten feet or more in height and a foot in diameter; and since they resist erosion somewhat better than the surrounding rock, they often stand out prominently against the bluffs. They have been considered the burrows of extinct rodents and also fossil algae, but the proof of the latter seems now to be well established.

Geological History of Sequoias—The history of the sequoias (big trees and redwoods) of California is a striking example of the fate of many animals and plants, and illustrates the difficulty of finding the cause of extinction in plants as well as animals. These big trees, which sometimes grow to a height of 325 feet, have a girth of 50 or 60 feet, and live to be 5000 years old, are now confined to the mountain slopes of California. In the Tertiary, the sequoias were common trees in the northern hemisphere, extending from Spitzbergen (78° north latitude) as far south as the middle of Italy, in Asia to the Sea of Japan, and over a large part of North America. The sequoias date back to the Cretaceous, where they were undoubtedly represented by several species. This is perhaps the most remarkable record in the whole history of vegetation. The sequoias are the giants of the conifers, the grandest representatives of the family, and the fact that, after spreading over the whole northern hemisphere and attaining to more than 20 specific forms, their decaying remnant should now be confined to one limited region in western America and to two species constitutes a sad memento of departed greatness. The small remnant of *Sequoia gigantea* (the big trees) still, however, towers above all competitors, as eminently the "big trees," but, had they and the allied species failed to escape the Tertiary continental submergences and the disasters of the Glacial Period, this grand genus would have been to us an extinct type. It is stated that the sequoias were so abundant in northwest Canada as to furnish much of the material for the great lignite beds of that region. That the climate of other parts of the world is still suited to the growth of these trees is shown by the fact that sequoias are now growing in England and around Lake Geneva in Switzerland from seeds carried there from California. Their extinction becomes more remarkable since it is known that under the most favorable conditions these giants probably live 5000 years or more, though few of the larger trees are more than half as old.

Diatoms—The earliest specimens known with certainty to be diatoms are from the Jurassic. This group did not become common until the Cretaceous, or abundant until the Tertiary. A stratum 50 feet thick in Bohemia is almost entirely composed of diatoms; about Richmond, Virginia, there is a deposit 30 feet thick and many miles in extent; and deposits of diatomaceous earth are common in other parts of the Coastal Plain. Thick beds also occur in the California Tertiary. Diatom deposits are now forming in the Yellowstone National Park, where they cover many square miles in the vicinity of active and extinct hot spring vents of the park, and are often 3 feet, 4 feet, and sometimes 5 to 6 feet thick. In the Tertiary, diatom deposits were formed in sluggish streams, in lakes, on the sea bottom, and in hot springs, as they are now in Nevada and California. The Cretaceous and

Tertiary species very closely resemble living forms.

Exceptional Preservation of Plants—The preservation of such delicate parts of plants as flowers, catkins of the oak, pollen grains, as well as fungi, in the amber of Oligocene trees, gives us almost as definite knowledge of some of the Oligocene plants as if they were living today.

In certain parts of Europe, Oligocene mineral springs covered with their deposits whatever organic remains they touched and thus buried them. After a time, the organic matter decayed, sometimes leaving perfect molds of their forms. When these molds are properly filled, casts even of fossil flowers and insects are sometimes obtained.

Plant Localities in North America—About 500 species of Miocene plants are known in North America, but the deposits in which they are found are small and widely separated. One at Brandon, Vermont, consists of a small, pocket-like deposit of lignite, at one time worked for fuel, which has yielded a large number of fossil fruits, among which walnuts and hickory nuts have been identified, although most of them are of unknown or doubtful affinity. In Colorado (Florissant) the deposits of a Miocene lake have afforded large numbers of plants, such as alders, oaks, narrow-leafed cotton woods, pines, roses, thistles, asters, and Virginia creepers. Mixed with these are others of a more southern type, such as the holly, smoke tree, sweet gum, and persimmon. The vegetation of the Pliocene was probably almost identical with that in the adjoining regions today, but so few plant fossils have been found that no definite statements can be made.

Climate

Difficulty in Determining Tertiary Climates—The determination of the climates of the Tertiary is complicated by a mingling of plants whose relatives are no longer found associated, some being at present restricted to tropical or subtropical regions and others to temperate zones. Since closely related modern species sometimes live under very different climatic conditions, it will readily be seen that the presence of plants in the Tertiary related to species now living only in subtropical regions, for example, does not necessarily prove that these early species lived under similar conditions, but is certainly strong evidence in favor of such a supposition; especially when it can be shown that the plant associations were then the same as now; e.g., breadfruit trees, cycads, and many ferns grew in association in Greenland (72° N.) as they now do in the tropics. This mingling of what are now tropical and temperate region types has led investigators to very different conclusions. For example, the Miocene climate of Colorado, because of the presence of genera now living in Colorado, is believed by one investigator (Cockerell) to have been in no sense tropical, while another (Knowlton), because of the presence of West Indian genera, believes that the climate was not unlike that of certain parts of the West Indies of today. Certain trees, such as palms, are generally agreed to indicate warm, tropical, or subtropical conditions, even when they grew in forests with the maple, elm, and other temperate region plants. When, however, the remains of insects, reptiles, or mammals occur in deposits with plants, the total evidence often becomes conclusive.

Eocene—The vegetation suffered so little change between the Upper Cretaceous and the Eocene that it is evident the climates of the two epochs were not unlike. In the early Eocene (Fort Union), a cool to mild temperate climate, with a much greater rainfall than now, prevailed over the Dakotas, Wyoming,

Montana, and as far north as the Mackenzie River in Canada, as is shown by the remains of the walnut, hickory, viburnum, grape, elm, poplar, sequoia, and yew, and by the presence of numerous, often thick, beds of lignite. In the earliest Eocene, the vegetation of Greenland, Iceland, and Spitzbergen included alders, magnolias, lindens, poplars, and birches, indicating a climate similar to that of south temperate France and California at the present time. In the later Eocene, palms flourished in southern England; and the waters were tenanted by crocodiles and giant sea snakes, indicating a climate like that of tropical America today, and warmer than in the western interior of North America.

Oligocene—The climate of the Oligocene in Europe appears to have been slightly cooler than during the Eocene. The occurrence of palms in the Baltic region, however, indicates a temperature such as now prevails in Spain and Italy. Nothing is known of the grasses, and the development of the teeth of the mammals does not afford a positive proof of their presence. The presence of crocodiles in the Oligocene deposits of South Dakota implies a climate such as is now found in Florida.

Miocene—Although the vegetation is similar to that of the Oligocene, there is evidence of a gradual lowering of the temperature in the Miocene; palms ceased to exist north of the Alps; and towards the end of the epoch there was a lowering of the temperature in the Arctics. In Colorado Miocene deposits no palms have been found, but the presence of figs, which now do not grow north of the coast of the southern states, and of two genera of trees which are now confined to the tropics (Weinmania) indicates that the climate of this mountainous region was more equable, moister, and somewhat warmer than now, although it is probable that on the Great Plains arid conditions prevailed. A layer of fan-palm leaves a foot in thickness in a formation of this epoch in northern Washington points to almost tropical conditions in that region. The occurrence of breadfruit trees associated with temperate region trees in the Middle Miocene of Oregon indicates a somewhat warmer climate there than now.

Pliocene—The gradual cooling of the climate of Europe continued in the Pliocene, during which epoch there was a slow southward movement of the northern forest trees and a disappearance of delicate tropical types. Towards the very end of the Pliocene there was a marked lowering of the temperature and perhaps the beginning of glaciation on the higher mountains. In the English Pliocene, the proportion of Arctic shells rises from five percent in the oldest to over sixty percent in the youngest beds.

The disappearance of rhinoceroses and the browsing types of horses and camels (those that lived largely on the leaves of shrubs and trees), as well as the existence of great herds of land tortoises on the Great Plains of North America, is perhaps proof of arid conditions in the western interior. The disappearance on the Pacific coast of warm temperate plants, as well as the character of the marine and freshwater invertebrates, indicates a change to colder conditions. Evidence is at hand showing that Japan was colder during the Pliocene than during the Glacial Period (Yokohama). This has again suggested the possibility of a "wandering pole" (pg. 485) to account for the Glacial Period.

Effects of Isolation and Migration

During the Age of Mammals, the seas were at times so expanded as to isolate large areas of land, while at others they were so restricted that continents now separated were then united (pgs. 457, 458). The mammalian life of Cuba seems to have been derived from a few species that were carried there on natural rafts. Other islands were doubtless populated in the same way. When the isolation was prolonged, the evolution of the animal life of the various continents took place independently. When the lands were again united, widespread

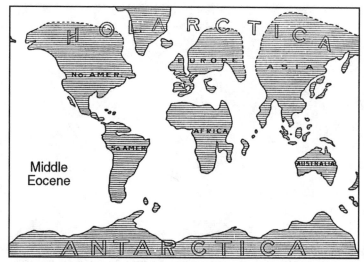

Map of the world in the Middle Eocene, showing the isolation of the continents and the conditions favorable for the development of provincial faunas.

The separation of Africa and South America from their neighboring continents caused the animals of the former to develop independently, while the union of Asia to North America permitted intermigration between these continents.

The continents, with the exception of South America, were broadly united during the Middle and Upper Miocene, permitting widespread migrations.

migrations took place. This isolation and later establishment of land connections occurred several times during the Tertiary. The proof of the separation or reunion of great land areas is based chiefly upon the dissimilarity in the one case, and the similarity in the other, of the life of the past.

The effect of isolation upon the animal life depends somewhat upon the size of the region and the diversity of the topography. If the topography, climate, and vegetation are varied, a diversified mammalian fauna will arise to take advantage of every opportunity of securing food; and the body, limbs, and feet will become adapted to a great variety of conditions; some will become adapted for burrowing, some for life in the water, some for rapid motion, and some for tree life. The larger the region and the more diverse the conditions, the greater will be the variety of mammals that will result. When after long periods of isolation animals which, because of the physical conditions under which they lived or because of the fierceness of the competition with other forms, had become especially fitted for life, were able to migrate to other regions where for various reasons evolution had not been so effective in producing such successful types; the better fitted quickly possessed the new regions, either forcing the former inhabitants into subordinate positions or causing their extinction.

No better example of the effect of such isolation can be found than in Australia today, where only mammals of a low type (marsupials) occur. These Australian mammals are very different from those of other parts of the world, but are related to those that lived in Europe in the Mesozoic; types which with a few exceptions have long since been extinct in other continents. The natural inference is that Australia was isolated during the whole of the Tertiary epoch and that, because of the non-interference of the higher mammals, the marsupials have been able to develop there along their own lines, producing the kangaroo, the wombat, and the other animals peculiar to that continent.

The effect of isolation and migration on animal life is especially well shown in the history of Tertiary mammals.

Eocene Invasion—The first, and perhaps most important Tertiary migration occurred at the very beginning of that epoch, as is shown (1) by the sudden appearance of true mammals which, though simple in structure, were already somewhat diversified, and (2) by their similarity in all parts of the world where found. The land connection, or connections, which permitted this migration from a center whose location is at present unknown, disappeared, perhaps by subsidence, and the continents of the Old and the New World were apparently again separated by broad seas for a long period of time (pg. 457, top), permitting the life of the isolated regions to develop independently during the Eocene. A comparison of the life of the Middle and Upper Eocene shows that the odd-toed, hoofed mammals (perissidactyls) of Europe and North America differed in many respects, and that, although horses developed on the two continents, they were markedly dissimilar. The same dissimilarity is shown in the carnivores and rodents. During this period of isolation three new families made their appearance in America, the camels, oreodonts, and armadillos; and the most striking of the Rocky Mountain Eocene mammals (Amblypoda) were probably extinct in Europe before the Upper Eocene, but did not become extinct in America until near the close of the Upper Eocene. The resemblance that existed between the mammals of the two continents is only that of descent from similar ancestors.

Oligocene Invasion—The simultaneous appearance in Europe and North America (pg. 457, bottom) of new families of mammals of a decidedly more modern type than those of the Eocene, points strongly to their evolution in some region separated from both continents for a long period, and united to them at approximately the same time by renewed land connections. It should not be inferred from the above that the intermigration was so great as to make the life of the two continents identical at the beginning of the Oligocene, for this was far from being the case.

Following the period of land connections at the beginning of the Oligocene, the Old and the New World were again isolated and independent evolution was permitted.

Miocene African Invasion—The similarity of the life of Europe and of North America indicates that these continents, as well as the East Indies, were united during the Miocene (pg. 458); but the dissimilarity of that of South America and Australia shows that these southern continents were separated from the others. The appearance in the Lower Miocene of Africa and Europe of ancestral elephants whose development had been taking place in Africa or some adjoining region, during the earlier portion of the Tertiary, shows that these continents of the Old World were then united, for the first time perhaps, since the early Eocene. The mastodons and rhinoceroses migrated to America at this time, and other tribes unquestionably came with them. They are believed to have reached here by way of the Alaskan land connection. These strangers had little effect on the life of the New World, as is shown by the development in America of its own pigs, oreodonts, deer, antelopes, camels, horses, etc. Although the continental connections were well-established between Europe, Asia, Africa, and North America in the Miocene and continued throughout the epoch, permitting the spread into North America of mastodons, rhinoceroses, and probably other mammals, many of the important races of the New World—such as the camels, llamas, ancestral horses, and ancestral American deer—were confined to this continent; and animals equally characteristic of the Old World—true rhinoceroses, African and Asiatic monkeys, bears, lynxes, foxes, hyenas, true antelopes—are not known to have migrated to North America at this time.

Pliocene South American Invasion and Intermigration between the Old and New Worlds—With the connection of South America and North America in the Pliocene, all of the continents of the world with the exception of Australia were united, and widespread migration occurred.

South America seems to have been isolated from the rest of the world from early Eocene times until the Pliocene. Its fauna, before the Pliocene connection with North America was established, was comprised of (1) marsupials resembling, in a marked degree, those of Australia today, and (2) true mammals differing greatly from those that had been developing in North America. The explanation of this peculiar fauna is probably to be found (1) in the absence of all true carnivores (the cat and dog family having, so far as is known, failed to send any Eocene representatives there), and (2) in the small variety of ancestral forms from which the fauna developed. This small number of ancestral true mammals indicates that the Eocene Central American connection had been of brief duration. These South American ancestral forms came from North America, or from Australia by way of the Antarctic Continent, or from both.

During the period of separation, several families of strange hoofed mammals were evolved to take advantage of the varied physical conditions, some of which (Litopterna) were odd-toed, with bodily proportions resembling those of the horse and llama. In this family, the third toe was always the largest, and in some species the evolution of the foot had been carried to the one-toed stage, producing a foot similar to that of the horse. They were, however, inferior in brain and teeth to the even and odd-toed herbivores of North America.

The most remarkable development occurred in the sloth tribe (edentates), in which huge forms, elephantine in size but of different proportions, some of which were covered with armor, were numerous and conspicuous.

Rodents of the porcupine type and monkeys were abundant during the Tertiary and are living in South America today.

With the joining of North America and South America, an intermigration of animals from the two continents began. Horses, llamas, deer, mastodons, tapirs, members of the cat and dog family, and others invaded South America; and at the same time the giant sloths and other South American forms moved north. The result was to be expected. Not all of the families of North American mammals found a home in South America, either because they did not migrate there or because the conditions were unfavorable for their existence; but, as a rule, these immigrants from the north in which brain and limb had been highly developed as a result of the severe struggle with highly specialized carnivorous enemies, as well as because of the competition with other herbivorous forms, soon crowded out the more conspicuous but less highly developed indigenous animals. Today the most conspicuous South American animals are those whose ancestors reached that continent in the Pliocene, although many characteristic ancient forms, such as the armadillo, are numerous.

Not only did the North American mammals invade South America but a similar invasion in the opposite direction was going on at the same time. These immigrants, however, failed to establish themselves permanently, although they probably lived on in their new home for some time, as the presence of their remains as far north as Oregon shows. Their extinction before the Pleistocene was due either to the competition with the higher forms or to the cold of the Glacial Period, but probably to the former.

Duration of the Tertiary—Because of the fact that the Tertiary rocks are, with the exception of the Pleistocene, the last deposited, some evidence of the duration of the period is at hand which is not available in estimating the length of former periods. The most important means which can be employed are the following:

(1) Biologists are generally agreed that the time necessary for the evolution of modern mammals from the generalized ancestors of the early Eocene was very long. For example, the highly specialized modern horse could not have been evolved from the little Eohippus with his four-toed foot, simple teeth, and carnivorous-like body, in tens of thousands, or hundreds of thousands of years; but a much greater length of time must have been required.

(2) Again, as in other periods, we can gain some idea of the vastness of Tertiary time by a consideration of the mountain ranges which had their birth and principal growth during the period. At the beginning of the Tertiary, Switzerland was probably a comparatively flat plain where the lofty peaks of the Alps now stand; and the grandest mountains of the world, the Himalayas, were not raised until about the middle of the Miocene. It is impossible to state how long, in years, this great deformation required, but it is evident that an almost inconceivable length of time was necessary.

(3) Since the close of the Eocene the Grand Canyon region has been elevated 11,000 feet, and the Colorado River has been able to carve its way through limestone, sandstone, and granite to a depth of 6500 feet.

(4) The most approved method of estimating geological time

is by the maximum thickness of the sediments deposited during the period. Upon this basis the duration of the period has been variously estimated at from 3,000,000 to 4,000,000 years, with the former estimate more generally accepted than the latter.

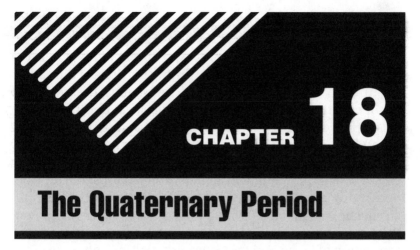

CHAPTER 18

The Quaternary Period

The last great period of the earth's history—the Quaternary—may be considered as beginning with the initiation of extensive sheets of ice in the northern hemisphere. It is divided as follows:

Quaternary
- *Recent, Post-Glacial or Human.* Since the disappearance of the continental ice sheets.
- *Pleistocene* (Greek, *pleistos*, most, and *kainos*, recent) *or Glacial.* Extending from the beginning of glaciation until the final disappearance of continental glaciers.

Changes at the Close of the Tertiary

Three important changes at the close of the Tertiary should be noted.

(1) **Elevation**—The later Pliocene and early Quaternary was a time of elevation, during which the continents stood higher than now, and broad land connections existed, permitting migration between the continents. On the Atlantic coast the Pleistocene elevation has been variously estimated at from a few hundred to a few thousand feet. Evidence is at hand of an elevation of 1500 feet in southern California, of 3000 to 6000 feet in the Sierra Nevadas, 1500 feet in Oregon, and of an even greater raising of the land in British Columbia. In the West

Indies, Panama, and South America, observations point to a higher level of the land than now during portions of the epoch. An upward movement increased the height and extent of Europe, so that at the time of maximum elevation Great Britain was a portion of Europe, and Europe was united to Africa by broad land connections across Gibraltar and through Italy by way of Sicily. After a time, elevation ceased, and a subsidence began which first separated Ireland from Wales and later from Scotland, and finally isolated Great Britain. The land connections with Sicily, which joined Europe and Africa, and that at Gibraltar also disappeared. It seems probable that the separation of Japan and the Philippine archipelago occurred in post-glacial times.

(2) **Glaciation**—The gradual refrigeration of the climate at the close of the Tertiary culminated in the Pleistocene. The cause, or causes, which produced this marked decrease in temperature will be discussed later.

The lowering of the temperature resulted in the accumulation of snow and ice to form great ice sheets, several hundreds to several thousands of feet thick, which spread over 6,000,000 to 8,000,000 square miles of the earth's surface, especially in the northern hemisphere. Although this is an important event in the world's history, it should be remembered that it is not unique, since extensive glaciation occurred in the Permian as well as in Pre-Cambrian times, and possibly at other periods. It should also be noted that these earlier ice invasions have not been considered of sufficient importance to form a basis for a further subdivision of these earlier periods. Why, then, is the separation into Tertiary and Quaternary made? It is because the event was so recent (geologically) that the evidences of glaciation are widespread and conspicuous, since sufficient time has not yet elapsed to obliterate them by erosion and weathering, and also because the indirect effect upon man has been of great importance.

(3) **Changes in Life**—The least important change between the periods is in the life. As far as this is concerned, the Quaternary might almost equally well be considered a continuation of the Pliocene. At the beginning of the period nearly all living species of mollusks were in existence, and most of the species of living mammals, but during the Pleistocene there was a gradual disappearance of many mammals, such, for example, as the mammoth, mastodon, woolly rhinoceros, and saber-toothed tiger. After the Tertiary there was no longer a mingling of tropical and subtropical plants with temperate and Arctic plants; but, apparently as a result of the migrations forced on them by the climate, they became adapted to special habitats.

Distribution of the Ice Sheets

The great event of the Pleistocene was the accumulation of vast continental glaciers.

1. Other Continents—As a result of the increasing cold, the whole of northern Europe (pg. 468) was buried under an ice sheet of great thickness, which filled up the basins of the Baltic and North seas and spread over Scotland and the greater part of England. The Alps and Pyrenees were covered by great snow fields and glaciers which stretched over the neighboring lowlands, and even the island of Corsica had glaciers. In the southern hemisphere and in the tropics, glaciers appear to have existed where none are now, and our present-day glaciers are insignificant remnants of those of Pleistocene times. One interesting exception to the general glaciation of the northern portion of the Old World was the absence of glaciers in Siberia, where, even at the present, a portion of the country has a mean temperature of five degrees, the soil is permanently frozen to a depth of several hundreds of feet, and Arctic conditions prevail over large areas. The absence of glaciers is now, and probably was then, due to the deficient precipitation, which makes an accumulation of snow impossible.

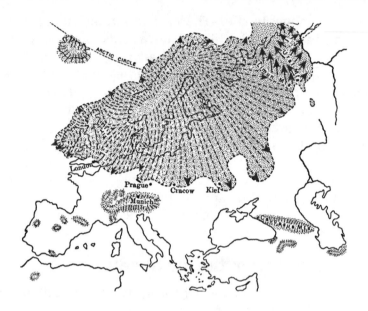

Map of Europe during glacial times. The area covered by glaciers is shaded, and the direction of the movement of the ice is indicated by arrows.

2. North America—North America (pg. 469) was more extensively affected by glaciation than any other part of the world, about 4,000,000 square miles being covered by ice at the time of its greatest extension. Two peculiar features are especially striking in the distribution of the North American ice sheets: (1) the greatest extent of ice was in the low regions of the northeast, instead of in the high mountains of the west and northwest; and (2) the northeastern portion of the continent, rather than the northern, was the scene of maximum glaciation, even Alaska being largely free from ice.

The great ice sheets moved in all directions from three great centers and probably to some extent from other smaller ones, as is proved by the direction of the striations on the underlying rocks, and the courses along which the boulders were carried. These three great centers of radiation were: (1) that situated in Labrador (the Labradorean); (2) that just west of Hudson Bay (the Keewatin); and (3) that in the western mountains (the Cordilleran). The greatest extent of the Labradorean ice sheet was to the southeast, where it stretched 1600 miles south of the center. There was also a movement north from this center, but it is not known to have been nearly so extensive. This ice sheet, in its greatest expansion, crossed the Ohio River into Kentucky, and extended into southern Illinois.

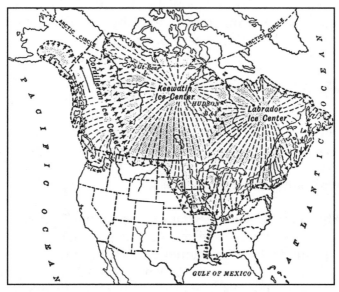

Map of North America, showing the area covered by ice at the stage of maximum glaciation, and the centers from which the ice moved. The arrows show the directions of ice movement.

The Keewatin ice sheet extended almost as far southward as the Labradorean, its front at one time being in Kansas and Michigan, about 1500 miles from its center. The movement of the Keewatin ice sheet is remarkable since, beginning in a low, flat region, which is now semiarid, the ice moved upgrade into the United States. This is more astonishing when we find that the Cordilleran sheet, starting from the lofty mountains of western North America, apparently failed to move beyond their foothills. The Cordilleran ice sheet should, perhaps, be considered as the product of the confluence of mountain glaciers, spreading out as they reached lower and less rugged ground, much as do some of the Alaskan glaciers today.

Besides these great centers of ice movement, large local glaciers accumulated on the mountains of the western United States, where they were vigorous for many years, as is shown by the cirques, moraines, rock basins, and other evidences of glaciation common throughout the high mountains of the west. This is well seen on the topographic maps of Montana, Wyoming, Colorado, and neighboring states.

Development of the Ice Sheets

The great ice sheets (with the probable exception of the Cordilleran center) did not begin as mountain glaciers which by their coalescence became one great glacier, but were the result of the gradual accumulation of snow in the north, due to the lowering of the temperature.

Thickness of Ice Sheets at Center—It has been held that the great ice sheets were several miles thick at the various centers, from which points they gradually thinned toward the margins. A study of existing Greenland and Antarctic glaciers shows that such is not now the case, but that the thickness of the ice not far from the margin is practically the same as that of the interior, the surface of which is a comparatively level plain.

The slope of the sides of the tongues of ice that reached down ravines of the Allegheny River somewhat beyond the margin of the main sheet varied from 100 to 130 feet a mile, and the average slope of the ice lobe of the Hudson valley has been estimated to have been 25 to 30 feet a mile. Even the smaller of these figures would make an enormous thickness for the ice sheets at the centers, if the slopes were uniform. Since the ice in Illinois is known to have reached 1500 to 1600 miles south of the center of accumulation, an average slope of 25 feet a mile would, on this basis, give a thickness of about eight miles at the center. It is probable, however, that the slope was not nearly so steep some distance back from the margin. When this is taken in connection with the fact that the thickness of the ice near the margin was probably approximately the same as that at the center, a much less depth is obtained than on the former estimates. Upon any basis, however, the thickness must have been great. In New England, for example, the ice was so thick that it passed over the Green Mountains where they are 3000 to 5000 feet high, in a course diagonal to their general direction, showing that such a mountain chain made scarcely a ripple on the surface.

Glacial and Interglacial Stages

The Glacial Period was made up of a number of advances of the ice and corresponding recessions when the ice either entirely, or largely, disappeared from the northern hemisphere. The duration of the various glacial stages was long, and that of the interglacial stages so extended as to permit trees and plants to clothe again the glaciated regions. The proofs of distinct ice sheets (below), separated by long intervals, when the land was free from glaciers and even warmer than at present in the same regions, are conclusive. The drift of the earlier glacial stages, where it has not been covered by more recent drift, differs from the latter in a number of particulars: (1) the drainage of the surface is mature, being in this respect in marked contrast to

Diagram showing the proof of successive ice sheets. Resting upon the lower till sheet and underlying the upper are ancient peat bogs (solid black). The erosion of the older (lower) drift where exposed is much further advanced than that of the younger (upper). The composition of the drift sheets also differs.

the immature drainage of the most recent drift, with its lakes and marshes; (2) the boulders of the older drift sheets are found to be different from those of later drifts, showing that they were brought by ice sheets that moved in a somewhat different direction; (3) the boulders of the older drift are, moreover, much weathered, so that even granite boulders can sometimes be crumbled in the hand, and the clay is deeply oxidized and the lime largely leached out. (4) Deeply eroded surfaces, covered with peat or ancient soils (pg. 473) occur, underlying more recent drift. The most satisfying proof is found in two sections, one in Iowa and the other near Toronto, in which stratified deposits containing fossils rest upon old, weathered till and are, in turn, overlain by younger drift.

The recognized glacial stages in America and Europe are:

North America		Germany
Post Glacial		Iron Bronze Neolithic
Wisconsin	Later Fourth Interglacial Earlier	Würm
(?)Iowa	(?) Interglacial (Peorian) (There is some doubt as to the interpretation of the Iowan.) Third Interglacial (Sangamon)	Third Interglacial Paleolithic man in Europe
Illinoian		Riss
	Second Interglacial (Yarmouth)	Second Interglacial Earliest remains of Paleolithic man in Europe
Kansan		Mindel Heidelberg man (Europe)
	First Interglacial (Aftonian)	First Interglacial
Subaftonian	(Jerseyan)	Günz

Characteristics of Former Drift Sheets—The different stages varied in duration, in the character of the material deposited, and in the extent of the glaciation. The deposits of the Kansan ice sheet, for example, contain little stratified drift, indicating that stream action was of little importance. This is surprising, since the melting of great masses of ice naturally carries with it the idea of flooded streams and great deposits of stratified drift, such as resulted from the last ice sheet. The effect of the last recrudescence of glaciation in the Wisconsin stage is best known, since its deposits cover the earlier drifts. It seems probable, moreover, that the surface of the last drift had a relief stronger, originally, than that of any of the former ice sheets. One marked feature is the lobate form of the moraines which occur in a succession of crescentic belts (below).

Map showing the moraines and the direction of the ice movement (shown by arrows) of the last continental ice sheet. The lobate character of the moraines is very pronounced. The "driftless area" was not covered by the ice.

History of the Great Lakes

There is general agreement that the Great Lakes were not in existence immediately previous to glacial times. This belief is based upon the fact that the region now occupied by them had been subjected to erosion so long that any lakes which might at some time have been in existence must have been destroyed by filling or the cutting down of their outlets. It is probable that in preglacial times the region in the vicinity of the Great Lakes was not unlike that of central and eastern Tennessee, Kentucky, and southern Indiana, where the weak rock formations are marked by lowlands, and the more resistant by highlands.

Preglacial Drainage—The precise course of the preglacial drainage of this region is yet to be determined, but the evidence indicates that the St. Lawrence River was not as now the channel through which it flowed to the ocean. At that time, indeed, the head of the St. Lawrence River may have been in the vicinity of the Thousand Islands. Well borings in Michigan and Indiana have revealed the fact that ancient, drift-filled valleys of great depth lead towards the south, giving strong support to the supposition that the preglacial drainage was southward to the Ohio and Mississippi, instead of to the east. In fact, at the present time, a lowering of the land a few feet at the southern end of Lake Michigan would turn the drainage to the Mississippi Valley.

Origin of the Basins—The basins of the Great Lakes are lowlands which have been modified in several ways: (1) by drift deposition which has not only blocked up the valleys leading south, but has also increased the height of the divides by terminal moraines; (2) by glacial erosion, though whether the glaciers accomplished more than the removal of the weathered rock and soil is a mooted question; all will concede, however, a deepening to this extent at least; (3) by a depression of the region at the north as a result of the weight of the ice. Since the disappearance of the ice sheets the land at the north has been slowly rising, as is shown by the beach lines of the ancient lakes which are now higher at the north than they are further south in some cases 400 feet or more. The surfaces of the lakes are held up by rock and drift barriers to levels several hundreds of feet above their rock beds, while the bottoms of all the lakes, except Lake Erie, are below sea level.

Great Lakes Stages—The Great Lakes had their inception when the ice sheet had retreated across the higher land which turns some of the water to the north and some to the south. The melting waters being prevented from flowing north, accumulated between it and the ice front, gradually enlarging upon the further recession of the ice. At first there were doubtless many small lakes which had temporary outlets to the south. As the ice retreated still further, these lakes coalesced into larger ones. The brief and incomplete history of the Great Lakes which is given below has been learned from a study of the beaches, sand bars, deltas, and outlets made at former lake levels.

The history of the Great Lakes may be considered as beginning after the ice had retreated to such an extent that a large lake (glacial Lake Maumee) came into existence near the western end of what is now Lake Erie, and which emptied through the Wabash River into the Mississippi River. This may, for convenience, be called the first stage (pg. 476, top) in the history. In the second stage, the ice front had retreated to such an extent that a greatly expanded Lake Erie (glacial Lake Warren) was formed which emptied through the Illinois River into the Mississippi River. Upon a further retreat of the ice, a third stage (pg. 476, bottom) was inaugurated, with the resulting formation of three Great Lakes with three outlets; Lake Superior (glacial Lake Duluth) discharging over the divide at Duluth through the St. Croix into the Mississippi River, Lake Michigan (glacial Lake Chicago) through the Illinois River as before, and lakes Erie and Huron (glacial Lake Lundy) through the Mohawk River in New

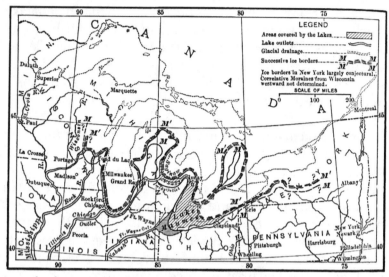

An early (*first*) stage in the history of the Great Lakes.

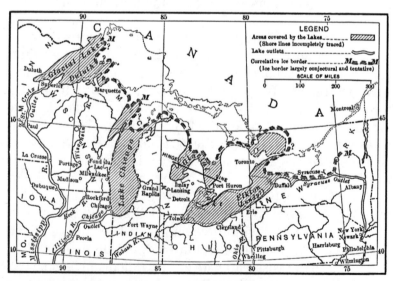

An early (*third*) stage, in which three Great Lakes with separate outlets were formed by the further retreat of the ice front.

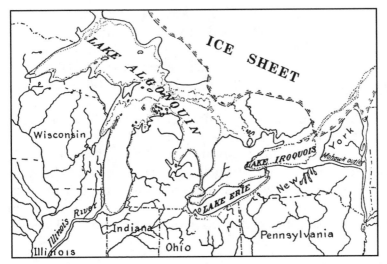

In this (*fourth*) stage, the Great Lakes had their greatest area, forming Lake Algonquin. The outlets were the Mohawk and Illinois rivers.

York to the Hudson River. In the fourth stage (above) the region of the Great Lakes was entirely uncovered with the exception of the St. Lawrence River; at this time the present lakes Michigan, Superior, and Huron were greatly expanded to form a lake which covered a greater area than that occupied by all of the present Great Lakes. This lake (named Lake Algonquin) discharged through the Mohawk River to the Hudson and probably also for a time through the old Chicago outlet. During an early part of the stage Lake Huron probably emptied into Lake Erie, but later, when the ice front had melted back farther to the north, drained through the Trent River in Canada to Lake Ontario (named Lake Iroquois), reducing Lake Erie to such an extent that the amount of water flowing over Niagara falls was probably not greater than that now pouring over the American falls. As the land was uplifted in the north (as the weight of the ice was removed), the drainage of Lake Huron was again discharged over Niagara Falls. The fifth stage (pg. 478) began with the opening of an eastern passage along the ice border into the Ottawa valley, which lowered the surface of the lakes (forming

Lake Nipissing). The drainage passed through this outlet until the elevation of the land on the north was sufficient to send the waters into their present course. The beach lines of this fifth stage rise at the north (are higher at the north than at the south), showing an uplift of 100 feet at the head of the Ottawa River since it was abandoned.

There seems little doubt that at each advance and retreat of the ice, during the different stages, lakes were formed in somewhat the same position as at the close of the last ice (Wisconsin) invasion. The proof of such lakes is not abundant, but is indicated by the sandy character of the drift in the moraines at the south end of Lake Michigan, the sand having probably been obtained by the ice from the deposits of a former Lake Michigan.

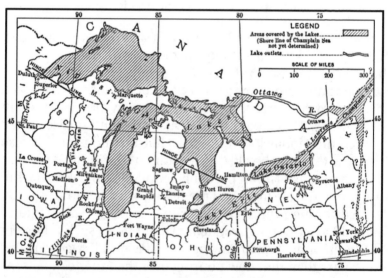

A (*fifth*) stage with the drainage through the Ottawa River. A lowering of the region resulted in the extension of the sea into the St. Lawrence valley and Lake Champlain.

The Champlain Subsidence—The fifth stage (pg. 478) of the Great Lakes was coincident with a great subsidence of the northeastern Atlantic coast, which permitted the sea to spread over the St. Lawrence valley, Lake Ontario, Lake Champlain, and the Hudson River, thus making New England an island.

The sediments carried into these bodies of water at that time contain marine shells and even the skeletons of whales, one of which was found in a Lake Champlain terrace, and another in the Ottawa valley. The terraces near Montreal which are 600 feet above the sea, those of Lake Champlain which are 500 feet at the northern end and 400 feet or less at the southern end, and those of Maine which are 200 or more feet in height, show both the amount of sinking in that epoch and the differential uplift since then.

Other Pleistocene Lakes

Lake Agassiz—The Red River valley of Manitoba, North Dakota, and Minnesota, so remarkable for its fertility as well as for its flatness, is the result of a glacial dam which prevented the usual drainage through the Red River to the north, and produced a great lake (Lake Agassiz) which discharged by way of the Minnesota River into the Mississippi. On its bottom the silts carried in by the streams were deposited, making a surface as flat perhaps as any on earth. Upon the retreat of the ice, drainage to the north was permitted, and the lake disappeared. At its greatest extent this lake had a larger area than that of the present Great Lakes combined.

Lake Bascom—In rugged regions the ice sheet formed many temporary lakes, a rather remarkable example of which is to be found in northwestern Massachusetts, where a lake (Lake Bascom) first stood at an elevation of 1100 feet above the sea, the level of the lake being determined by the pass (col) through which the water was discharged. As the ice retreated, lower passes, approximately 1000, 900, 700, and 600 feet above the sea, were found; and the lake was finally drained when the present outlet—the Hoosic River —was uncovered.

Great Basin Lakes—In some semiarid regions not covered by the ice sheet the climate of the Pleistocene seems to have been moister than at present and previous to glacial times. This is brought out by a study of the Great Basin region of Utah and Nevada. Great Salt Lake is a small remnant of a Pleistocene lake (Lake Bonneville), which was many times larger (below), discharging at one stage by a northern outlet to the Pacific. This lake (Lake Bonneville) at its maximum was 1000 feet deep and covered an area of 17,000 square miles. Terraces marking former levels of the lake are conspicuous features of the landscape, as seen from Salt Lake City and other portions of the basin. A return to an arid climate caused the shrinkage of the lake to its present area and maximum depth of only 50 feet. The salt lake which doubtless formerly existed there in preglacial times was gradually freshened as the lake level was raised, and became entirely fresh when the excess water flowed through the outlet. When the climate again became drier, the lake shrank, and all of the soluble salts of the larger lake, as well as those brought into the basin since that time, have accumulated to form the

Map showing the position of Lake Bonneville on the east and Lake Lahontan on the west. The relative size of Great Salt Lake is indicated.

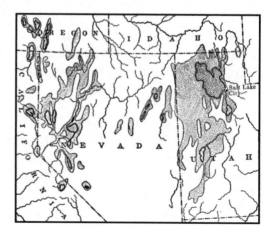

present exceedingly saline waters.

Further west in the same basin were other lakes (Lake Lahontan) which, however, were not as large as Lake Bonneville, although of considerable extent.

Loess

In the Mississippi basin, especially in Illinois, Iowa, Nebraska, and states to the south, are extensive areas of a deposit (loess) intermediate between fine sand and clay. The fact that it contains angular, undecomposed particles of calcite, dolomite, feldspar, hornblende, mica, and magnetite indicates that loess was derived from the finely ground rock flour of the glaciers. Pebbles, with the exception of lime and iron concretions, are absent, except at the base of the deposits. The most striking characteristic of loess is its ability to form vertical cliffs, a feature which can best be seen in railroad cuts and along stream courses.

One peculiarity of its distribution is its independence of topography. Its thickness seldom exceeds 50 feet, while 10 feet is more common. Loess occurs on the drift, between drift sheets, and even beyond the limit of the drift sheets.

The question of the origin of this widespread deposit has given rise to much discussion, but to a large extent the aqueous theory, i.e., that loess is a deposit that was laid down in standing water, has been replaced by the eolian. According to the latter, glacial streams, heavily loaded with rock flour, spread silt upon their flood plains, exposing it to the action of the winds, which caught it up and redeposited it on the adjacent uplands, where after its deposition it was held by the vegetation. A similar deposit is forming today on the western plains, where loess-like dust is held by the grasses and is slowly building up portions of the surface. If the above explanation is correct, the presence of such extensive areas of loess indicates aridity during some of the glacial or interglacial stages, since if the climate was moist, the action of the winds would be inconsiderable.

Duration

The difficulty of arriving at a definite conclusion as to the length of the Glacial Period, as expressed in years, is seen when the elements upon which such estimates must be based are analyzed. These are (1) the weathering and erosion of drift; (2) the time necessary for the climatic changes between the glacial and interglacial stages; (3) the amount of vegetable growth in interglacial stages; (4) the time necessary for the immigration of plants and animals. These all show that a long period of time must have elapsed, but afford little basis for an estimate in terms of years. (5) The time required for the advance and retreat of the ice sheets, however, affords something of a clue, since the rate of the advance and retreat of existing glaciers is known; but such estimates, at best, are subject to wide variations, depending upon the rate used as a basis. This is well shown in the figures given by different investigators, which vary from 100,000 years, to 500,000 to 1,000,000 years. The former estimate, however, is evidently much too small.

The time which has elapsed since the beginning of the retreat of the last ice sheet is better known, because of other lines of evidence. These are (1) the time required for the retreat of the ice, and (2) the time necessary for the excavation of the Niagara (pg. 483) and St. Anthony gorges, the present rate of the recession of these falls being known. When allowances are made for fluctuation in the volume of the rivers at different times after the ice had retreated so as to permit the streams to flow over their new courses, it is seen that 10,000 to 50,000 years must have elapsed since the cutting of the Niagara gorge began. For the recession of the St. Anthony Falls between 12,000 and 16,000 years seem necessary. To these estimates must be added the time required for the retreat of the ice from its terminal moraine to the Niagara and Minnesota rivers. For this 10,000 to 30,000 years more should be added. This, then, would give 20,000 to 80,000 years since the beginning of the retreat of the last ice sheet. It is evident upon comparing the weathering of

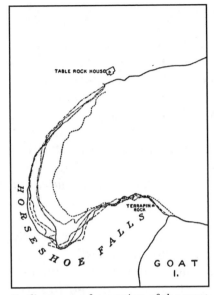

Outline map of a portion of the crest line of Niagara Falls, showing the recession of the brink during various intervals since 1842.

the Wisconsin drift with that of older drifts that the time which has elapsed since the last ice sheet is a small fraction of some of the interglacial stages. This has led to the suggestion that perhaps we are now in an interglacial stage. It is interesting in this connection to note that the extent of the area at present covered by glaciers is one tenth that of the maximum glaciation.

Marine postglacial clays in Sweden have furnished an interesting basis for determining the length of postglacial time. These clays have been deposited in regular layers, with different colors and composition, the same succession being repeated time after time. The layers laid down in summer are brown, due to oxidation, and thicker; those laid down in the autumn are darker as a result of the greater amount of organic matter, and thinner. Counting these layers (much as the age of a tree is determined by the rings), it is estimated that Stockholm was covered with ice only nine thousand years ago, and that the glaciers withdrew at a rate of 800 feet a year. The ice is believed to have receded from Ragunda, Sweden, only 7000 years ago.

Causes of Glaciation

Numerous theories have been offered to account for the refrigeration of the climate which resulted in the Glacial Period, each of which has elements of probability, but none of which, as at present worked out, is perfect. In the consideration of all of the theories discussed below, it should be borne in mind that the appearance of an ice sheet does not necessarily imply an extremely low average temperature. It has been estimated that a fall of 3°F in the average temperature of the Scottish Highlands, and a fall of 12° in the Laurentian region of Canada would result in a glacial period for these regions.

1. Elevation—The explanation of glaciation which naturally suggests itself is that the refrigeration was due to a great elevation of the land in the northern hemisphere, which so reduced the temperature that snow accumulated to form glaciers as it does now on high mountains. Those who hold this theory point to the evidence of an elevation of several thousands of feet as shown by the fiords on northern coasts. The objections to the theory are (1) that it is probable maximum elevation and maximum glaciation did not coincide in time; (2) that the elevation was not as great as once supposed; (3) that glaciation not only occurred in the northern hemisphere, but that mountain glaciers throughout the world were more extensive than now. (4) Moreover, this hypothesis would require a great elevation for the glacial stages and a corresponding depression for the interglacial.

2. Astronomical—A theory which at one time had wide acceptance was offered by James Croll and is known as "Croll's hypothesis." It is based upon (1) the variation in the eccentricity of the earth's orbit as a result of which the relative length of the summer and winter seasons changes (pg. 485). When the eccentricity is greatest the earth is 14,000,000 miles farther from the sun in the one season than in the other. (2) By means of the precession of the equinoxes the winter of the northern hemisphere, which now occurs when the sun is nearest the earth (perihelion) (below, *A*), is gradually, in 10,500 years, brought around so as to occur when the earth is farthest from the sun (aphelion) (below, *B*). The combined effect of (1) maximum eccentricity and (2) the precession of the equinoxes is to make the winters 22 days longer and 20° colder, and the summers 22 days shorter and hotter than now. The cold of the northern hemisphere would be further intensified by the diverting of some of the ocean currents to the south as the "heat equator" moved south. In the Atlantic Ocean, if the heat equator were farther south, the equatorial current would be turned southward by the wedge-shaped eastern coast of South America. The lowering of the temperature would be further increased if elevation occurred at the same time.

Some of the objections to the theory are (1) that the various ice invasions were not of equal duration as the theory requires; (2) that the duration of each glacial stage was greater than 10,500

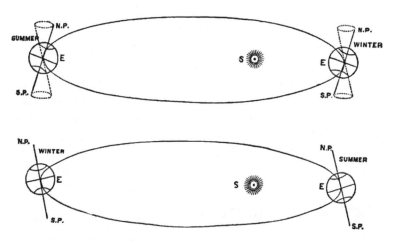

Diagram illustrating the astronomical theory of glaciation. *A*, diagram showing the relative positions of the earth and sun when the northern summer occurs in aphelion. This is the condition now. *B*, diagram showing the relative positions of the sun and earth when the northern summer occurs in perihelion. This condition favors glaciation, since the winters are longer and colder.

years, in some cases many times greater; (3) that during the Pleistocene, glaciation was greater in equatorial regions than now, where, according to the theory, there should have been little change of temperature.

3. Atmospheric Hypothesis—A hypothesis based upon the varying amounts of carbon dioxide and water vapor in the atmosphere has been favorably received, especially in America. Carbon dioxide and water vapor act much as does the glass in a greenhouse—i.e., they form a thermal blanket which prevents the radiation of much of the heat derived from the sun. When the amount of one or both of these gases is diminished, the radiation increases, and the climate becomes colder. During periods of great elevation and continental extension, erosion would be greatly increased, and the withdrawal of carbon dioxide from the air would be rapid. This would be the case since carbon dioxide is consumed in large quantities by rocks in weathering. Such consumption, under conditions favorable to great erosion, may be in excess of the supply. Also, at times of great land extension the water surfaces are relatively small, and since less evaporation occurs there is diminution of the water vapor in the air. The elevation of the land at the close of the Tertiary favored the consumption of carbon dioxide, and the contraction of the oceans furnished less water vapor to the atmosphere than formerly. By variations in the consumption of carbon dioxide, especially in its absorption and escape from the ocean, the hypothesis attempts to explain the periodicity of glaciation. Localization is attributed to the two great areas of permanent low pressure in proximity to which the ice sheet developed.

Effects of Glaciation

Glaciation benefited some regions and was harmful to others, but the former effects were, on the whole, greater than the latter. Among other benefits may be mentioned (1) waterfalls and rapids, which afford valuable water power; (2)

lakes which not only afford means of transportation, but so ameliorate the climate as to permit the raising of fruits, such as peaches and grapes, which otherwise would not thrive. In addition to this, they beautify the region where they occur, affording for the city dwellers many attractive places for rest and recreation. (3) The bays of New England have for the most part been modified by glaciation. (4) Kames, eskers, and delta deposits furnish gravel for roads and for concrete where rocks suitable for these purposes were absent in preglacial times. (5) Deposits of clay suitable for the manufacture of brick are abundant in old glacial lakes and valleys. (6) Soils are sometimes more fertile and sometimes more sterile as a result of glaciation, but the balance is in favor of the former. The mixing of soils from different regions by glaciers is often beneficial, especially when they contain fine fragments of fresh rock which, upon weathering, furnish a constant supply of plant food. On the other hand, large areas overspread by glacial sand and gravel are comparatively worthless, and hilly regions are often covered with boulders. Regions such as central Kentucky and the valley of Virginia, for example, would probably be injured were an ice sheet to pass over them, while others, such as the Piedmont of Virginia, might be benefited.

Life of the Pleistocene

As the dinosaurs culminated in the Jurassic and Cretaceous, so the mammals attained their greatest size and variety in the Pliocene and Pleistocene. The early and mid-Pleistocene life of North America is the grandest and most varied assemblage of the entire Cenozoic Period on our continent. It lacks the rhinoceroses of Europe, but possesses the mastodons, in addition to an array of elephants more varied and quite as majestic as those of the Old World. Great herds of large llamas and camels are interspersed with enormous troops of horses. Tapirs roam through the forests. True cattle (Bos) are not present, but imposing and varied species of bison are widely distributed. An element entirely lacking in Europe is that of the varied types of giant sloths, which were scattered all over the country, as well as the great armored glyptodonts in the south. Preying upon these animals are not only saber-toothed cats, but true cats, rivaling the modern lion and tiger in size.

Our record of the life of the Tertiary and previous periods has been obtained largely from marine, lake, and blown-sand deposits. Deposits of these kinds also contain Pleistocene fossils; but, since Pleistocene animals were in existence but a comparatively short time ago, their remains are also found in superficial deposits, such as river terraces, peat bogs, frozen soils, ice cliffs, and cave deposits (below), which, being easily destroyed by erosion, are seldom found in the older formations.

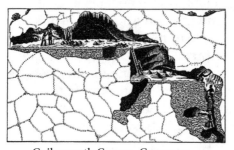
Gailenreuth Cavern, Germany.

Marine Pleistocene deposits are not common, since the subsidence which followed the emergent condition of that epoch buried most of the sediments of the time beneath the sea. This is unfortunate, since, if a complete marine record were extant, we should have, as the ice sheet advanced and retreated, a succession of faunas and floras; the temperate life changing to arctic as the ice sheet advanced, and this, in turn, being replaced by the temperate, or even subtropical life (if the climate changed to that extent), when the ice again retreated. During each advance and retreat of the ice sheet a migration of the life to and fro would, theoretically, be recorded. Unfortunately, however, no such series of deposits has been discovered, either because ideal conditions did not exist in any one place, or because the deposits are inaccessible.

Interglacial Deposits—Fortunately two deposits are known which were laid down during interglacial periods. In one of these, near Toronto, Canada, the fossil beds were deposited upon the eroded surface of boulder clay and were, in turn, eroded to some extent before the re-advance of the ice sheet which covered them with a layer of drift. The lower portion (Don) of this deposit yields plants which show that the climate was warmer in this interglacial stage than at present on Lake Ontario, being similar to that of Virginia today. This is indicated by the presence of the Judas tree (Cercis), the Osage orange (Maclura) and the papaw (Asimina). Besides these more typically southern trees, there are maples, spruces, oaks, elms, and hickories. In the upper portion (Scarborough beds) the fossil flora indicates a colder climate, showing that the ice sheet was again advancing.

A second occurrence of interglacial deposit (Aftonian) is in Iowa and yields the bones of a number of animals: elephants and mastodons, giant beavers (Castoroides), camels, horses, some dwarf and some perhaps larger than the domestic horse, and all of extinct species, giant sloths (Megalonyx), etc. This deposit is of special importance since it gives some idea of the life of the first interglacial period and furnishes a clue to the age of the fossil deposits south of the limit of the ice sheet, since, if the fauna in any of these are the same as that of the above, it is probable that they lived at the same time.

North and South Migrations During Glacial and Interglacial Times—We have seen that warm temperate plants, such as the papaw and Judas tree, lived in southern Canada during at least one interglacial period. The discovery of a fossil tamarack (larch) in Georgia, 480 miles below its present limit, shows that the Pleistocene climate of the southern states was colder at certain times than at present, and for a period sufficiently long to permit trees to grow. Walruses of a northern type lived on the coast of Georgia, and caribou and moose ranged into Pennsylvania and Ohio.

An interesting suggestion explaining the present migratory habit of birds is that it is due to the reduction in the temperature of the Arctic regions during late Tertiary and Pleistocene times. During the earlier Tertiary the comparatively uniform climate of the world would not necessitate any extended periodic movements, but the cold of the Glacial Period must have enforced prolonged migration, and periodic migration developed later. During the waning ice period the areas offering a congenial home to a great multitude of birds became greatly extended, from which, however, they were driven by semi-arctic winters to seek favorable winter haunts farther southward.

Deposits beyond the Ice Sheets or Protected from Them— (1) Caverns have been the greatest source of our knowledge of the mammalian life of the Pleistocene, the bones found in them having been brought there by floods or by beasts of prey which used the caverns as their lairs. The term "cavern" is used in the broad sense to include true caves, sink holes, and fissures. A section of the cave of Gailenreuth (pg. 488), showing the bones embedded in clay and the whole covered by a stalagmitic crust, is typical of many European bone caverns. Sometimes, however, there is more than one stalagmitic crust. The number of individuals and species represented by the bones found in some of these caves is remarkable. In the Gailenreuth cave the remains of 800 cave bears were found; from one in Sicily 20 tons of hippopotamus bones were taken. A cave in Pennsylvania (Port Kennedy), 60 to 70 feet deep, has yielded 64 species of mammals, of which 40 are extinct, among them being giant sloths, tapirs, mastodons, and saber-toothed tigers. In a California cave, horses camels, ground sloths, mastodons, and other extinct forms have been identified.

Caves were doubtless inhabited by the land animals of the Tertiary and other periods, but as caves have a relatively short life, they and their contents are rarely preserved in the older formations.

(2) Marshy ground in the vicinity of springs has often

Pleistocene tar pool near Los Angeles, California, with entrapped animals. The elephant and wolves were caught, and the saber-toothed tiger is about to suffer the same fate.

preserved many fossils, since at such places carnivores frequently kill their prey when the latter are coming down to drink; and in times of drought the animals of the region congregate about the water holes, where they often die in large numbers. Many also are doubtless mired in wet seasons. One of the most famous of such deposits (Big Bone Lick, Kentucky) has yielded 100 specimens of mastodon, 20 specimens of elephant, as well as bisons, musk oxen, and other animals.

(3) An asphalt deposit (above) not far from Los Angeles, California is remarkable, not only because it is unusual, but also because of the number of specimens preserved in it. In the early stages of the accumulation of the asphalt, the gummy surface apparently acted as a trap for unwary animals; where there were pools of water, aquatic birds of many kinds were entrapped in the soft tar about their margins, while land birds and smaller mammals were ensnared in attempting to reach the water.

Sloths, mammoths, horses, camels, saber-toothed tigers, together with many birds, among which is a fossil peacock, are only a few of the numerous species already identified.

(4) Wind-blown sand and volcanic dust have covered and preserved skeletons which have since been uncovered and studied.

Difficulty is experienced in determining to what portion of the Pleistocene the deposits not found between sheets of drift belong, but four "zones," or subdivisions, of which certain animals are characteristic, have been recognized.

Deposits on the Last Drift—The peat bogs which rest upon the drift of the last ice sheet (Wisconsin) have yielded a number of mastodon and mammoth skeletons. One mastodon found at Newburg, New York, had its legs bent under the body and the head thrown up, evidently in the very position in which it was mired. The teeth were still filled with the half-chewed remnants of its food which consisted of twigs of spruce, fir, and other trees.

Vegetation—The vegetation of the Pleistocene is a continuation of that of the Tertiary, and aside from the extinction of a few species little change is noticed. Some minor effects of glaciation on vegetation are, however, interesting. It is found, for example, that the number of genera of trees in Europe (33 genera) is much smaller than in eastern North America (66 genera). A study of a glacial map of Europe and North America suggests an explanation. It is seen that at the time of maximum glaciation in Europe the vegetation was confined between the front of the great ice sheet on the north and the expanded mountain glaciers from the east-west ranges (Pyrenees and Alps) at the south. As a result, the less hardy plants were killed, leaving a flora rather poor in species. In eastern North America, on the other hand, the mountain ranges have a north-south direction; and the broad plains offered few obstacles to the migration, back and forth, of plants and animals. In western North America where north and south migration was more difficult, 31 genera of trees are found as contrasted with 66 genera in the east.

On the summits of some mountains, arctic plants and insects are found whose presence is difficult to explain except on the assumption that, as the ice retreated northward, they followed the front closely moving up the sides of the mountains as the ice retreated; and that they were stranded there when the ice disappeared from the region.

Attention has been called to the fact that, apparently as a result of the oscillations of climate which marked the Pleistocene, plants were forced to special adaptations and habitats, with the result that there is now little mingling of tropical and subtropical types such as was the case in the Tertiary.

Mammoths and Mastodons—Perhaps the most characteristic Pleistocene mammals were the mammoth and the mastodon. The mammoths (Columbian and Imperial) are true elephants, and the mastodon is closely related. They have the same general appearance; the most conspicuous differences being (1) in the teeth, which in the mastodon have large, transverse ridges, but in the mammoth (as in the living elephant) are made up of many plates of enamel, alternating with cement and dentine (pg. 431); (2) in the forehead, which is low in the mastodon and high and bulging in the mammoth; and (3) in the shorter and more massive legs of the mastodon.

The largest specimens of the mammoth (Elephas primigenius) exceed in size that of any elephant now living, but their average height was probably not much greater. The mastodon was somewhat smaller than the mammoth. Both were covered with long hair, with probably an undercoating of fine wool. It is known that the mammoth had this additional protection against the cold, but it is not definitely known that the mastodon was thus protected.

The mastodon was abundant in America, possibly as abundant as the buffalo; 413 specimens of mammoths and mastodons have been reported from North America, of which 330 are mastodons. The mastodon ranged over the whole of North America. Both lived in America after the disappearance of the ice sheets, as is proved by the burial of their remains in peat bogs on top of the latest drift. The finding of charcoal (perhaps the result of lightning) beneath a mastodon skeleton in New York, and charcoal and pottery at the same level in the same bog, suggests the possibility that man and the mastodon were contemporaneous in America. Drawings upon ivory and sketches on cave walls of the mammoth prove, without question, that in Europe, man had seen mammoths. Carcasses of the mammoth have been found frozen in the ice in Siberia, where they were so perfectly preserved in this cold storage that the body of one, at least, furnished food for dogs, and perhaps even for man, several thousands of years after its death.

The distribution of the elephant tribe in the New World at that time proves that land connections must have existed between Asia and North America, and between North and South America.

Edentates—This class is found abundantly in the Tertiary formations of South America, and towards its close developed into gigantic and highly specialized forms which were among the largest animals of that continent. The Pliocene land connections between North and South America permitted some of these large creatures to immigrate to North America, where they lived into the Pleistocene. They never reached the Old World, and in South America their living relatives are small and inconspicuous. Some of the Megatherium (Greek, *megas*, large, and *therion*, a beast) tribe (Megatherium, Mylodon, Megalonyx) which reached this continent attained the bulk of a rhinoceros. What strikes one most in examining the skeleton of a Megatherium (pg. 495) is its pyramidal shape, the hind legs being massive as compared with the fore, and the backbone rapidly enlarging toward the hind quarters. The Megatherium lived upon leaves and twigs, and when standing on its hind legs could, if necessary, use its tail as the third leg of a tripod, leaving the fore limbs free to pull down branches or even trees of considerable size.

Restoration of the gigantic sloth, *Megatherium.*

The edentates roamed over South America, and some members of the tribe (Megalonyx and Mylodon) over a large portion of the United States. The finding in Patagonia of a large piece of skin (of Grypotherium) covered with hair, whose edges showed the marks of tools and seems to have been stripped off the carcass by man, indicates that some members of the tribe were alive a comparatively short time ago. But the evidence that man was contemporaneous with the giant sloths in North America is not conclusive.

Another edentate of very different appearance, which is distantly related to the armadillo, is Glyptodon (Greek, *glyptos*, carved, and *odont*-, tooth) (pg. 496) The body was covered with a bony shell, similar in appearance to that of a tortoise, but made up of a large number of small, polygonal bones united to form an immovable armature, so that this sloth has been called the tortoise armadillo. Not only was the body protected, but the tail also was surrounded by bony plates, and the top of the head was similarly armored. The animals grew to be 15 to 16 feet long. In their migrations they reached Texas and Florida.

The great armored sloth, *Glyptodon*.

Pleistocene Carnivores—One of the most characteristic animals of the early Pleistocene was the saber-toothed tiger (pg. 491) (Machairodus), so named because of the enormously developed, sharp edged, upper canine teeth which in some species extended 10 inches beyond the jaw. An examination shows that if the jaw were constructed like that of other carnivores, a time would come in the evolution of the great canine teeth when biting would be impossible. A more careful study of the jaws, however, reveals the fact that the lower jaw could be dropped straight down, thus permitting the animal to use the full length of its teeth in stabbing its prey. The tribe has had a long history, small ancestral forms with moderate canine teeth being known from the Oligocene.

Restoration of a saber-toothed tiger.

Although perhaps the most powerful of the carnivorous animals of the Pliocene and Pleistocene, they nevertheless became extinct early in the Pleistocene in Europe, and disappeared from the New World before the close of the epoch, their place being taken by existing carnivores, such as the lion, tiger, and leopard.

Horses, Camels, etc.—Horses roamed over the plains of North America in great herds and were of great variety, but became fewer and fewer until all had disappeared before the close of the Glacial Period. Some (*Equus giganteus*) had teeth exceeding in size those of the largest modern horses, while others (*Equus tau*) were more diminutive than any other true horses living or extinct.

True camels, as well as llamas, were abundant in portions of the United States in the early Pleistocene, living at least as far north in the United States as Nebraska (Hay Springs) and within the Arctic circle in the Yukon territory. They probably became extinct on this continent before the close of the epoch.

Bisons of many species roamed over North America during the Pleistocene, some of which were of great size, if the horns can be taken as a measure, one pair of horns measuring more than six feet from tip to tip. Wolves, musk oxen, bears, arid rodents (among which is a giant beaver, Castoroides), were also present.

Although Europe was not invaded by the sloth tribe, it was, nevertheless, the meeting place of many animals, those of the tropics and those of the Arctic regions coming at different times and even mingling at others.

Birds—The Pleistocene birds of Europe and America were not of exceptional size, nor did they differ to any important degree from those now living, but in New Zealand and Madagascar, gigantic flightless birds were abundant during that epoch. The name moa includes, in a general way, 20 to 25 species of these New Zealand birds, the largest of which stood 10 feet high, or from two to three feet higher than an ostrich, while the smallest were about the size of a turkey. In all of these, wings are entirely wanting. The development of flightless birds on these islands seems to be the indirect result of the absence of carnivorous enemies. With an abundance of food throughout the year and no powerful enemies, the New Zealand Pleistocene birds had not the usual incentives to flight. Under such conditions, some of them increased in bodily size until flight was impossible (25 or 30 pounds seems to have been the limit of the weight of flying animals). Once the power of flight was lost, the larger and more powerful the bird the better was the chance of its preservation as long as food was abundant, and great size resulted. A change in climatic conditions, however, was fatal to these bulky birds, since having lost the power of flight they were unable to migrate, and were, therefore, forced to depend upon the food of the islands. Their extinction appears to have been due partly to the cold of the Glacial Period and partly to man who, it is thought, completed their extermination about 500 years ago.

Prehistoric Man

The record of prehistoric man and his ancestors is a matter of geological as well as of anthropological investigation. Our knowledge of the presence, although not of the evolution, of prehistoric man is far more complete than of other animals, because the source of information is not confined to his bones, but is obtained also from the implements of stone which he made and the discovery of which is especially likely, as they were frequently lost in fishing or in the chase. Moreover, since they are practically indestructible, they are preserved after the skeletons of their makers have been destroyed.

It is a mooted question whether or not man existed in America at an early period, but in Europe, the remains of prehistoric man have been found in situations which prove beyond question their antiquity.

A brief classification based on the evolution of human implements in Europe is as follows:

	Iron Age	The Present.
Recent	Bronze Age	Implements of bronze as well as of stone. Some tribes passed directly from the Stone to the Iron Age.
	Stone Age	Neolithic (Greek, *neos*, new, and *lithos*, stone). Implements of stone, often polished and with ground edges.
Pleistocene	Prehistoric in the Old World	Paleolithic (Greek, *paleos*, old, and *lithos*, stone). Stone implements, rough with chipped edges but never ground.
Tertiary		Eolithic (Greek, *eos*, dawn, and *lithos*, stone). Dawn of the Stone Age. Implements so crude that it is often difficult to distinguish them from those made by accident.

Eolithic—In Pliocene and Miocene deposits and, it is asserted, even in those of the Oligocene, extremely rude flints, called eoliths, have been found. Although rough and crude, they often show one part shaped as if to be held in the hand, while the other part appears to have been designed for cutting. It has long been a question whether these flints were the result of accident or were made by a "tool-making animal," either very early man or a prehuman type given to shaping implements. If the flints did not occur in deposits earlier than the Pleistocene, the question might be answered more certainly, but since they are found in beds laid down more than a million years ago, the difficulty is increased.

The discovery, near Heidelberg, Germany, of a lower jaw of a very low type in early Pleistocene deposits said to contain eoliths, is important, since it gives a clue to the makers of these flints. This lower jaw is massive, with an essentially human set of teeth, its most noticeable feature being the absence of a chin projection. In other words, it is the jaw of an anthropoid (manlike) ape with the dentition of a man. As compared with the oldest Paleolithic skulls (Neanderthal), this one is of a much lower type. It is possible, therefore, that the eoliths of the later Tertiary were made by some tool-making ape.

A creature (*Pithecanthropus erectus*) whose fragmentary remains have been found in Pleistocene deposits of Java, associated with the bones of extinct animals, may also have been a member of a race which made eoliths. These remains consist of a skull cap, two molar teeth, and a diseased thigh bone, and are remarkable because of the combination of ape and human characters. The skull differs from that of an ape, its brain capacity being about twice that of an ape of equal bodily size. The brain capacity of an ape's skull is, on an average, 500 cubic centimeters; of the skull of this so-called ape-man (*Pithecanthropus erectus*) 850 cubic centimeters; of an average man 1400 to 1500 cubic centimeters. The skulls of aborigines of Tasmania have an average of only 1199 cubic centimeters. This skull, then, as regards capacity, occupies an intermediate position between the large apes and man. Moreover, the forehead is low and the frontal ridge prominent, and the characteristic features are, in general, intermediate between those of the lowest man and the highest apes. The teeth are human, with certain apelike characters, and the thigh bone is considered to be intermediate.

Paleolithic Man—Although at first merely chipped into shape and never ground at the edges or polished, the Paleolithic stone implements (pg. 501, *A*) indicate that their makers had a much greater intelligence and skill than that possessed by the tool-making animals of Eolithic times. The works of Paleolithic man are found principally in caves and in river gravels, often associated with the bones of extinct animals and occasionally with the bones of man himself. It seems to be well established that Paleolithic man, together with other southern animals, reached western Europe during one of the interglacial periods, probably during the second.

On the left, a Paleolithic implement; on the right, a Neolithic implement.

The relative age of Paleolithic human relics can often be determined by a study of the fauna with which they are associated. The oldest relics are found with elephants (*Elephas antiquus*) more ancient than the mammoth, very old rhinoceroses (*Rhinoceros merckii*), and hippopotamuses (*Hippopotamus amphibius*). In the next oldest stage the mammoth, woolly rhinoceros, cave bear, cave hyena, and other extinct animals are common. The last stage occurred at the close of the Glacial Period, at which time reindeer were crossing Europe in great numbers; and their remains often occur with those of man, giving it the name of "Reindeer stage."

We are assisted in our conception of Paleolithic man by a study of the recently extinct aborigines of Tasmania, who were, though recent, a true Paleolithic, or perhaps a degenerate race. Their clothing consisted of skins thrown over the shoulders, and they protected themselves from the rain by daubing themselves with grease and ocher. They had no fixed place of abode; and even in winter, a screen of bark served as a shelter. Their implements were few and simple, and were made of wood and stone, the latter being fashioned by striking off chips from one flake with another. Cooking by boiling was unknown; and their sea food consisted of shellfish, as they knew nothing of fishing with a hook. Their survival until the present was due to their isolated position.

The skulls and skeletons of the older Paleolithic men of Europe show that they were savages of the lowest type (below, *B*, bottom), with low foreheads and rather large though not highly organized brains. They were small in stature (five feet, three inches in average height), with knees that were bent slightly forward, giving them a carriage that was not fully erect.

Their skill as hunters is shown by the great quantities of bones of animals about their ancient camps; the environs of one such camp having yielded the fragments of at least 100,000 horses. Other animals, such as the mammoth, rhinoceros,

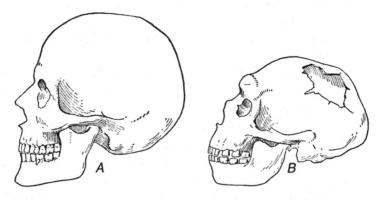

Skulls of modern and older Paleolithic man. The contrast in forehead, brow, teeth, chin, and shape of skull is very marked.

Thigh bone of modern man (shaded), and of older Paleolithic man (outline).

bison, and reindeer were also eaten. The presence of charcoal in their caves shows that Paleolithic men knew how to produce fire by friction, and that they probably roasted the flesh upon which they largely subsisted. They apparently knew nothing of agriculture and had no domestic animals, not even a dog. The stone arrows, lance heads, and hatchets, as has been said, were never ground at the edges nor polished. In some caves implements made of bone, such as arrows, harpoons, fishhooks, awls for piercing skins, and needles are not uncommon. The love of adornment is proved by the occurrence of numerous perforated teeth and shells which were doubtless strung into necklaces. The artistic skill displayed in carvings on bone and ivory (below), sketches on mammoth tusks (bottom), as well as pictures on the walls (pg. 504) of their caves, is unexpected, being superior to that possessed by any other primitive men ancient or modern. Indeed, in our own time few people not artists can equal some of the art of Paleolithic man. Although there is no perspective composition in the pictures, the drawing is excellent and the proportions and postures are unusually good. Sketches were made in red and black (pg. 504), as well as outline drawings in black. The artists chose almost exclusively the large animals of the time, the bison, mammoth, reindeer,

Carvings on bone made by Paleolithic man.

Paleolithic carving of a mammoth.

horse, boar, and rhinoceros. Man for some reason was seldom portrayed. During the Paleolithic, the making of flint implements was gradually perfected, and before its close bone and horn implements of highly useful and artistic forms were made.

Paleolithic man had a crude form of religion, and the dead were buried ceremoniously. There appears to have been a division of labor, some men devoting themselves to hunting, some to flint making, and some to art, although it is hardly probable that the specialists in any of these groups were not often employed in other work.

Neolithic Man—Man of Neolithic times was not

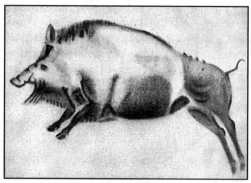

Paleolithic painting of a boar.

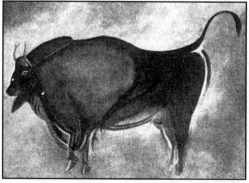

Paleolithic painting, in red and black, of a bull.

contemporaneous with the great extinct mammals, with the exception of the Irish elk. Their remains have been found in caves, cemeteries, and river deposits, in peat bogs, and lake bottoms (pile dwellings), and in shell mounds. Neolithic implements and weapons are often ground at the edge (pg. 501, *B*) and more or less polished and finely finished, and are frequently of graceful design. With some doubtful exceptions, Paleolithic man in western Europe seems to have been suddenly replaced by Neolithic man, who brought with him not only greater skill in the manufacture of implements, but domesticated animals, such as the dog, horse, sheep, goat, and hog. Moreover, he was acquainted with agriculture, as grains and the seeds of fruits, as well as dried fruits, show. Spinning, weaving, and pottery making were also practiced.

An important part of our knowledge of Neolithical man comes from the lake dwellings in Switzerland and Sweden. These dwellings were on piles driven into shallow lakes, and were connected with the shore by drawbridges which could be withdrawn in case of attack.

The Age of Stone gradually merges into the Bronze Age, as that, in turn, merges into the Age of Iron, but since these last two stages belong to protohistorical and historical times, they are outside our province. The Age of Stone did not come to an end throughout the world at the same time. The natives of the New World, Australia, and the islands of the Pacific were in the Neolithic Age, and those of Tasmania in the Paleolithic Age, when discovered by Europeans; and some isolated tribes are today still using stone implements.

Man in North America—No conclusive proof of the presence of man in North America during the Pleistocene has yet been offered. Indeed, it is doubtful if Paleolithic man ever lived on this continent.

Earlier investigators were led to assign a greater antiquity to many human relics than subsequent study has shown to be possible. Such errors were the result of over-enthusiasm and a failure to take into consideration all of the elements of the problem, some of which are the following.

(1) The presence in river gravels of rude flints has led to the conclusion that they were made by Paleolithic man. The danger in such a conclusion lies in the fact that, in the shaping of a stone tool, the maker sometimes loses the half-finished stone and often rejects others early in his work because of some imperfection or unfavorable quality in the stone. As a consequence, many unfinished stone implements are left, especially along river courses where the pebbles from which the implements were made occur.

(2) If a stone implement is found buried to a great depth, the thickness of the overlying deposit has often been taken as a measure of its age. Such data are very uncertain, since during floods a river may scour out deep holes in its bed, and within a few weeks, or months, completely fill the excavation. The Missouri, for example, scours out its bed to a depth of 40 feet or more during floods, and soon fills it again. It will readily be seen, therefore, that the finding of a flint implement in river gravels at a depth of 40 feet might not indicate any greater antiquity for it than for one on the surface.

(3) The age of flints in talus slopes is uncertain, since what was in the top portion of the cliff naturally becomes part of the base of the talus as the cliff crumbles back.

(4) The antiquity of stone implements found beneath layers of stalagmite has often been overstated, because a too low rate of deposition was used in the estimate. This, however, has not been a source of difficulty in America, since cave deposits are rare on this continent.

(5) The admixture of human remains with those of extinct animals is not necessarily a proof of their contemporaneousness, since it may have been due to accidental causes, such as human burial in deposits containing extinct animals, or to the washing out from older deposits of the bones of extinct animals and their redeposition with those of recent species.

It will readily be seen from the above that the error in

determining the age of human relics will usually be in the direction of too great antiquity. From the similarity in physical appearance of the aborigines of North and South America, it seems probable that the original inhabitants of the New World were immigrants who came from Asia and spread over the Americas after they had become differentiated in Asia, but long enough ago to permit of the development in their new home of the many languages and dialects now spoken by the Indians.

Birthplace of Man—The location of the original home of man was a matter of speculation even by the ancients and is still in doubt, but some suggestions have recently been made which are worthy of consideration. Science points, unlike the biblical account, to Africa as the birthplace of man. It is evident that the Americas were not populated by man until comparatively recent geological times and that Paleolithic and Neolithic man probably migrated to Europe from some other continent. It is a suggestive fact that all of our domesticated animals, with the exception of the llama, the vicuna, and the turkey, had their origin in Asia, and that they are the most highly specialized of their kinds. Moreover, possibly all of the cereals, with the exception of maize, are of Asiatic origin. Man was born in Africa, and attained elemental civilization in Asia because there was the place of all others upon the earth where evolution, in general, of organic life reached its highest development in late Cenozoic times. The loss of man's hairy covering is evidence of his origin in a temperate, or cold temperate climate, where he found clothing necessary to protect himself from the inclemencies of the weather.

Effect of the Advent of Man—The appearance of man was one of the greatest events in the whole history of the world, not only because, for the first time, brute strength and agility were at a disadvantage in a struggle with higher intelligence, but also because of the changes which he directly, or indirectly, caused, not only in the life of the world, but also in the very topography of the earth itself. (1) Man has directly caused and is still causing the rapid disappearance of many animals: such as the bison, the moa, seal, whale, furbearing animals, and the big game of Africa and Asia. (2) Indirectly, by the introduction of animals and plants into new regions, he has accomplished as great, or even greater, changes in life. The introduction of the mongoose into Cuba, which soon destroyed not only the snakes, but the birds that nested on the ground, has almost revolutionized the fauna of that island. The rabbits brought to Australia have overrun that continent, with a marked effect on the indigenous life. The various insects introduced into North America by man are changing the flora of this country. Many other examples might be added. (3) His work has not, however, been entirely destructive to life. Animals and plants on the verge of extinction have been preserved. The ginkgo tree, for example, would have been to us an extinct species if man had not preserved it by cultivation. (4) Not only has his contact with the life of the world been important, but his indirect effect upon inanimate nature has been stupendous. The cutting and burning of forests in certain regions has resulted in the rapid erosion of large areas, and the pulverization of the soil in plowing has permitted rainwash to carry away the best of the soil. By deforestation alone a single lumber merchant may in 50 years deprive the human race of soil that required thousands of years to form. Another effect which will eventually greatly lessen the fertility of the soil is the enormous and irrecoverable loss of phosphates in the sewerage of cities. As the result of these and many other effects of man's supremacy, the earth has suffered a vastly greater change in the past few hundred years than in many thousands of years in the most destructive periods of the past.

Future Habitability of the Earth

Two statements were made of old about the future of the earth: one that the climate will become progressively cooler until it will be unsuited for the existence of plants and animals; the other, that the earth will eventually be consumed by fire. The former statement is based on the assumption that the heat of the sun is diminishing, and that, since the earth depends upon it for its heat, a cooling of the sun will cause refrigeration. There is no question but that this would be the case were the sun's heat to decrease, but no such change appears likely based on what we know about the life cycle of stars similar to our sun.

The statement that the earth will eventually be consumed by fire appears to correspond to the pattern recognized as predicting that of our sun's future. No such catastrophe, however, is believed to be forthcoming soon. The sun has approximately five billion more years of hydrogen fuel left, and though its changes well before its own end may result in the earth's destruction, even those changes are almost an inconceivably long way off by human standards. It seems safe, consequently, to predict that for many years—perhaps millions—the sun will continue to provide the essential ingredients for Earth's supply of life. What happens on Earth, however, can not be considered as reliable.

When it is remembered that man has come up from the cave and the stone hammer in the past 50,000 or 75,000 years, and that in the past 100 years the greatest achievements of science have been accomplished, so that man today lives under conditions radically different from those of his ancestors of a few generations past, it would seem that the evolution which will take place will change profoundly the human race, if not interfered with. The progress of evolution does not, however, have a free course since, as never before in the history of animal life, the less competitive do not disappear in the struggle for existence, but instead the life of the physically and mentally disabled is lengthened through the aid of medical science and charity. The future will, hopefully, bring solution for more and more of mankind's vital problems, and the evolution of the human race may then be expected to continue, with the development of a type of man much superior to that now on earth.

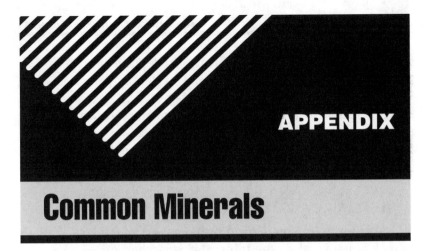

APPENDIX

Common Minerals

Every student of geology should be able to recognize the common minerals by sight and know their approximate chemical composition. In order to determine minerals without the aid of chemical tests, one must depend upon their physical properties. Of these the color, streak, hardness, specific gravity, and crystalline form are important.

The color sometimes varies greatly in the same mineral, but nevertheless often affords a strong clue to its identity. The color of a mineral is often due to the inclusion of foreign matter, such as iron oxide and organic matter, but some minerals, such as the carbonate of copper, malachite, vary slightly.

Each mineral has a characteristic hardness and this quality often affords an easy means of positive identification. The scale of hardness in common use is: 1, Talc; 2, Gypsum; 3, Calcite; 4, Fluorite; 5, Apatite; 6, Orthoclase; 7, Quartz; 8, Topaz; 9, Corundum; 10, Diamond. Minerals with a hardness of 1 and 2 can be scratched with the fingernail. If a mineral will barely scratch a copper coin, it may be considered as about 3 in hardness; if it fails to scratch glass, its hardness is less than 5; if it scratches glass but fails to scratch quartz, its hardness is between 5 and 7. A knife point is almost indispensable in determining hardness, since with a little practice, the hardness of all minerals between 1 and 6 can be readily determined.

The streak or mark that a mineral makes on a hard white substance, such as a piece of unglazed porcelain, is often important in distinguishing between minerals. The color of the streak is the same as that of the fine powder.

When a mineral breaks or cleaves in definite directions so as to form plane surfaces, it is said to have a cleavage. Since cleavage is caused by the separation along and between layers of molecules, it occurs only in crystals. The thin leaves of mica are formed by the splitting of the mineral along cleavage planes.

The relative weight of a mineral, or its specific gravity, is often an important aid in determining a mineral by its physical properties.

Iron Minerals

Magnetite (magnetic iron ore), Fe_3O_4—Color, black. A black streak is made when the mineral is scratched on a hard white surface. The hardness is slightly greater than steel (H=6). It is always attracted by a magnet and is sometimes capable itself of lifting particles of iron and steel. It is a valuable ore of iron. The Adirondack iron ore is largely magnetite.

Hematite (red iron ore), Fe_2O_3—The color is black to brick red. Streak, red. Slightly harder than steel (H=6). Occurs in compact masses composed of micalike flakes, in an earthy form, and in thin crystals set on edge. It is the most widely used iron ore in North America, the most famous localities of which are in the Lake Superior region and in Alabama.

Limonite (brown hematite, iron hydroxide), $2\,Fe_2O_3 \cdot 3\,H_2O$—The color is usually dark brown, but is sometimes yellow. The streak is yellow. The hardness of compact kinds is slightly less than steel (H=5). The ocher which occurs with limonite is composed of clay and limonite in a finely divided condition. Limonite is really iron rust and is formed from the hydration of many iron minerals, and consequently occurs in many situations and is widespread. It is an ore of excellent quality, but is little used in this country because of the more abundant and more easily mined hematite. It is common in New England and the Appalachians.

Siderite (spathic ore), $FeCO_3$—The color is gray on freshly broken surfaces; surfaces exposed to the weather, even for a few weeks, are brown. The streak is white, or nearly so. It can be easily scratched with a knife (H=4). It occurs commonly in masses which show shiny, bent, cleavage surfaces. Siderite effervesces with warm hydrochloric acid, giving off carbon dioxide. It is an ore of iron which, however, is little used in the United States.

Pyrite or Iron Pyrites (fool's gold), FeS_2—The color is brass-yellow when fresh, but oxidizes on the outside to brown limonite. It is harder than steel (H=6.5). It occurs in veins and is disseminated throughout many igneous and sedimentary rocks. It often occurs in cubical crystals or in crystalline masses. The yellow stains on rocks are often due to the weathering of grains of pyrite. Pyrite is a common mineral. It is not used as an ore of iron, but is used in the manufacture of sulphuric acid.

Pyrrhotite (magnetic pyrites), $Fe_{11}S_{12}$—The color is bronze-yellow when fresh, but weathers readily to brown on the outside. It is darker than pyrite. Pyrrhotite is softer than pyrite and can easily be scratched with a knife (H=4). Small fragments are attracted by a magnet. Pyrrhotite is of little use in itself but, since it often bears nickel, it is mined for that metal. The most valuable deposits occur in Canada, but large quantities are found in Vermont, Pennsylvania, and elsewhere.

Zinc Minerals

Sphalerite (blend, black jack, jack, zinc sulphide), ZnS—The color varies from yellow to brown-black. When a fragment is crushed, the pieces look like resin. This resinous luster can usually be seen whenever a specimen is fractured. The streak is light yellow. Sphalerite is softer than steel (H=3.5) and occurs in crystals or in masses with well-developed cleavage faces. It is an important ore of zinc and is often associated with lead and silver ores. It is extensively mined in Missouri and is of common occurrence in smaller quantities elsewhere.

Calcium Minerals

Calcite (calc spar), $CaCO_3$—The color, when pure, is white or colorless, but when impurities are present, the color depends upon the foreign substance; yellow, green, gray, salmon, lavender, and other colors are common. Calcite is much softer than glass (H=3). It is readily distinguished from other minerals by its strong rhomboidal cleavage, its hardness, and its effervescence with acids. It is one of the most widespread and abundant minerals. There are a number of varieties, including dogtooth spar, so-called because of the shape of the crystals; marble, a crystalline rock composed of large and small grains of calcite; Mexican onyx, an agatelike rock formed by successive layers of lime deposited from solution in a cavity.

Dolomite (pearl spar), CaMg $(CO_3)_2$—The color is usually white or with a yellow tint. It is softer than steel (H=3.5). Dolomite is distinguished from calcite, which it resembles, by its curved cleavage surfaces, its pearly luster, and its lack of effervescence with cold hydrochloric acid. It occurs in distinct crystals and forms thick strata of limestone. It is a common vein mineral.

Gypsum, $CaSO_4 \cdot 2 H_2O$—This mineral is colorless or white unless tinted by impurities. It is softer than calcite and can be scratched with the finger nail (H=2). Gypsum occurs in veins or beds; in crystals, or compact, rocklike masses. The most important variety is selenite, a crystalline gypsum with a perfect cleavage, thin leaves of which may be split off and resemble those of mica. They differ from the latter in their inelasticity and vertical cleavage. Alabaster is a compact, fine-grained, usually translucent gypsum, used in making ornaments and statuary. Satin spar is a fibrous gypsum which has somewhat the appearance of satin. It is occasionally used in the manufacture of cheap jewelry. Rock gypsum is compact and rock like. It is used for plaster of Paris. Gypsum is common in many portions of North America, but is especially abundant in New York, Iowa, Michigan, and Ohio.

Fluorite (fluorspar, blue john), CaF_2—The color is commonly blue or green, but is occasionally white or yellow. It is slightly harder than calcite and can be scratched with a knife (H=4). Its principal use is as a flux in reducing iron, but it is used to some extent for ornamental purposes. It occurs in clear, cubical crystals, and in masses. Fluorite is mined in Illinois and Kentucky.

Apatite (asparagus stone, phosphate rock, calcium phosphate)— The color is usually green or reddish brown. It is harder than fluorite and cannot easily be scratched with a knife (H=5). After being treated with sulphuric acid, it becomes a valuable fertilizer. It is found in many parts of North America, but the most valuable deposits occur in Canada.

Copper Minerals

Chalcopyrite (copper pyrites), $CuFeS_2$—The color is a deeper yellow than pyrite. Chalcopyrite can be easily scratched with a knife (H=3.5) and this character alone easily distinguishes it from pyrite, but not from pyrrhotite. The bluish tarnish of chalcopyrite is also distinctive. Since it is not attracted by a magnet, it is easily distinguishable from pyrrhotite. It is a valuable and widespread ore of copper and is mined in many of the Western States.

Malachite (green copper carbonate), $(CuOH)_2 CO_3$—The color is bright green. The color, hardness, which is less than that of steel (H=3.5), and its effervescence with acids readily distinguish it from other minerals. Its principal use in the United States is as an ore of copper, although in Europe the compact varieties have long been much sought after for vases, table tops, and mosaics.

Lead Minerals

Galenite (galena), PbS—The color is lead gray. Its softness (H=2.5), its high specific gravity which is greater than that of iron, and its strong cubical cleavage make it one of the most easily recognizable minerals. It occurs in masses and as cubical crystals. Galena is valuable as an ore of lead, as well as for the silver which it usually carries.

Silica Minerals

Quartz and its Varieties, SiO_2—When pure, quartz is colorless or white, but in no other mineral do the colors vary so widely; red, pink, yellow, brown, green, blue, lavender, and black, in fact almost every conceivable color is found in quartz. Quartz is harder than steel and scratches glass (H=7). It is the commonest of minerals. It makes up most of the sand of the seashore; it occurs as a rock in the forms of sandstone and quartzite, and is a prominent part of many other important rocks, as granite and gneiss. It is readily distinguished from other minerals by its hardness and its lack of cleavage. The crystals are six-sided (hexagonal). The principal varieties are: rock crystal, as the clear quartz crystals are called, which is used for making "pebble lenses," "Japanese balls," and other objects; amethyst, purple crystalline quartz which is cut for gem stones; rose quartz, which is light pink or rose color; milky, smoky, and yellow quartz, named because of their color. Chalcedony is a translucent variety with a waxy luster which varies greatly in color. Agate is a banded chalcedony in which the bands are variously colored. Flint and chert are gray to black translucent or opaque quartz masses which occur in chalk and limestone. Jasper is similar to flint in appearance, but is usually red, black, white, or yellow.

Silicate Minerals

Orthoclase Feldspar (potash feldspar), $KAlSi_3O_8$—The color is usually white, gray, or flesh. The hardness is about that of steel (H=6). The mineral cleaves readily, the cleavage planes being at right angles to each other. Orthoclase feldspar is an important constituent of granite and sometimes occurs in large crystals. Pure feldspar is used to make the glaze on porcelain.

Labradorite Feldspar (lime feldspar)—The color is dark gray, often with blue, green, and red iridescence. It is slightly harder than steel (H=6). The cleavage planes are often striated and are not at right angles to each other as in orthoclase. It is an important constituent of some igneous rocks. It is used to a limited extent for ornamental purposes.

Muscovite Mica (isinglass, white mica), $H_2KAl_3(SiO_4)_3$—It is usually transparent or gray. It can be scratched with the fingernail (H=2). The most distinctive characters of muscovite are its ability to be cleaved into thin leaves, its hardness, the elasticity of its leaves, and its color. It is used in stove doors, for insulation in electrical apparatus, and, when ground, as a lubricant.

Biotite Mica (black mica), a complex silicate—With the exception of the color and chemical composition, biotite has the same characters as muscovite.

Chlorite, a complex silicate—The color is usually dark green. It can be easily scratched with the fingernail (H=1-2). It occurs in dark green masses in which the flakes are usually so small they are difficult to distinguish. Chlorite occurs commonly in metamorphic rocks.

Talc, a hydro-magnesian silicate—The color is white, greenish, or gray. It is readily distinguished by its soapy feel (H=1), in which it differs from gypsum. Talc commonly occurs in plates or leaves like mica. It occasionally occurs in beds 15 or more feet in thickness. It is ground to make talcum powder and has many other uses, such as a filler for paper, a lubricant, and an adulterant. Large deposits of talc occur in New York, Massachusetts, North Carolina, and other states.

Serpentine, a hydro-magnesian silicate—The color is usually green or yellow, and the hardness is less than that of steel (H=usually about 3). There are two principal varieties: massive serpentine, a compact mineral with a greasy or waxy luster, and asbestos or chrysotile, a fibrous variety. The massive serpentine is polished for table tops and other ornamental purposes; the asbestos is used in the manufacture of fire-proof articles, such as theater curtains, coverings of steam pipes and boilers, and for firemen's suits. The province of Quebec is the great center for asbestos.

Hornblende, a silicate of several elements—The color is commonly black, and the hardness about that of steel (H=5.6). The most distinctive character of hornblende is its occurrence, usually, in slender, flat crystals, the larger angles of the crystals being about 124 degrees. A fibrous variety known as hornblende asbestos has much the same appearance, and is used for the same purpose, as serpentine asbestos. Hornblende is a constituent of some igneous rocks.

Augite, a silicate of several elements—The color is black or dark green and the hardness about that of steel (H=5-6). It usually occurs in short, thick crystals. It is a rock-making mineral of wide distribution and is an important constituent of "trap."

Olivine (chrysolite, peridot), an iron magnesium silicate— The color is usually yellowish green, and the hardness that of quartz (H=6.5-7). It is an important constituent of some igneous rocks. Large, clear crystals are cut for gem stones.

Garnet, variable silicates of various bases—The color is commonly red or black, but brown and green garnets also occur. The hardness is that of quartz (H=7). Garnets usually occur in crystals with 12 similar faces (dodecahedrons or trapezohedrons) and are found embedded in metamorphic rocks of various kinds. Garnets are crushed and manufactured into sandpaper, and fine, clear specimens of good color are cut for gem stones.